WORLD HEALTH ORGANIZATION
INTERNATIONAL AGENCY FOR RESEARCH ON CANCER

IARC Monographs on the Evaluation of Carcinogenic Risks to Humans

VOLUME 86

Cobalt in Hard Metals and Cobalt Sulfate, Gallium Arsenide, Indium Phosphide and Vanadium Pentoxide

This publication represents the views and expert opinions
of an IARC Working Group on the
Evaluation of Carcinogenic Risks to Humans,
which met in Lyon,

7–14 October 2003

2006

IARC MONOGRAPHS

In 1969, the International Agency for Research on Cancer (IARC) initiated a programme on the evaluation of the carcinogenic risk of chemicals to humans involving the production of critically evaluated monographs on individual chemicals. The programme was subsequently expanded to include evaluations of carcinogenic risks associated with exposures to complex mixtures, life-style factors and biological and physical agents, as well as those in specific occupations.

The objective of the programme is to elaborate and publish in the form of monographs critical reviews of data on carcinogenicity for agents to which humans are known to be exposed and on specific exposure situations; to evaluate these data in terms of human risk with the help of international working groups of experts in chemical carcinogenesis and related fields; and to indicate where additional research efforts are needed.

The lists of IARC evaluations are regularly updated and are available on Internet: http://monographs. iarc.fr/

This project was supported by Cooperative Agreement 5 UO1 CA33193 awarded by the United States National Cancer Institute, Department of Health and Human Services. Additional support has been provided since 1992 by the United States National Institute of Environmental Health Sciences.

This publication was made possible, in part, by a Cooperative Agreement between the United States Environmental Protection Agency, Office of Research and Development (USEPA-ORD) and the International Agency for Research on Cancer (IARC) and does not necessarily express the views of USEPA-ORD.

Published by the International Agency for Research on Cancer,
150 cours Albert Thomas, 69372 Lyon Cedex 08, France
©International Agency for Research on Cancer, 2006

Distributed by WHO Press, World Health Organization, 20 Avenue Appia, 1211 Geneva 27, Switzerland (tel.: +41 22 791 3264; fax: +41 22 791 4857; e-mail: bookorders@who.int).

IARC Library Cataloguing in Publication Data

Cobalt in Hard Metals and Cobalt Sulfate, Gallium Arsenide, Indium Phosphide and Vanadium Pentoxide/IARC Working Group on the Evaluation of Carcinogenic Risks to Humans (2006 : Lyon, France)
(IARC monographs on the evaluation of carcinogenic risks to humans ; v. 86)

1. Arsenic – adverse effects   2. Carcinogens   3. Cobalt – adverse effects   4. Gallium – adverse effects   5. Indium – adverse effects   6. Metals   7. Vanadium compounds – adverse effects
8. Tungsten – adverse effects   I. IARC Working Group on the Evaluation of Carcinogenic Risks to Humans   II. Series

ISBN 92 832 1286 X   (NLM Classification: W1)
ISSN 1017-1606

Cover: Some metallic compounds evaluated in this volume are used in integrated circuit boards which represent a new use of these metals.

Cover design by Georges Mollon, IARC

CONTENTS

NOTE TO THE READER

The term 'carcinogenic risk' in the *IARC Monographs* series is taken to mean the probability that exposure to an agent will lead to cancer in humans.

Inclusion of an agent in the *Monographs* does not imply that it is a carcinogen, only that the published data have been examined. Equally, the fact that an agent has not yet been evaluated in a monograph does not mean that it is not carcinogenic.

The evaluations of carcinogenic risk are made by international working groups of independent scientists and are qualitative in nature. No recommendation is given for regulation or legislation.

Anyone who is aware of published data that may alter the evaluation of the carcinogenic risk of an agent to humans is encouraged to make this information available to the Unit of Carcinogen Identification and Evaluation, International Agency for Research on Cancer, 150 cours Albert Thomas, 69372 Lyon Cedex 08, France, in order that the agent may be considered for re-evaluation by a future Working Group.

Although every effort is made to prepare the monographs as accurately as possible, mistakes may occur. Readers are requested to communicate any errors to the Unit of Carcinogen Identification and Evaluation, so that corrections can be reported in future volumes.

IARC WORKING GROUP ON THE EVALUATION
OF CARCINOGENIC RISKS TO HUMANS:
COBALT IN HARD METALS AND COBALT SULFATE,
GALLIUM ARSENIDE, INDIUM PHOSPHIDE
AND VANADIUM PENTOXIDE

Lyon, 7–14 October 2003

LIST OF PARTICIPANTS

Members

Mario Altamirano-Lozano, Unidad de Investigación en Genética y Toxicología Ambiental (UNIGEN), Facultad de Estudios Superiores-Zaragoza, Battalla del 5 de mayo esq. Fuerte de Loreto Col. Ejercito de Oriente, C.P. 09230 Mexico, DF, Mexico

Detmar Beyersmann, Department of Biology & Chemistry, Fachbereich 2, University of Bremen, Leobener Strasse NW2, Raum B2230, 28359 Bremen, Germany (*Chairman*)

Dean E. Carter, Department of Pharmacology & Toxicology, Center for Toxicology, College of Pharmacy, University of Arizona, 1703 E. Mabel, Tucson, AZ 85721, USA (*unable to attend*)

Bruce A. Fowler, Senior Research Advisor, ATSDR/CDC, 1600 Clifton Road NE, MS E-29, Atlanta, GA 30333, USA (*Subgroup Chair: Other Relevant Data*)

Bice Fubini, Department of Inorganic, Physical & Material Chemistry and Interdepartmental Center 'G. Scansetti' for Studies on Asbestos and other Toxic Particulates, Facoltà di Farmacia, Università degli Studi di Torino, Via P. Giuria 7, 10125 Torino, Italy

Janet Kielhorn, Fraunhofer Institute of Toxicology & Experimental Medicine, Department of Chemical Risk Assessment, Nikolai-Fuchs-Strasse 1, 30625 Hannover, Germany

Micheline Kirsch-Volders, Laboratorium voor Cellulaire Genetica, Faculteit Wetenschappen, Vrije Universiteit Brussel, Pleinlaan 2, 1050 Brussels, Belgium

Jan Kucera, Nuclear Physics Institute, 250 68 Rez near Prague, Czech Republic

Yukinori Kusaka, Department of Environmental Health, School of Medicine, Fukui Medical University, Matsuoka-cho, Fukui 910-1193, Japan (*Subgroup Chair: Exposure Data*)

Gerard Lasfargues, Médecine et Santé au Travail, Faculté de Médecine, 2 bis Bd Tonnelé, B.P. 3223, 37325 Tours Cedex, France

Dominique Lison, Industrial Toxicology & Occupational Medicine Unit, Catholic University of Louvain, Clos Chapelle-aux-Champs 30, 1200 Brussels, Belgium

Inge Mangelsdorf, Chemikalienbewertung, Fraunhofer-Institut fur Toxikologie und Experi-
mentelle Medizin, Nikolai-Fuchs-Strasse 1, 30625 Hannover, Germany (*unable to
attend*)

Damien McElvenny, Epidemiology & Medical Statistics Unit, Health & Safety Executive,
Room 246, Magdalen House, Stanley Precinct, Bootle, Merseyside L20 3QZ, United
Kingdom (*Subgroup Chair: Cancer in Humans*)

Benoit Nemery, Professor of Toxicology and Occupational Medicine, Laboratorium voor
Pneumologie, K.U. Leuven, Herestraat 49, 3000 Leuven, Belgium

Joseph Roycroft, National Toxicology Program, National Institute of Environmental Health
Sciences, 79 Alexander Drive, Research Triangle Park, NC 27709, USA (*Subgroup
Chair: Cancer in Experimental Animals*)

Magnus Svartengren, Division of Occupational Medicine, Department of Public Health
Sciences, Karolinska Institutet, Norrbacka, 171 76 Stockholm, Sweden

Invited specialists

Ted Junghans, Technical Resources International Inc., 6500 Rock Spring Drive, Suite 650,
Bethesda, MD 20817-1197, USA

Steve Olin, ILSI Risk Science Institute, One Thomas Circle, NW, 9th Floor, Washington,
DC 20005-5802, USA

Roger Renne, Battelle Toxicology Northwest, 902 Battelle Bd, PO Box 999, Richland,
WA 99352, USA

Representatives

David G. Longfellow, Cancer Etiology Branch, Division of Cancer Biology, National
Cancer Institute, 6130 Executive Blvd, Suite 5000, MSC7398, Rockville, MD 20892-
7398, USA

Kyriakoula Ziegler-Skylakakis, European Commission, DG Employment D/5, Bâtiment
Jean Monnet, Plateau du Kirchberg, 2920 Luxembourg, Grand Duchy of Luxembourg

IARC Secretariat

Robert Baan, Unit of Carcinogen Identification and Evaluation (*Co-Rapporteur, Subgroup
on Other Relevant Data*)

Vincent Cogliano, Unit of Carcinogen Identification and Evaluation (*Head of Programme*)

Fatiha El Ghissassi, Unit of Carcinogen Identification and Evaluation (*Co-Rapporteur,
Subgroup on Other Relevant Data*)

Tony Fletcher, Unit of Environmental Cancer Epidemiology

Marlin Friesen, Unit of Nutrition and Cancer

Yann Grosse, Unit of Carcinogen Identification and Evaluation (*Responsible Officer;
Rapporteur, Subgroup on Cancer in Experimental Animals*)

Nikolai Napalkov[1]
Béatrice Secretan, Unit of Carcinogen Identification and Evaluation (Rapporteur, Subgroup on Exposure Data)
Kurt Straif, Unit of Carcinogen Identification and Evaluation (Rapporteur, Subgroup on Cancer in Humans)
Zhao-Qi Wang, Unit of Gene–Environment Interactions
Rosamund Williams (Editor)

Post-meeting scientific assistance
Catherine Cohet

Technical assistance
Sandrine Egraz
Martine Lézère
Jane Mitchell
Elspeth Perez

---

[1] Present address: Director Emeritus, Petrov Institute of Oncology, Pesochny-2, 197758 St Petersburg, Russia

PREAMBLE

IARC MONOGRAPHS PROGRAMME ON THE EVALUATION OF CARCINOGENIC RISKS TO HUMANS

PREAMBLE

1. BACKGROUND

In 1969, the International Agency for Research on Cancer (IARC) initiated a programme to evaluate the carcinogenic risk of chemicals to humans and to produce monographs on individual chemicals. The *Monographs* programme has since been expanded to include consideration of exposures to complex mixtures of chemicals (which occur, for example, in some occupations and as a result of human habits) and of exposures to other agents, such as radiation and viruses. With Supplement 6 (IARC, 1987a), the title of the series was modified from *IARC Monographs on the Evaluation of the Carcinogenic Risk of Chemicals to Humans* to *IARC Monographs on the Evaluation of Carcinogenic Risks to Humans*, in order to reflect the widened scope of the programme.

The criteria established in 1971 to evaluate carcinogenic risk to humans were adopted by the working groups whose deliberations resulted in the first 16 volumes of the *IARC Monographs series*. Those criteria were subsequently updated by further ad-hoc working groups (IARC, 1977, 1978, 1979, 1982, 1983, 1987b, 1988, 1991a; Vainio *et al.*, 1992).

2. OBJECTIVE AND SCOPE

The objective of the programme is to prepare, with the help of international working groups of experts, and to publish in the form of monographs, critical reviews and evaluations of evidence on the carcinogenicity of a wide range of human exposures. The *Monographs* may also indicate where additional research efforts are needed.

The *Monographs* represent the first step in carcinogenic risk assessment, which involves examination of all relevant information in order to assess the strength of the available evidence that certain exposures could alter the incidence of cancer in humans. The second step is quantitative risk estimation. Detailed, quantitative evaluations of epidemiological data may be made in the *Monographs*, but without extrapolation beyond the range of the data available. Quantitative extrapolation from experimental data to the human situation is not undertaken.

The term 'carcinogen' is used in these monographs to denote an exposure that is capable of increasing the incidence of malignant neoplasms; the induction of benign neoplasms may in some circumstances (see p. 19) contribute to the judgement that the expo-sure is carcinogenic. The terms 'neoplasm' and 'tumour' are used interchangeably.

Some epidemiological and experimental studies indicate that different agents may act at different stages in the carcinogenic process, and several mechanisms may be involved. The aim of the *Monographs* has been, from their inception, to evaluate evidence of carci-nogenicity at any stage in the carcinogenesis process, independently of the underlying mechanisms. Information on mechanisms may, however, be used in making the overall evaluation (IARC, 1991a; Vainio *et al.*, 1992; see also pp. 25–27).

The *Monographs* may assist national and international authorities in making risk assessments and in formulating decisions concerning any necessary preventive measures. The evaluations of IARC working groups are scientific, qualitative judgements about the evidence for or against carcinogenicity provided by the available data. These evaluations represent only one part of the body of information on which regulatory measures may be based. Other components of regulatory decisions vary from one situation to another and from country to country, responding to different socioeconomic and national priorities. **Therefore, no recommendation is given with regard to regulation or legislation, which are the responsibility of individual governments and/or other international organizations.**

The *IARC Monographs* are recognized as an authoritative source of information on the carcinogenicity of a wide range of human exposures. A survey of users in 1988 indi-cated that the *Monographs* are consulted by various agencies in 57 countries. About 2500 copies of each volume are printed, for distribution to governments, regulatory bodies and interested scientists. The Monographs are also available from IARC*Press* in Lyon and via the Marketing and Dissemination (MDI) of the World Health Organization in Geneva.

3.   SELECTION OF TOPICS FOR MONOGRAPHS

Topics are selected on the basis of two main criteria: (a) there is evidence of human exposure, and (b) there is some evidence or suspicion of carcinogenicity. The term 'agent' is used to include individual chemical compounds, groups of related chemical compounds, physical agents (such as radiation) and biological factors (such as viruses). Exposures to mixtures of agents may occur in occupational exposures and as a result of personal and cultural habits (like smoking and dietary practices). Chemical analogues and compounds with biological or physical characteristics similar to those of suspected carcinogens may also be considered, even in the absence of data on a possible carcino-genic effect in humans or experimental animals.

The scientific literature is surveyed for published data relevant to an assessment of carcinogenicity. The IARC information bulletins on agents being tested for carcino-genicity (IARC, 1973–1996) and directories of on-going research in cancer epide-miology (IARC, 1976–1996) often indicate exposures that may be scheduled for future meetings. Ad-hoc working groups convened by IARC in 1984, 1989, 1991, 1993 and

1998 gave recommendations as to which agents should be evaluated in the IARC Monographs series (IARC, 1984, 1989, 1991b, 1993, 1998a,b).

As significant new data on subjects on which monographs have already been prepared become available, re-evaluations are made at subsequent meetings, and revised monographs are published.

4. DATA FOR MONOGRAPHS

The *Monographs* do not necessarily cite all the literature concerning the subject of an evaluation. Only those data considered by the Working Group to be relevant to making the evaluation are included.

With regard to biological and epidemiological data, only reports that have been published or accepted for publication in the openly available scientific literature are reviewed by the working groups. In certain instances, government agency reports that have undergone peer review and are widely available are considered. Exceptions may be made on an ad-hoc basis to include unpublished reports that are in their final form and publicly available, if their inclusion is considered pertinent to making a final evaluation (see pp. 25–27). In the sections on chemical and physical properties, on analysis, on production and use and on occurrence, unpublished sources of information may be used.

5. THE WORKING GROUP

Reviews and evaluations are formulated by a working group of experts. The tasks of the group are: (i) to ascertain that all appropriate data have been collected; (ii) to select the data relevant for the evaluation on the basis of scientific merit; (iii) to prepare accurate summaries of the data to enable the reader to follow the reasoning of the Working Group; (iv) to evaluate the results of epidemiological and experimental studies on cancer; (v) to evaluate data relevant to the understanding of mechanism of action; and (vi) to make an overall evaluation of the carcinogenicity of the exposure to humans.

Working Group participants who contributed to the considerations and evaluations within a particular volume are listed, with their addresses, at the beginning of each publication. Each participant who is a member of a working group serves as an individual scientist and not as a representative of any organization, government or industry. In addition, nominees of national and international agencies and industrial associations may be invited as observers.

6. WORKING PROCEDURES

Approximately one year in advance of a meeting of a working group, the topics of the monographs are announced and participants are selected by IARC staff in consultation with other experts. Subsequently, relevant biological and epidemiological data are collected by the Carcinogen Identification and Evaluation Unit of IARC from recognized sources of information on carcinogenesis, including data storage and retrieval systems such as MEDLINE and TOXLINE.

For chemicals and some complex mixtures, the major collection of data and the preparation of first drafts of the sections on chemical and physical properties, on analysis, on production and use and on occurrence are carried out under a separate contract funded by the United States National Cancer Institute. Representatives from industrial associations may assist in the preparation of sections on production and use. Information on production and trade is obtained from governmental and trade publications and, in some cases, by direct contact with industries. Separate production data on some agents may not be available because their publication could disclose confidential information. Information on uses may be obtained from published sources but is often complemented by direct contact with manufacturers. Efforts are made to supplement this information with data from other national and international sources.

Six months before the meeting, the material obtained is sent to meeting participants, or is used by IARC staff, to prepare sections for the first drafts of monographs. The first drafts are compiled by IARC staff and sent before the meeting to all participants of the Working Group for review.

The Working Group meets in Lyon for seven to eight days to discuss and finalize the texts of the monographs and to formulate the evaluations. After the meeting, the master copy of each monograph is verified by consulting the original literature, edited and prepared for publication. The aim is to publish monographs within six months of the Working Group meeting.

The available studies are summarized by the Working Group, with particular regard to the qualitative aspects discussed below. In general, numerical findings are indicated as they appear in the original report; units are converted when necessary for easier comparison. The Working Group may conduct additional analyses of the published data and use them in their assessment of the evidence; the results of such supplementary analyses are given in square brackets. When an important aspect of a study, directly impinging on its interpretation, should be brought to the attention of the reader, a comment is given in square brackets.

7.   EXPOSURE DATA

Sections that indicate the extent of past and present human exposure, the sources of exposure, the people most likely to be exposed and the factors that contribute to the exposure are included at the beginning of each monograph.

Most monographs on individual chemicals, groups of chemicals or complex mixtures include sections on chemical and physical data, on analysis, on production and use and on occurrence. In monographs on, for example, physical agents, occupational exposures and cultural habits, other sections may be included, such as: historical perspectives, description of an industry or habit, chemistry of the complex mixture or taxonomy. Monographs on biological agents have sections on structure and biology, methods of detection, epidemiology of infection and clinical disease other than cancer.

For chemical exposures, the Chemical Abstracts Services Registry Number, the latest Chemical Abstracts primary name and the IUPAC systematic name are recorded; other synonyms are given, but the list is not necessarily comprehensive. For biological agents, taxonomy and structure are described, and the degree of variability is given, when applicable.

Information on chemical and physical properties and, in particular, data relevant to identification, occurrence and biological activity are included. For biological agents, mode of replication, life cycle, target cells, persistence and latency and host response are given. A description of technical products of chemicals includes trade names, relevant specifications and available information on composition and impurities. Some of the trade names given may be those of mixtures in which the agent being evaluated is only one of the ingredients.

The purpose of the section on analysis or detection is to give the reader an overview of current methods, with emphasis on those widely used for regulatory purposes. Methods for monitoring human exposure are also given, when available. No critical evaluation or recommendation of any of the methods is meant or implied. The IARC published a series of volumes, *Environmental Carcinogens: Methods of Analysis and Exposure Measurement* (IARC, 1978–93), that describe validated methods for analysing a wide variety of chemicals and mixtures. For biological agents, methods of detection and exposure assessment are described, including their sensitivity, specificity and reproducibility.

The dates of first synthesis and of first commercial production of a chemical or mixture are provided; for agents which do not occur naturally, this information may allow a reasonable estimate to be made of the date before which no human exposure to the agent could have occurred. The dates of first reported occurrence of an exposure are also provided. In addition, methods of synthesis used in past and present commercial production and different methods of production which may give rise to different impurities are described.

Data on production, international trade and uses are obtained for representative regions, which usually include Europe, Japan and the United States of America. It should not, however, be inferred that those areas or nations are necessarily the sole or major sources or users of the agent. Some identified uses may not be current or major applications, and the coverage is not necessarily comprehensive. In the case of drugs, mention of their therapeutic uses does not necessarily represent current practice, nor does it imply judgement as to their therapeutic efficacy.

Information on the occurrence of an agent or mixture in the environment is obtained from data derived from the monitoring and surveillance of levels in occupational environments, air, water, soil, foods and animal and human tissues. When available, data on the generation, persistence and bioaccumulation of the agent are also included. In the case of mixtures, industries, occupations or processes, information is given about all agents present. For processes, industries and occupations, a historical description is also given, noting variations in chemical composition, physical properties and levels of occupational exposure with time and place. For biological agents, the epidemiology of infection is described.

Statements concerning regulations and guidelines (e.g., pesticide registrations, maximal levels permitted in foods, occupational exposure limits) are included for some countries as indications of potential exposures, but they may not reflect the most recent situation, since such limits are continuously reviewed and modified. The absence of information on regulatory status for a country should not be taken to imply that that country does not have regulations with regard to the exposure. For biological agents, legislation and control, including vaccines and therapy, are described.

8.   STUDIES OF CANCER IN HUMANS

(a)   Types of studies considered

Three types of epidemiological studies of cancer contribute to the assessment of carcinogenicity in humans — cohort studies, case–control studies and correlation (or ecological) studies. Rarely, results from randomized trials may be available. Case series and case reports of cancer in humans may also be reviewed.

Cohort and case–control studies relate the exposures under study to the occurrence of cancer in individuals and provide an estimate of relative risk (ratio of incidence or mortality in those exposed to incidence or mortality in those not exposed) as the main measure of association.

In correlation studies, the units of investigation are usually whole populations (e.g. in particular geographical areas or at particular times), and cancer frequency is related to a summary measure of the exposure of the population to the agent, mixture or exposure circumstance under study. Because individual exposure is not documented, however, a causal relationship is less easy to infer from correlation studies than from cohort and case–control studies. Case reports generally arise from a suspicion, based on clinical experience, that the concurrence of two events — that is, a particular exposure and occurrence of a cancer — has happened rather more frequently than would be expected by chance. Case reports usually lack complete ascertainment of cases in any population, definition or enumeration of the population at risk and estimation of the expected number of cases in the absence of exposure. The uncertainties surrounding interpretation of case reports and correlation studies make them inadequate, except in rare instances, to form the sole basis for inferring a causal relationship. When taken together with case–control and cohort studies, however, relevant case reports or correlation studies may add materially to the judgement that a causal relationship is present.

Epidemiological studies of benign neoplasms, presumed preneoplastic lesions and other end-points thought to be relevant to cancer are also reviewed by working groups. They may, in some instances, strengthen inferences drawn from studies of cancer itself.

(b)   Quality of studies considered

The Monographs are not intended to summarize all published studies. Those that are judged to be inadequate or irrelevant to the evaluation are generally omitted. They may be mentioned briefly, particularly when the information is considered to be a useful supplement to that in other reports or when they provide the only data available. Their inclusion does not imply acceptance of the adequacy of the study design or of the analysis and interpretation of the results, and limitations are clearly outlined in square brackets at the end of the study description.

It is necessary to take into account the possible roles of bias, confounding and chance in the interpretation of epidemiological studies. By 'bias' is meant the operation of factors in study design or execution that lead erroneously to a stronger or weaker association than in fact exists between disease and an agent, mixture or exposure circumstance. By 'confounding' is meant a situation in which the relationship with disease is made to appear stronger or weaker than it truly is as a result of an association between the apparent causal factor and another factor that is associated with either an increase or decrease in the incidence of the disease. In evaluating the extent to which these factors have been minimized in an individual study, working groups consider a number of aspects of design and analysis as described in the report of the study. Most of these considerations apply equally to case–control, cohort and correlation studies. Lack of clarity of any of these aspects in the reporting of a study can decrease its credibility and the weight given to it in the final evaluation of the exposure.

Firstly, the study population, disease (or diseases) and exposure should have been well defined by the authors. Cases of disease in the study population should have been identified in a way that was independent of the exposure of interest, and exposure should have been assessed in a way that was not related to disease status.

Secondly, the authors should have taken account in the study design and analysis of other variables that can influence the risk of disease and may have been related to the exposure of interest. Potential confounding by such variables should have been dealt with either in the design of the study, such as by matching, or in the analysis, by statistical adjustment. In cohort studies, comparisons with local rates of disease may be more appropriate than those with national rates. Internal comparisons of disease frequency among individuals at different levels of exposure should also have been made in the study.

Thirdly, the authors should have reported the basic data on which the conclusions are founded, even if sophisticated statistical analyses were employed. At the very least, they should have given the numbers of exposed and unexposed cases and controls in a case–control study and the numbers of cases observed and expected in a cohort study. Further tabulations by time since exposure began and other temporal factors are also important. In a cohort study, data on all cancer sites and all causes of death should have been given, to reveal the possibility of reporting bias. In a case–control study, the effects of investigated factors other than the exposure of interest should have been reported.

Finally, the statistical methods used to obtain estimates of relative risk, absolute rates of cancer, confidence intervals and significance tests, and to adjust for confounding should have been clearly stated by the authors. The methods used should preferably have been the generally accepted techniques that have been refined since the mid-1970s. These methods have been reviewed for case–control studies (Breslow & Day, 1980) and for cohort studies (Breslow & Day, 1987).

(c)   Inferences about mechanism of action

Detailed analyses of both relative and absolute risks in relation to temporal variables, such as age at first exposure, time since first exposure, duration of exposure, cumulative exposure and time since exposure ceased, are reviewed and summarized when available. The analysis of temporal relationships can be useful in formulating models of carcino-genesis. In particular, such analyses may suggest whether a carcinogen acts early or late in the process of carcinogenesis, although at best they allow only indirect inferences about the mechanism of action. Special attention is given to measurements of biological markers of carcinogen exposure or action, such as DNA or protein adducts, as well as markers of early steps in the carcinogenic process, such as proto-oncogene mutation, when these are incorporated into epidemiological studies focused on cancer incidence or mortality. Such measurements may allow inferences to be made about putative mecha-nisms of action (IARC, 1991a; Vainio et al., 1992).

(d)   Criteria for causality

After the individual epidemiological studies of cancer have been summarized and the quality assessed, a judgement is made concerning the strength of evidence that the agent, mixture or exposure circumstance in question is carcinogenic for humans. In making its judgement, the Working Group considers several criteria for causality. A strong asso-ciation (a large relative risk) is more likely to indicate causality than a weak association, although it is recognized that relative risks of small magnitude do not imply lack of causality and may be important if the disease is common. Associations that are replicated in several studies of the same design or using different epidemiological approaches or under different circumstances of exposure are more likely to represent a causal relation-ship than isolated observations from single studies. If there are inconsistent results among investigations, possible reasons are sought (such as differences in amount of exposure), and results of studies judged to be of high quality are given more weight than those of studies judged to be methodologically less sound. When suspicion of carcino-genicity arises largely from a single study, these data are not combined with those from later studies in any subsequent reassessment of the strength of the evidence.

If the risk of the disease in question increases with the amount of exposure, this is considered to be a strong indication of causality, although absence of a graded response is not necessarily evidence against a causal relationship. Demonstration of a decline in risk after cessation of or reduction in exposure in individuals or in whole populations also supports a causal interpretation of the findings.

Although a carcinogen may act upon more than one target, the specificity of an association (an increased occurrence of cancer at one anatomical site or of one morphological type) adds plausibility to a causal relationship, particularly when excess cancer occurrence is limited to one morphological type within the same organ.

Although rarely available, results from randomized trials showing different rates among exposed and unexposed individuals provide particularly strong evidence for causality.

When several epidemiological studies show little or no indication of an association between an exposure and cancer, the judgement may be made that, in the aggregate, they show evidence of lack of carcinogenicity. Such a judgement requires first of all that the studies giving rise to it meet, to a sufficient degree, the standards of design and analysis described above. Specifically, the possibility that bias, confounding or misclassification of exposure or outcome could explain the observed results should be considered and excluded with reasonable certainty. In addition, all studies that are judged to be methodologically sound should be consistent with a relative risk of unity for any observed level of exposure and, when considered together, should provide a pooled estimate of relative risk which is at or near unity and has a narrow confidence interval, due to sufficient population size. Moreover, no individual study nor the pooled results of all the studies should show any consistent tendency for the relative risk of cancer to increase with increasing level of exposure. It is important to note that evidence of lack of carcinogenicity obtained in this way from several epidemiological studies can apply only to the type(s) of cancer studied and to dose levels and intervals between first exposure and observation of disease that are the same as or less than those observed in all the studies. Experience with human cancer indicates that, in some cases, the period from first exposure to the development of clinical cancer is seldom less than 20 years; studies with latent periods substantially shorter than 30 years cannot provide evidence for lack of carcinogenicity.

9. STUDIES OF CANCER IN EXPERIMENTAL ANIMALS

All known human carcinogens that have been studied adequately in experimental animals have produced positive results in one or more animal species (Wilbourn *et al.*, 1986; Tomatis *et al.*, 1989). For several agents (aflatoxins, 4-aminobiphenyl, azathioprine, betel quid with tobacco, bischloromethyl ether and chloromethyl methyl ether (technical grade), chlorambucil, chlornaphazine, ciclosporin, coal-tar pitches, coal-tars, combined oral contraceptives, cyclophosphamide, diethylstilboestrol, melphalan, 8-methoxypsoralen plus ultraviolet A radiation, mustard gas, myleran, 2-naphthylamine, nonsteroidal estrogens, estrogen replacement therapy/steroidal estrogens, solar radiation, thiotepa and vinyl chloride), carcinogenicity in experimental animals was established or highly suspected before epidemiological studies confirmed their carcinogenicity in humans (Vainio *et al.*, 1995). Although this association cannot establish that all agents and mixtures that cause cancer in experimental animals also cause cancer in humans, nevertheless, **in the absence of adequate data on humans, it is biologically plausible and prudent to regard agents and mixtures for which there is *sufficient evidence* (see p. 24) of carcinogenicity in experimental animals as if they presented a carcinogenic risk to humans**. The possibility that a given agent may cause cancer through a species-specific mechanism which does not operate in humans (see p. 27) should also be taken into consideration.

The nature and extent of impurities or contaminants present in the chemical or mixture being evaluated are given when available. Animal strain, sex, numbers per group, age at start of treatment and survival are reported.

Other types of studies summarized include: experiments in which the agent or mixture was administered in conjunction with known carcinogens or factors that modify carcinogenic effects; studies in which the end-point was not cancer but a defined precancerous lesion; and experiments on the carcinogenicity of known metabolites and derivatives.

For experimental studies of mixtures, consideration is given to the possibility of changes in the physicochemical properties of the test substance during collection, storage, extraction, concentration and delivery. Chemical and toxicological interactions of the components of mixtures may result in nonlinear dose–response relationships.

An assessment is made as to the relevance to human exposure of samples tested in experimental animals, which may involve consideration of: (i) physical and chemical characteristics, (ii) constituent substances that indicate the presence of a class of substances, (iii) the results of tests for genetic and related effects, including studies on DNA adduct formation, proto-oncogene mutation and expression and suppressor gene inactivation. The relevance of results obtained, for example, with animal viruses analogous to the virus being evaluated in the monograph must also be considered. They may provide biological and mechanistic information relevant to the understanding of the process of carcinogenesis in humans and may strengthen the plausibility of a conclusion that the biological agent under evaluation is carcinogenic in humans.

(a)   Qualitative aspects

An assessment of carcinogenicity involves several considerations of qualitative importance, including (i) the experimental conditions under which the test was per-formed, including route and schedule of exposure, species, strain, sex, age, duration of follow-up; (ii) the consistency of the results, for example, across species and target organ(s); (iii) the spectrum of neoplastic response, from preneoplastic lesions and benign tumours to malignant neoplasms; and (iv) the possible role of modifying factors.

As mentioned earlier (p. 11), the *Monographs* are not intended to summarize all published studies. Those studies in experimental animals that are inadequate (e.g., too short a duration, too few animals, poor survival; see below) or are judged irrelevant to the evaluation are generally omitted. Guidelines for conducting adequate long-term carcinogenicity experiments have been outlined (e.g. Montesano *et al.*, 1986).

Considerations of importance to the Working Group in the interpretation and evaluation of a particular study include: (i) how clearly the agent was defined and, in the case of mixtures, how adequately the sample characterization was reported; (ii) whether the dose was adequately monitored, particularly in inhalation experiments; (iii) whether the doses and duration of treatment were appropriate and whether the survival of treated animals was similar to that of controls; (iv) whether there were adequate numbers of animals per group; (v) whether animals of each sex were used; (vi) whether animals were allocated randomly to groups; (vii) whether the duration of observation was adequate; and (viii) whether the data were adequately reported. If available, recent data on the incidence of specific tumours in historical controls, as well as in concurrent controls, should be taken into account in the evaluation of tumour response.

When benign tumours occur together with and originate from the same cell type in an organ or tissue as malignant tumours in a particular study and appear to represent a stage in the progression to malignancy, it may be valid to combine them in assessing tumour incidence (Huff *et al.*, 1989). The occurrence of lesions presumed to be pre-neoplastic may in certain instances aid in assessing the biological plausibility of any neoplastic response observed. If an agent or mixture induces only benign neoplasms that appear to be end-points that do not readily progress to malignancy, it should nevertheless be suspected of being a carcinogen and requires further investigation.

(b)   Quantitative aspects

The probability that tumours will occur may depend on the species, sex, strain and age of the animal, the dose of the carcinogen and the route and length of exposure. Evidence of an increased incidence of neoplasms with increased level of exposure strengthens the inference of a causal association between the exposure and the development of neoplasms.

The form of the dose–response relationship can vary widely, depending on the particular agent under study and the target organ. Both DNA damage and increased cell division are important aspects of carcinogenesis, and cell proliferation is a strong determinant of dose–response relationships for some carcinogens (Cohen & Ellwein, 1990). Since many chemicals require metabolic activation before being converted into their reactive intermediates, both metabolic and pharmacokinetic aspects are important in determining the dose–response pattern. Saturation of steps such as absorption, activation, inactivation and elimination may produce nonlinearity in the dose–response relationship, as could saturation of processes such as DNA repair (Hoel *et al.*, 1983; Gart *et al.*, 1986).

(c)   *Statistical analysis of long-term experiments in animals*

Factors considered by the Working Group include the adequacy of the information given for each treatment group: (i) the number of animals studied and the number examined histologically, (ii) the number of animals with a given tumour type and (iii) length of survival. The statistical methods used should be clearly stated and should be the generally accepted techniques refined for this purpose (Peto *et al.*, 1980; Gart *et al.*, 1986). When there is no difference in survival between control and treatment groups, the Working Group usually compares the proportions of animals developing each tumour type in each of the groups. Otherwise, consideration is given as to whether or not appropriate adjustments have been made for differences in survival. These adjustments can include: comparisons of the proportions of tumour-bearing animals among the effective number of animals (alive at the time the first tumour is discovered), in the case where most differences in survival occur before tumours appear; life-table methods, when tumours are visible or when they may be considered 'fatal' because mortality rapidly follows tumour development; and the Mantel-Haenszel test or logistic regression, when occult tumours do not affect the animals' risk of dying but are 'incidental' findings at autopsy.

In practice, classifying tumours as fatal or incidental may be difficult. Several survival-adjusted methods have been developed that do not require this distinction (Gart *et al.*, 1986), although they have not been fully evaluated.

10.   OTHER DATA RELEVANT TO AN EVALUATION OF CARCINOGENICITY AND ITS MECHANISMS

In coming to an overall evaluation of carcinogenicity in humans (see pp. 25–27), the Working Group also considers related data. The nature of the information selected for the summary depends on the agent being considered.

For chemicals and complex mixtures of chemicals such as those in some occupational situations or involving cultural habits (e.g. tobacco smoking), the other data considered to be relevant are divided into those on absorption, distribution, metabolism and excretion; toxic effects; reproductive and developmental effects; and genetic and related effects.

Concise information is given on absorption, distribution (including placental transfer) and excretion in both humans and experimental animals. Kinetic factors that may affect the dose–response relationship, such as saturation of uptake, protein binding, metabolic activation, detoxification and DNA repair processes, are mentioned. Studies that indicate the metabolic fate of the agent in humans and in experimental animals are summarized briefly, and comparisons of data on humans and on animals are made when possible. Comparative information on the relationship between exposure and the dose that reaches the target site may be of particular importance for extrapolation between species. Data are given on acute and chronic toxic effects (other than cancer), such as organ toxicity, increased cell proliferation, immunotoxicity and endocrine effects. The presence and toxicological significance of cellular receptors is described. Effects on reproduction, teratogenicity, fetotoxicity and embryotoxicity are also summarized briefly.

Tests of genetic and related effects are described in view of the relevance of gene mutation and chromosomal damage to carcinogenesis (Vainio *et al.*, 1992; McGregor *et al.*, 1999). The adequacy of the reporting of sample characterization is considered and, where necessary, commented upon; with regard to complex mixtures, such comments are similar to those described for animal carcinogenicity tests on p. 18. The available data are interpreted critically by phylogenetic group according to the end-points detected, which may include DNA damage, gene mutation, sister chromatid exchange, micro-nucleus formation, chromosomal aberrations, aneuploidy and cell transformation. The concentrations employed are given, and mention is made of whether use of an exogenous metabolic system *in vitro* affected the test result. These data are given as listings of test systems, data and references. The data on genetic and related effects presented in the *Monographs* are also available in the form of genetic activity profiles (GAP) prepared in collaboration with the United States Environmental Protection Agency (EPA) (see also Waters *et al.*, 1987) using software for personal computers that are Microsoft Windows® compatible. The EPA/IARC GAP software and database may be downloaded free of charge from *www.epa.gov/gapdb*.

Positive results in tests using prokaryotes, lower eukaryotes, plants, insects and cultured mammalian cells suggest that genetic and related effects could occur in mammals. Results from such tests may also give information about the types of genetic effect produced and about the involvement of metabolic activation. Some end-points described are clearly genetic in nature (e.g., gene mutations and chromosomal aberra-tions), while others are to a greater or lesser degree associated with genetic effects (e.g. unscheduled DNA synthesis). In-vitro tests for tumour-promoting activity and for cell transformation may be sensitive to changes that are not necessarily the result of genetic alterations but that may have specific relevance to the process of carcinogenesis. A critical appraisal of these tests has been published (Montesano *et al.*, 1986).

Genetic or other activity detected in experimental mammals and humans is regarded as being of greater relevance than that in other organisms. The demonstration that an agent or mixture can induce gene and chromosomal mutations in whole mammals indi-cates that it may have carcinogenic activity, although this activity may not be detectably expressed in any or all species. Relative potency in tests for mutagenicity and related effects is not a reliable indicator of carcinogenic potency. Negative results in tests for mutagenicity in selected tissues from animals treated *in vivo* provide less weight, partly because they do not exclude the possibility of an effect in tissues other than those examined. Moreover, negative results in short-term tests with genetic end-points cannot be considered to provide evidence to rule out carcinogenicity of agents or mixtures that act through other mechanisms (e.g. receptor-mediated effects, cellular toxicity with regenerative proliferation, peroxisome proliferation) (Vainio *et al.*, 1992). Factors that may lead to misleading results in short-term tests have been discussed in detail elsewhere (Montesano *et al.*, 1986).

When available, data relevant to mechanisms of carcinogenesis that do not involve structural changes at the level of the gene are also described.

The adequacy of epidemiological studies of reproductive outcome and genetic and related effects in humans is evaluated by the same criteria as are applied to epidemiological studies of cancer.

Structure–activity relationships that may be relevant to an evaluation of the carcinogenicity of an agent are also described.

For biological agents — viruses, bacteria and parasites — other data relevant to carcinogenicity include descriptions of the pathology of infection, molecular biology (integration and expression of viruses, and any genetic alterations seen in human tumours) and other observations, which might include cellular and tissue responses to infection, immune response and the presence of tumour markers.

11. SUMMARY OF DATA REPORTED

In this section, the relevant epidemiological and experimental data are summarized. Only reports, other than in abstract form, that meet the criteria outlined on p. 11 are considered for evaluating carcinogenicity. Inadequate studies are generally not summarized: such studies are usually identified by a square-bracketed comment in the preceding text.

(a)   *Exposure*

Human exposure to chemicals and complex mixtures is summarized on the basis of elements such as production, use, occurrence in the environment and determinations in human tissues and body fluids. Quantitative data are given when available. Exposure to biological agents is described in terms of transmission and prevalence of infection.

(b)   *Carcinogenicity in humans*

Results of epidemiological studies that are considered to be pertinent to an assessment of human carcinogenicity are summarized. When relevant, case reports and correlation studies are also summarized.

(c)   *Carcinogenicity in experimental animals*

Data relevant to an evaluation of carcinogenicity in animals are summarized. For each animal species and route of administration, it is stated whether an increased incidence of neoplasms or preneoplastic lesions was observed, and the tumour sites are indicated. If the agent or mixture produced tumours after prenatal exposure or in single-dose experiments, this is also indicated. Negative findings are also summarized. Dose–response and other quantitative data may be given when available.

(d)   *Other data relevant to an evaluation of carcinogenicity and its mechanisms*

Data on biological effects in humans that are of particular relevance are summarized. These may include toxicological, kinetic and metabolic considerations and evidence of DNA binding, persistence of DNA lesions or genetic damage in exposed humans. Toxicological information, such as that on cytotoxicity and regeneration, receptor binding and hormonal and immunological effects, and data on kinetics and metabolism in experimental animals are given when considered relevant to the possible mechanism of the carcinogenic action of the agent. The results of tests for genetic and related effects are summarized for whole mammals, cultured mammalian cells and nonmammalian systems.

When available, comparisons of such data for humans and for animals, and particularly animals that have developed cancer, are described.

Structure–activity relationships are mentioned when relevant.

For the agent, mixture or exposure circumstance being evaluated, the available data on end-points or other phenomena relevant to mechanisms of carcinogenesis from studies in humans, experimental animals and tissue and cell test systems are summarized within one or more of the following descriptive dimensions:

(i) Evidence of genotoxicity (structural changes at the level of the gene): for example, structure–activity considerations, adduct formation, mutagenicity (effect on specific genes), chromosomal mutation/aneuploidy

(ii) Evidence of effects on the expression of relevant genes (functional changes at the intracellular level): for example, alterations to the structure or quantity of the product of a proto-oncogene or tumour-suppressor gene, alterations to metabolic activation/inactivation/DNA repair

(iii) Evidence of relevant effects on cell behaviour (morphological or behavioural changes at the cellular or tissue level): for example, induction of mitogenesis, compensatory cell proliferation, preneoplasia and hyperplasia, survival of premalignant or malignant cells (immortalization, immunosuppression), effects on metastatic potential

(iv) Evidence from dose and time relationships of carcinogenic effects and interactions between agents: for example, early/late stage, as inferred from epidemiological studies; initiation/promotion/progression/malignant conversion, as defined in animal carcinogenicity experiments; toxicokinetics

These dimensions are not mutually exclusive, and an agent may fall within more than one of them. Thus, for example, the action of an agent on the expression of relevant genes could be summarized under both the first and second dimensions, even if it were known with reasonable certainty that those effects resulted from genotoxicity.

12.   EVALUATION

Evaluations of the strength of the evidence for carcinogenicity arising from human and experimental animal data are made, using standard terms.

It is recognized that the criteria for these evaluations, described below, cannot encompass all of the factors that may be relevant to an evaluation of carcinogenicity. In considering all of the relevant scientific data, the Working Group may assign the agent, mixture or exposure circumstance to a higher or lower category than a strict inter-pretation of these criteria would indicate.

(a)   Degrees of evidence for carcinogenicity in humans and in experimental animals and supporting evidence

These categories refer only to the strength of the evidence that an exposure is carcino-genic and not to the extent of its carcinogenic activity (potency) nor to the mechanisms involved. A classification may change as new information becomes available.

An evaluation of degree of evidence, whether for a single agent or a mixture, is limited to the materials tested, as defined physically, chemically or biologically. When the agents evaluated are considered by the Working Group to be sufficiently closely related, they may be grouped together for the purpose of a single evaluation of degree of evidence.

(i)   Carcinogenicity in humans

The applicability of an evaluation of the carcinogenicity of a mixture, process, occu-pation or industry on the basis of evidence from epidemiological studies depends on the variability over time and place of the mixtures, processes, occupations and industries. The Working Group seeks to identify the specific exposure, process or activity which is considered most likely to be responsible for any excess risk. The evaluation is focused as narrowly as the available data on exposure and other aspects permit.

The evidence relevant to carcinogenicity from studies in humans is classified into one of the following categories:

Sufficient evidence of carcinogenicity: The Working Group considers that a causal relationship has been established between exposure to the agent, mixture or exposure circumstance and human cancer. That is, a positive relationship has been observed between the exposure and cancer in studies in which chance, bias and confounding could be ruled out with reasonable confidence.

Limited evidence of carcinogenicity: A positive association has been observed between exposure to the agent, mixture or exposure circumstance and cancer for which a causal interpretation is considered by the Working Group to be credible, but chance, bias or confounding could not be ruled out with reasonable confidence.

Inadequate evidence of carcinogenicity: The available studies are of insufficient quality, consistency or statistical power to permit a conclusion regarding the presence or absence of a causal association between exposure and cancer, or no data on cancer in humans are available.

Evidence suggesting lack of carcinogenicity: There are several adequate studies covering the full range of levels of exposure that human beings are known to encounter, which are mutually consistent in not showing a positive association between exposure to the agent, mixture or exposure circumstance and any studied cancer at any observed level of exposure. A conclusion of 'evidence suggesting lack of carcinogenicity' is inevitably limited to the cancer sites, conditions and levels of exposure and length of observation covered by the available studies. In addition, the possibility of a very small risk at the levels of exposure studied can never be excluded.

In some instances, the above categories may be used to classify the degree of evidence related to carcinogenicity in specific organs or tissues.

(ii)   Carcinogenicity in experimental animals

The evidence relevant to carcinogenicity in experimental animals is classified into one of the following categories:

Sufficient evidence of carcinogenicity: The Working Group considers that a causal relationship has been established between the agent or mixture and an increased incidence of malignant neoplasms or of an appropriate combination of benign and malignant neoplasms in (a) two or more species of animals or (b) in two or more independent studies in one species carried out at different times or in different laboratories or under different protocols.

Exceptionally, a single study in one species might be considered to provide sufficient evidence of carcinogenicity when malignant neoplasms occur to an unusual degree with regard to incidence, site, type of tumour or age at onset.

Limited evidence of carcinogenicity: The data suggest a carcinogenic effect but are limited for making a definitive evaluation because, e.g. (a) the evidence of carcinogenicity is restricted to a single experiment; or (b) there are unresolved questions regarding the adequacy of the design, conduct or interpretation of the study; or (c) the agent or mixture increases the incidence only of benign neoplasms or lesions of uncertain neoplastic potential, or of certain neoplasms which may occur spontaneously in high incidences in certain strains.

Inadequate evidence of carcinogenicity: The studies cannot be interpreted as showing either the presence or absence of a carcinogenic effect because of major qualitative or quantitative limitations, or no data on cancer in experimental animals are available.

Evidence suggesting lack of carcinogenicity: Adequate studies involving at least two species are available which show that, within the limits of the tests used, the agent or mixture is not carcinogenic. A conclusion of evidence suggesting lack of carcinogenicity is inevitably limited to the species, tumour sites and levels of exposure studied.

(b)   Other data relevant to the evaluation of carcinogenicity and its mechanisms

Other evidence judged to be relevant to an evaluation of carcinogenicity and of sufficient importance to affect the overall evaluation is then described. This may include data on preneoplastic lesions, tumour pathology, genetic and related effects, structure–activity relationships, metabolism and pharmacokinetics, physicochemical parameters and analogous biological agents.

Data relevant to mechanisms of the carcinogenic action are also evaluated. The strength of the evidence that any carcinogenic effect observed is due to a particular mechanism is assessed, using terms such as weak, moderate or strong. Then, the Working Group assesses if that particular mechanism is likely to be operative in humans. The strongest indications that a particular mechanism operates in humans come from data on humans or biological specimens obtained from exposed humans. The data may be considered to be especially relevant if they show that the agent in question has caused changes in exposed humans that are on the causal pathway to carcinogenesis. Such data may, however, never become available, because it is at least conceivable that certain compounds may be kept from human use solely on the basis of evidence of their toxicity and/or carcinogenicity in experimental systems.

For complex exposures, including occupational and industrial exposures, the chemical composition and the potential contribution of carcinogens known to be present are considered by the Working Group in its overall evaluation of human carcinogenicity. The Working Group also determines the extent to which the materials tested in experimental systems are related to those to which humans are exposed.

(c)   Overall evaluation

Finally, the body of evidence is considered as a whole, in order to reach an overall evaluation of the carcinogenicity to humans of an agent, mixture or circumstance of exposure.

An evaluation may be made for a group of chemical compounds that have been evaluated by the Working Group. In addition, when supporting data indicate that other, related compounds for which there is no direct evidence of capacity to induce cancer in humans or in animals may also be carcinogenic, a statement describing the rationale for this conclusion is added to the evaluation narrative; an additional evaluation may be made for this broader group of compounds if the strength of the evidence warrants it.

The agent, mixture or exposure circumstance is described according to the wording of one of the following categories, and the designated group is given. The categorization of an agent, mixture or exposure circumstance is a matter of scientific judgement, reflecting the strength of the evidence derived from studies in humans and in experimental animals and from other relevant data.

Group 1 — The agent (mixture) is carcinogenic to humans.
The exposure circumstance entails exposures that are carcinogenic to humans.

This category is used when there is *sufficient evidence* of carcinogenicity in humans. Exceptionally, an agent (mixture) may be placed in this category when evidence of carcinogenicity in humans is less than sufficient but there is *sufficient evidence* of carcinogenicity in experimental animals and strong evidence in exposed humans that the agent (mixture) acts through a relevant mechanism of carcinogenicity.

Group 2

This category includes agents, mixtures and exposure circumstances for which, at one extreme, the degree of evidence of carcinogenicity in humans is almost sufficient, as well as those for which, at the other extreme, there are no human data but for which there is evidence of carcinogenicity in experimental animals. Agents, mixtures and exposure circumstances are assigned to either group 2A (probably carcinogenic to humans) or group 2B (possibly carcinogenic to humans) on the basis of epidemiological and experimental evidence of carcinogenicity and other relevant data.

Group 2A — The agent (mixture) is probably carcinogenic to humans.
The exposure circumstance entails exposures that are probably carcinogenic to humans.

This category is used when there is *limited evidence* of carcinogenicity in humans and *sufficient evidence* of carcinogenicity in experimental animals. In some cases, an agent (mixture) may be classified in this category when there is *inadequate evidence* of carcinogenicity in humans, *sufficient evidence* of carcinogenicity in experimental animals and strong evidence that the carcinogenesis is mediated by a mechanism that also operates in humans. Exceptionally, an agent, mixture or exposure circumstance may be classified in this category solely on the basis of *limited evidence* of carcinogenicity in humans.

Group 2B — The agent (mixture) is possibly carcinogenic to humans.
The exposure circumstance entails exposures that are possibly carcinogenic to humans.

This category is used for agents, mixtures and exposure circumstances for which there is *limited evidence* of carcinogenicity in humans and less than *sufficient evidence* of carcinogenicity in experimental animals. It may also be used when there is *inadequate evidence* of carcinogenicity in humans but there is *sufficient evidence* of carcinogenicity in experimental animals. In some instances, an agent, mixture or exposure circumstance for which there is *inadequate evidence* of carcinogenicity in humans but *limited evidence* of carcinogenicity in experimental animals together with supporting evidence from other relevant data may be placed in this group.

Group 3 — The agent (mixture or exposure circumstance) is not classifiable as to its carcinogenicity to humans.

This category is used most commonly for agents, mixtures and exposure circumstances for which the *evidence of carcinogenicity* is *inadequate* in humans and *inadequate* or *limited* in experimental animals.

Exceptionally, agents (mixtures) for which the *evidence of carcinogenicity* is *inadequate* in humans but *sufficient* in experimental animals may be placed in this category when there is strong evidence that the mechanism of carcinogenicity in experimental animals does not operate in humans.

Agents, mixtures and exposure circumstances that do not fall into any other group are also placed in this category.

Group 4 — The agent (mixture) is probably not carcinogenic to humans.

This category is used for agents or mixtures for which there is *evidence suggesting lack of carcinogenicity* in humans and in experimental animals. In some instances, agents or mixtures for which there is *inadequate evidence* of carcinogenicity in humans but *evidence suggesting lack of carcinogenicity* in experimental animals, consistently and strongly supported by a broad range of other relevant data, may be classified in this group.

13.   REFERENCES

Breslow, N.E. & Day, N.E. (1980) *Statistical Methods in Cancer Research*, Vol. 1, *The Analysis of Case–Control Studies* (IARC Scientific Publications No. 32), Lyon, IARC*Press*

Breslow, N.E. & Day, N.E. (1987) *Statistical Methods in Cancer Research*, Vol. 2, *The Design and Analysis of Cohort Studies* (IARC Scientific Publications No. 82), Lyon, IARC*Press*

Cohen, S.M. & Ellwein, L.B. (1990) Cell proliferation in carcinogenesis. *Science*, **249**, 1007–1011

Gart, J.J., Krewski, D., Lee, P.N., Tarone, R.E. & Wahrendorf, J. (1986) *Statistical Methods in Cancer Research*, Vol. 3, *The Design and Analysis of Long-term Animal Experiments* (IARC Scientific Publications No. 79), Lyon, IARC*Press*

Hoel, D.G., Kaplan, N.L. & Anderson, M.W. (1983) Implication of nonlinear kinetics on risk estimation in carcinogenesis. *Science*, **219**, 1032–1037

Huff, J.E., Eustis, S.L. & Haseman, J.K. (1989) Occurrence and relevance of chemically induced benign neoplasms in long-term carcinogenicity studies. *Cancer Metastasis Rev.*, **8**, 1–21

IARC (1973–1996) *Information Bulletin on the Survey of Chemicals Being Tested for Carcinogenicity/Directory of Agents Being Tested for Carcinogenicity*, Numbers 1–17, Lyon, IARC*Press*

IARC (1976–1996), Lyon, IARC*Press*

   Directory of On-going Research in Cancer Epidemiology 1976. Edited by C.S. Muir & G. Wagner

   Directory of On-going Research in Cancer Epidemiology 1977 (IARC Scientific Publications No. 17). Edited by C.S. Muir & G. Wagner

   Directory of On-going Research in Cancer Epidemiology 1978 (IARC Scientific Publications No. 26). Edited by C.S. Muir & G. Wagner

   Directory of On-going Research in Cancer Epidemiology 1979 (IARC Scientific Publications No. 28). Edited by C.S. Muir & G. Wagner

   Directory of On-going Research in Cancer Epidemiology 1980 (IARC Scientific Publications No. 35). Edited by C.S. Muir & G. Wagner

   Directory of On-going Research in Cancer Epidemiology 1981 (IARC Scientific Publications No. 38). Edited by C.S. Muir & G. Wagner

Directory of On-going Research in Cancer Epidemiology 1982 (IARC Scientific Publications No. 46). Edited by C.S. Muir & G. Wagner

Directory of On-going Research in Cancer Epidemiology 1983 (IARC Scientific Publications No. 50). Edited by C.S. Muir & G. Wagner

Directory of On-going Research in Cancer Epidemiology 1984 (IARC Scientific Publications No. 62). Edited by C.S. Muir & G. Wagner

Directory of On-going Research in Cancer Epidemiology 1985 (IARC Scientific Publications No. 69). Edited by C.S. Muir & G. Wagner

Directory of On-going Research in Cancer Epidemiology 1986 (IARC Scientific Publications No. 80). Edited by C.S. Muir & G. Wagner

Directory of On-going Research in Cancer Epidemiology 1987 (IARC Scientific Publications No. 86). Edited by D.M. Parkin & J. Wahrendorf

Directory of On-going Research in Cancer Epidemiology 1988 (IARC Scientific Publications No. 93). Edited by M. Coleman & J. Wahrendorf

Directory of On-going Research in Cancer Epidemiology 1989/90 (IARC Scientific Publications No. 101). Edited by M. Coleman & J. Wahrendorf

Directory of On-going Research in Cancer Epidemiology 1991 (IARC Scientific Publications No.110). Edited by M. Coleman & J. Wahrendorf

Directory of On-going Research in Cancer Epidemiology 1992 (IARC Scientific Publications No. 117). Edited by M. Coleman, J. Wahrendorf & E. Démaret

Directory of On-going Research in Cancer Epidemiology 1994 (IARC Scientific Publications No. 130). Edited by R. Sankaranarayanan, J. Wahrendorf & E. Démaret

Directory of On-going Research in Cancer Epidemiology 1996 (IARC Scientific Publications No. 137). Edited by R. Sankaranarayanan, J. Wahrendorf & E. Démaret

IARC (1977) *IARC Monographs Programme on the Evaluation of the Carcinogenic Risk of Chemicals to Humans*. Preamble (IARC intern. tech. Rep. No. 77/002)

IARC (1978) *Chemicals with Sufficient Evidence of Carcinogenicity in Experimental Animals —* IARC Monographs *Volumes 1–17* (IARC intern. tech. Rep. No. 78/003)

IARC (1978–1993) *Environmental Carcinogens. Methods of Analysis and Exposure Measurement*, Lyon, IARC*Press*

Vol. 1. Analysis of Volatile Nitrosamines in Food (IARC Scientific Publications No. 18). Edited by R. Preussmann, M. Castegnaro, E.A. Walker & A.E. Wasserman (1978)

Vol. 2. Methods for the Measurement of Vinyl Chloride in Poly(vinyl chloride), Air, Water and Foodstuffs (IARC Scientific Publications No. 22). Edited by D.C.M. Squirrell & W. Thain (1978)

Vol. 3. Analysis of Polycyclic Aromatic Hydrocarbons in Environmental Samples (IARC Scientific Publications No. 29). Edited by M. Castegnaro, P. Bogovski, H. Kunte & E.A. Walker (1979)

Vol. 4. Some Aromatic Amines and Azo Dyes in the General and Industrial Environment (IARC Scientific Publications No. 40). Edited by L. Fishbein, M. Castegnaro, I.K. O'Neill & H. Bartsch (1981)

Vol. 5. Some Mycotoxins (IARC Scientific Publications No. 44). Edited by L. Stoloff, M. Castegnaro, P. Scott, I.K. O'Neill & H. Bartsch (1983)

Vol. 6. N-Nitroso Compounds (IARC Scientific Publications No. 45). Edited by R. Preussmann, I.K. O'Neill, G. Eisenbrand, B. Spiegelhalder & H. Bartsch (1983)

Vol. 7. Some Volatile Halogenated Hydrocarbons (IARC Scientific Publications No. 68). Edited by L. Fishbein & I.K. O'Neill (1985)

Vol. 8. Some Metals: As, Be, Cd, Cr, Ni, Pb, Se, Zn (IARC Scientific Publications No. 71). Edited by I.K. O'Neill, P. Schuller & L. Fishbein (1986)

Vol. 9. Passive Smoking (IARC Scientific Publications No. 81). Edited by I.K. O'Neill, K.D. Brunnemann, B. Dodet & D. Hoffmann (1987)

*Vol. 10. Benzene and Alkylated Benzenes (*IARC Scientific Publications No. 85). Edited by L. Fishbein & I.K. O'Neill (1988)

Vol. 11. Polychlorinated Dioxins and Dibenzofurans (IARC Scientific Publications No. 108). Edited by C. Rappe, H.R. Buser, B. Dodet & I.K. O'Neill (1991)

Vol. 12. Indoor Air (IARC Scientific Publications No. 109). Edited by B. Seifert, H. van de Wiel, B. Dodet & I.K. O'Neill (1993)

IARC (1979) *Criteria to Select Chemicals for* IARC Monographs (IARC intern. tech. Rep. No. 79/003)

IARC (1982) *IARC Monographs on the Evaluation of the Carcinogenic Risk of Chemicals to Humans,* Supplement 4, *Chemicals, Industrial Processes and Industries Associated with Cancer in Humans* (IARC Monographs, Volumes 1 to 29), Lyon, IARCPress

IARC (1983) *Approaches to Classifying Chemical Carcinogens According to Mechanism of Action* (IARC intern. tech. Rep. No. 83/001)

IARC (1984) *Chemicals and Exposures to Complex Mixtures Recommended for Evaluation in IARC Monographs and Chemicals and Complex Mixtures Recommended for Long-term Carcinogenicity Testing* (IARC intern. tech. Rep. No. 84/002)

IARC (1987a) *IARC Monographs on the Evaluation of Carcinogenic Risks to Humans,* Supplement 6, *Genetic and Related Effects: An Updating of Selected* IARC Monographs *from Volumes 1 to 42,* Lyon, IARCPress

IARC (1987b) *IARC Monographs on the Evaluation of Carcinogenic Risks to Humans,* Supplement 7, *Overall Evaluations of Carcinogenicity: An Updating of* IARC Monographs *Volumes 1 to 42,* Lyon, IARCPress

IARC (1988) *Report of an IARC Working Group to Review the Approaches and Processes Used to Evaluate the Carcinogenicity of Mixtures and Groups of Chemicals* (IARC intern. tech. Rep. No. 88/002)

IARC (1989) *Chemicals, Groups of Chemicals, Mixtures and Exposure Circumstances to be Evaluated in Future IARC Monographs, Report of an ad hoc Working Group* (IARC intern. tech. Rep. No. 89/004)

IARC (1991a) *A Consensus Report of an IARC Monographs Working Group on the Use of Mechanisms of Carcinogenesis in Risk Identification* (IARC intern. tech. Rep. No. 91/002)

IARC (1991b) *Report of an ad-hoc* IARC Monographs *Advisory Group on Viruses and Other Biological Agents Such as Parasites* (IARC intern. tech. Rep. No. 91/001)

IARC (1993) *Chemicals, Groups of Chemicals, Complex Mixtures, Physical and Biological Agents and Exposure Circumstances to be Evaluated in Future* IARC Monographs, *Report of an ad-hoc Working Group* (IARC intern. Rep. No. 93/005)

IARC (1998a) *Report of an ad-hoc* IARC Monographs *Advisory Group on Physical Agents* (IARC Internal Report No. 98/002)

IARC (1998b) *Report of an ad-hoc* IARC Monographs *Advisory Group on Priorities for Future Evaluations* (IARC Internal Report No. 98/004)

McGregor, D.B., Rice, J.M. & Venitt, S., eds (1999) *The Use of Short and Medium-term Tests for Carcinogens and Data on Genetic Effects in Carcinogenic Hazard Evaluation* (IARC Scientific Publications No. 146), Lyon, IARC*Press*

Montesano, R., Bartsch, H., Vainio, H., Wilbourn, J. & Yamasaki, H., eds (1986) *Long-term and Short-term Assays for Carcinogenesis — A Critical Appraisal* (IARC Scientific Publications No. 83), Lyon, IARC*Press*

Peto, R., Pike, M.C., Day, N.E., Gray, R.G., Lee, P.N., Parish, S., Peto, J., Richards, S. & Wahrendorf, J. (1980) Guidelines for simple, sensitive significance tests for carcinogenic effects in long-term animal experiments. In: *IARC Monographs on the Evaluation of the Carcinogenic Risk of Chemicals to Humans*, Supplement 2, *Long-term and Short-term Screening Assays for Carcinogens: A Critical Appraisal*, Lyon, IARC*Press*, pp. 311–426

Tomatis, L., Aitio, A., Wilbourn, J. & Shuker, L. (1989) Human carcinogens so far identified. *Jpn. J. Cancer Res.*, **80**, 795–807

Vainio, H., Magee, P.N., McGregor, D.B. & McMichael, A.J., eds (1992) *Mechanisms of Carcinogenesis in Risk Identification* (IARC Scientific Publications No. 116), Lyon, IARC*Press*

Vainio, H., Wilbourn, J.D., Sasco, A.J., Partensky, C., Gaudin, N., Heseltine, E. & Eragne, I. (1995) Identification of human carcinogenic risk in IARC Monographs. *Bull. Cancer*, **82**, 339–348 (in French)

Waters, M.D., Stack, H.F., Brady, A.L., Lohman, P.H.M., Haroun, L. & Vainio, H. (1987) Appendix 1. Activity profiles for genetic and related tests. In: *IARC Monographs on the Evaluation of Carcinogenic Risks to Humans*, Suppl. 6, *Genetic and Related Effects: An Updating of Selected IARC Monographs from Volumes 1 to 42*, Lyon, IARC*Press*, pp. 687–696

Wilbourn, J., Haroun, L., Heseltine, E., Kaldor, J., Partensky, C. & Vainio, H. (1986) Response of experimental animals to human carcinogens: an analysis based upon the IARC Monographs Programme. *Carcinogenesis*, **7**, 1853–1863

GENERAL REMARKS ON THE SUBSTANCES CONSIDERED

This eighty-sixth volume of *IARC Monographs* considers cobalt (with or without tungsten carbide) in hard metals and cobalt sulfate, gallium arsenide, indium phosphide and vanadium pentoxide.

Most of the materials evaluated in this volume are poorly soluble solid materials that are deposited in particulate form in the lung, where they may be retained for long periods of time. In this respect, they should be considered as 'particulate toxicants', the toxic effects of which are regulated not only by their chemical composition but also by their particle size and surface properties.

Workers in the hard-metal industry can have significant exposures to metallic cobalt particles in general in the presence but occasionally in the absence of tungsten carbide. Cobalt and cobalt compounds were evaluated in volume 52 (1991) as being *possibly carcinogenic to humans (Group 2B)*, and the evidence of carcinogenicity in humans was *inadequate*. Since that time, new epidemiological studies of the hard-metal industry have been conducted in Sweden and in France and are evaluated here. Exposure to metallic cobalt is also prevalent in the cobalt production industry, and studies on that industry were also considered in the evaluation of cobalt. Because most data from the hard-metal industry deal with mixtures of cobalt and tungsten carbide, the Working Group also evaluated studies of tungsten miners, especially in China. Although these studies explored an association between exposure to silica and lung cancer and no data on exposure to tungsten were available, risks for lung cancer were nevertheless presented separately for tungsten miners. These were not increased compared with the reference population, but there is major potential for confounding by silica and other carcinogens in these studies.

No new studies in experimental animals were available for cobalt compounds used in the hard-metal industry. Nevertheless, this volume re-evaluates some of the experimental evidence for cobalt that was presented in the previous volume. The Working Group questioned the relevance of the routes of administration used in some of the animal carcinogenesis bioassays for the evaluation of carcinogenicity of cobalt metal and cobalt alloys. These included, for example, intramuscular injection into rats of cobalt metal powder or cobalt–chromium–molybdenum alloy, which produced sarcomas at the site of injection. The bioassays were reviewed again in this volume and the Working Group maintained the same conclusion as that reached in the previous monograph.

The semiconductor industry is a rapidly growing and changing industry that uses several compounds which have been evaluated as being potentially carcinogenic to humans. Inhalation studies by the National Toxicology Program have recently become available on two metal compounds used in this industry — gallium arsenide and indium phosphide. The available human epidemiological evidence from studies of the semiconductor industry is summarized and evaluated, although this is not extensive and is not particularly informative for the monographs on gallium arsenide and indium phosphide. Exposures to gallium arsenide and indium phosphide in the semiconductor industry may be very low, and other potential carcinogens present in this industry include trichloroethylene (*Group 2A*; IARC, 1995) and ultraviolet radiation (*Group 2A*; IARC, 1992).

In addition, there have been indications of adverse reproductive and developmental effects in workers in the semiconductor industry, although it has been suggested that these may be attributed in part to factors that are unrelated to employment in this industry. Therefore, more comprehensive epidemiological investigations of the semiconductor industry are needed.

Although they are not used in either the hard-metal or semiconductor industries, inhalation studies by the National Toxicology Program have recently become available on cobalt sulfate heptahydrate and vanadium pentoxide. Because the Working Group that convened to elaborate this volume had considerable expertise in metal carcinogenicity, it was considered advantageous to evaluate these compounds also. The evaluation of cobalt sulfate heptahydrate in this volume brings up to date the evaluations of cobalt compounds that appear in volume 52.

THE MONOGRAPHS

METALLIC COBALT PARTICLES
(WITH OR WITHOUT TUNGSTEN CARBIDE)

METALLIC COBALT PARTICLES (WITH OR WITHOUT TUNGSTEN CARBIDE)

1. Exposure Data

1.1 Chemical and physical data

1.1.1 *Nomenclature*

Metallic cobalt

Chem. Abstr. Serv. Reg. No.: 7440-48-4
Deleted CAS Reg. No.: 177256-35-8; 184637-91-0; 195161-79-6
Chem. Abstr. Name: Cobalt
IUPAC Systematic Name: Cobalt
Synonyms: C.I. 77320; Cobalt element; Cobalt-59

Cobalt sulfate heptahydrate

Chem. Abstr. Serv. Reg. No.: 10026-24-1
Chem. Abstr. Name: Sulfuric acid, cobalt(2+) salt (1:1), heptahydrate
IUPAC Systematic Name: Cobaltous sulfate heptahydrate
Synonyms: Cobalt monosulfate heptahydrate; cobalt(II) sulfate heptahydrate; cobalt(II) sulfate (1:1), heptahydrate

Tungsten carbide

Chem. Abstr. Serv. Reg. No.: 12070-12-1
Deleted CAS Reg. No.: 52555-87-0; 182169-08-0; 182169-11-5; 188300-42-7; 188300-43-8; 188300-44-9; 188300-45-0
Chem. Abstr. Name: Tungsten carbide
IUPAC Systematic Name: Tungsten carbide
Synonyms: Tungsten carbide (WC); tungsten monocarbide; tungsten monocarbide (WC)

1.1.2   *Molecular formulae and relative molecular mass*

Co   Relative atomic mass: 58.93
$CoSO_4.7H_2O$ Relative molecular mass: 281.10
WC   Relative molecular mass: 195.85

1.1.3   *Chemical and physical properties of the pure substance* (from Lide, 2003, unless otherwise specified)

Cobalt

  (a) *Description*: Hexagonal or cubic crystalline grey metal; exists in two allotropic modifications; both forms can exist at room temperature, although the hexagonal form is more stable than the cubic form (O'Neil, 2001)
  (b) *Boiling-point*: 2927 °C
  (c) *Melting-point*: 1495 °C
  (d) *Density*: 8.86 g/cm³
  (e) *Solubility*: Soluble in dilute acids; ultrafine metal cobalt powder is soluble in water at 1.1 mg/L (Kyono *et al.*, 1992)

Cobalt sulfate heptahydrate

  (a) *Description*: Pink to red monoclinic, prismatic crystals (O'Neil, 2001)
  (b) *Melting-point*: 41 °C, decomposes
  (c) *Density*: 2.03 g/cm³
  (d) *Solubility*: Soluble in water; slightly soluble in ethanol and methanol (O'Neil, 2001)

Tungsten carbide

  (a) *Description*: Grey hexagonal crystal
  (b) *Boiling-point*: 6000 °C (Reade Advanced Materials, 1997)
  (c) *Melting-point*: 2785 °C
  (d) *Density*: 15.6 g/cm³
  (e) *Solubility*: Insoluble in water; soluble in nitric and hydrofluoric acids

1.1.4   *Technical products and impurities*

Cobalt-metal and tungsten carbide powders are produced widely in high purity for use in the hard-metal industry, in the manufacture of superalloys and for other applications. [Superalloys are alloys usually based on group VIIIA elements (iron, cobalt, nickel) developed for elevated temperature use, where relatively severe mechanical stressing is encountered and where high surface stability is frequently required (Cobalt Development Institute, 2003).] Specifications of cobalt-metal powders are closely controlled to meet the requirements of particular applications. Commercial cobalt-metal powders are available in purities ranging from 99% to $\geq$ 99.999% in many grades, particle size ranges and forms; commercial tungsten carbide powders are available in purities ranging from 93% to 99.9%, also in many grades, particle size ranges and forms. Tables 1 and 2 show the specifications for selected cobalt-metal and tungsten-carbide powder products.

1.1.5 *Analysis*

(*a*) *Biological monitoring*

The presence of cobalt in samples of whole blood, plasma, serum and urine is used as a biological indicator of exposure to cobalt (Ichikawa *et al.*, 1985; Ferioli *et al.*, 1987; Angerer *et al.*, 1989). Soluble cobalt compounds are readily absorbed and excreted in the urine (see Section 4.1) and therefore urinary cobalt is considered a good indicator of exposure to these, but not to insoluble cobalt compounds (Cornelis *et al.*, 1995).

For an accurate determination of cobalt concentration in body fluids, it is necessary to use blood collection devices which do not themselves produce detectable amounts of cobalt. All containers must be washed with high purity acids. Urine samples may be acidified with high purity nitric acid and stored at 4 °C for one week, or at –20 °C for longer periods, prior to analysis (Minoia *et al.*, 1992; Cornelis *et al.*, 1995).

(*b*) *Analytical methods for workplace air and biological monitoring*

Analytical methods used until 1988 for the determination of cobalt in air particulates (for workplace air monitoring) and in biological materials (for biological monitoring) have been reviewed in a previous monograph on cobalt and its compounds (IARC, 1991). These methods are primarily flame and graphite-furnace atomic absorption spectrometry (F-AAS, and GF-AAS, respectively) and inductively coupled plasma atomic emission spectrometry (ICP-AES). Minor applications of electrochemical methods, namely adsorption voltametry, differential pulse anodic stripping voltametry and neutron activation analysis (NAA) for the determination of cobalt in serum have also been mentioned (IARC, 1991; Cornelis *et al.*, 1995).

Inductively coupled plasma mass spectrometry (ICP-MS) has become more widely available since the early 1990s, and is increasingly used for multi-elemental analysis of human blood, serum or urine, including determination of cobalt concentrations in these body fluids (Schmit *et al.*, 1991; Moens & Dams, 1995; Barany *et al.*, 1997; Sariego Muñiz *et al.*, 1999, 2001).

(*c*) *Reference values for occupationally non-exposed populations*

Normal concentrations of cobalt in the body fluids of healthy individuals are uncertain. Cornelis *et al.* (1995) give a range of 0.1–1 µg/L for cobalt concentrations in urine. Results obtained in national surveys of healthy adults yielded a mean cobalt concentration in urine of 0.57 µg/L in a population sample in Italy (Minoia *et al.*, 1990) and of 0.46 µg/L in a population sample in the United Kingdom of Great Britain and Northern

Table 1. Specifications for selected technical cobalt-metal powder products

Minimum % cobalt	Maximum %[a] contaminants permitted	Grade/particle size/crystal structure	Country of production	Reference
99.85	C, 0.02; S, 0.001; P, 0.01; Fe, 0.015	Not stated	India	Jayesh Group (2003)
>99.95	C, 0.0015–0.002; Cu, <0.0005; H, <0.0005; Fe, <0.001; Pb, <0.0002; Ni, 0.03–0.05; N, <0.0001; O, <0.005; Si, <0.0003; S, 0.0002–0.035; Zn, 0.0001–0.0002	Electrolytic and S-type/ 25 mm cut squares	Canada	Falconbridge (2002)
99.9	Bi, <0.00002; C, 0.0025; Cu, 0.0001; H, 0.0002; Fe, 0.0004; Pb, 0.0003; Ni, 0.095; N, 0.0004; O, 0.005; Se, <0.00002; S, 0.0005; Zn, 0.0008	Electrolytic rounds/button-shaped pieces circa 35 mm in diameter and circa 5 mm thick	Canada	Inco Ltd (2003)
99.999	[mg/kg] Cu, Cd, Pb, Cr, Al, Ag, Na, Sb, W, Li, Mg, Mn, Mo, Si, Ti, Cl, K, Ca and Ni, <1; Fe, <2; Zn and As, <5; S, <10; C, <20	Shiny silver-grey cathode plates/hexagonal	Belgium	Umicore Specialty Metals (2002)
99.5	Ni, 0.05; Fe, 0.11; Mn, 0.01; Cu, 0.007; Pb, <0.001; Zn, 0.003; Si and Ca, 0.04; Mg, 0.02; Na, 0.005; S, 0.01; C, 0.025; O_2, 0.30	Coarse particle/400 or 100 mesh/50% hexagonal, 50% cubic	Belgium	Umicore Specialty Metals (2002)
99.8	Ni, 0.15; Ag, 0.02; Fe, 0.003; Mg, Mn and Cu, <0.0005; Zn and Na, 0.001; Al, Ca and Si, <0.001; Pb, <0.002; S, 0.006; C, 0.07; O_2, 0.5	5M powder/3.3–4.7 μm/ 90% hexagonal, 10% cubic	Belgium	Umicore Specialty Metals (2002)

Table 1 (contd)

Minimum % cobalt	Maximum %[a] contaminants permitted	Grade/particle size/crystal structure	Country of production	Reference
99.88	Ni, 0.05; Fe, 0.005; Mg, Mn, Pb and S, < 0.001; Ca, Cu and Zn, 0.003; Si, < 0.002; Na 0.002; C, 0.015; O_2, 0.35	Extra fine powder/1.2–1.5 μm/70% hexagonal, 30% cubic	Belgium	Umicore Specialty Metals (2002)
99.8	Ni, 0.10; Ag, 0.12; Al, Fe, Na and Pb, < 0.001; Cu, Mg and Mn, < 0.0005; Zn, 0.0011; Ca, 0.0013; Si, < 0.003; S, 0.005; C, 0.22; O_2, 0.8	Half micron powder/0.55 μm/80% hexagonal, 20% cubic	Belgium	Umicore Specialty Metals (2002)
99.7	[mg/kg] C, 1000; Ni and Cl, 500; Fe and Ca, 70; Na, 60; Mg, 30; Cu and Zn, 20; Al, Mn, Pb and S, < 10; Si, < 20; O_2, 0.8%	Submicron-size powder/0.8 μm/85% hexagonal, 15% cubic	Belgium	Umicore Specialty Metals (2002)
99.8	Ni, 0.15; Ag, 0.12; Fe and Na, 0.001; Al, Cu, Mg and Mn, < 0.0005; Zn, 0.0013; Ca, 0.0015; Pb, < 0.002; Si, < 0.001; S, 0.006; C, 0.18; O_2, 0.7	Ultrafine powder/0.9 μm/90% hexagonal, 10% cubic	Belgium	Umicore Specialty Metals (2002)
> 99.8	[mg/kg] Ca, Fe and Si, < 100; Ni, < 400–1000; O_2, < 0.8%	Extrafine powder/1.05–1.45 μm	France	Eurotungstene Metal Powders (2003)
99.80	C and Ni, 0.20; Ag, 0.15; Fe, 0.02; Cu, 0.005; S, 0.01; O, 0.80	Ultrafine powder/0.9–8.0 μm	Luxembourg	Foxmet SA (2003)
99.8	[mg/kg] Ni, 600; C, 300; Fe, 100; Cu and S, 50; O, 0.50%	Extrafine powder/1.40–3.90 μm	Luxembourg	Foxmet SA (2003)

Table 1 (contd)

Minimum % cobalt	Maximum %[a] contaminants permitted	Grade/particle size/crystal structure	Country of production	Reference
99.20	[mg/kg] Ni and Fe, 1000; Ca, 750; C and S, 300; O, 0.50%	Fine powder-400 mesh/4.2–14.0 µm	Luxembourg	Foxmet SA (2003)
99.90	Ni, 0.30; C, 0.10; Fe and S, 0.01; Cu, 0.001; O, 0.60	Fine powder-5M/4.0 µm	Luxembourg	Foxmet SA (2003)
99.80	Ni, 0.05; C, 0.10; Fe, 0.003; S, 0.03; Cu, 0.002	Coarse powder-'S' grade/75–600 µm	Luxembourg	Foxmet SA (2003)
99.8	[mg/kg] C, 1000; S, 350; Ni, 200; Fe, 35; Cu and Zn, 15	Coarse powder-'DGC' grade/45–600 µm	Luxembourg	Foxmet SA (2003)
Not stated	Not stated	Coarse powder-100 & 400 mesh; battery grade briquette; extrafine powder (standard & high density); submicron (0.8 µm) powder	USA	OM Group (2003)

[a] Unless stated otherwise

Table 2. Specifications for selected technical tungsten-carbide (WC) powder products

Minimum % WC	Maximum %[a] contaminants permitted	Grade/particle size	Country of production	Reference
Not stated	Total C, 6.11–6.16; free C, 0.03; [mg/kg] Al, Cr and Na, 10; Ca and Ni, 20; Co, Cu, K, Mg and Mn, 5; Mo, 50; Si and Fe, 30	100–200 mesh 0.7–20.0 μm	Israel	Metal-Tech Ltd (2003)
93–94	Total C, 6; free C, 0.04	Mesh size, 200	India	Jayesh Group (2003)
99.70–99.90	Total C, 6.08–6.29; free C, 0.05–0.16; Fe, 0.02; Mo, 0.01	Standard grade/0.7–12 μm	Japan	Japan New Metals Co. Ltd (2003)
Not stated	Total C, 6.05–6.25; free C, 0.10; Fe, 0.05; Mo, 0.02; Cr, 1; V, 1	Fine grade/0.45–0.75 μm	Japan	Japan New Metals Co. Ltd (2003)
99.8	Total C, 6.13; free C, 0.10; Fe, 0.05; Mo, 0.02	Standard grade/0.7–7.1 μm	Japan	A.L.M.T. Corp. (2003)
99.8	Total C, 6.13; free C, 0.05; Fe, 0.02; Mo, 0.02	Coarse grade/2.5–16 μm	Japan	A.L.M.T. Corp. (2003)
Not stated	Total C, 6.15–6.20; free C, 0.15–0.25; Fe, 0.02; Mo, 0.02	Ultrafine grade/0.1–0.70 μm	Japan	A.L.M.T. Corp. (2003)
Not stated	Total C, 6.11–6.18; free C, < 0.08; [mg/kg] Al and Ca, < 10; Cr, < 40; Fe, < 200; K, Mg and Na, < 15; Mo, < 50; Ni, < 25; Si, < 40; V, 1400–2000; O_2, < 0.16–0.25%	0.6–1.1 μm (doped with 0.2% VC)	France	Eurotungstene Metal Powders (2003)
Not stated	Combined C, 6.05 min.; free C, 0.08; O_2, 0.025–0.030	2.6–5.5 μm	France	Eurotungstene Metal Powders (2003)

Table 2 (contd)

Minimum % WC	Maximum %[a] contaminants permitted	Grade/particle size	Country of production	Reference
Not stated	Total C, 3.9–4.2; free C, 0.1; Fe, 0.4	Fused powders (eutectic mixture of WC and W_2C)/ < 45–450 μm	France	Eurotungstene Metal Powders (2003)
Not stated	Not stated	DS/0.45–2.5 μm MAS/5.0–50 μm HC/2.5–14 μm DR/3–10 μm MA/4–12 μm	Germany	Starck (2003)
99.7	Total C, 6.13; free C, 0.06; [mg/kg] Fe and Mo, 250; Co, 100; Cr, 75; Ca, Ni and Si, 50; Al, 25; Na, 20; Cu, 15	Fine grade powder/0.9–6.3 μm	Luxembourg	Foxmet SA (2003)
Not stated	Total C, 3.90–4.20 ; free C, 0.10; Fe, 0.40; O, 0.10	Fused powder/0–150 μm	Luxembourg	Foxmet SA (2003)
80–88% WC & 12–20% Co	Not stated	pre-alloyed WC/Co powder/ 0–300 μm	Luxembourg	Foxmet SA (2003)
10–50% WC & 50–90% Co	Not stated	Ready-mixed powder	Luxembourg	Foxmet SA (2003)
Not stated	Total C, 6.08–6.24; free C, 0.05; Fe, 0.03; Mo and Nb, 0.15; Ta, 0.1; Ti, 0.20	Macrocrystalline powder/0–420 μm	USA	Kennametal (2003)

Table 2 (contd)

Minimum % WC	Maximum %[a] contaminants permitted	Grade/particle size	Country of production	Reference
Not stated	Not stated	Conventional carburized powder/0.8–4.8 μm Cast carbide vacuum-fused powder/44–2000 μm Chill cast carbide/37–420 μm) Sintered WC/Co hard metal/44–2000 μm	USA	Kennametal (2003)

[a] Unless stated otherwise

Ireland (White & Sabbioni, 1998). Significant differences between concentrations of cobalt in the urine of men and women (median values of [0.22] and [0.39 µg/L], respectively) were reported by Kristiansen *et al.* (1997).

Concentrations of cobalt in blood and serum are expected to be at the lower end of the 0.1–1 µg/L range (Versieck & Cornelis, 1980); a median cobalt concentration in serum of 0.29 µg/L was determined by Iyengar and Woittiez (1988). In an Italian population, Minoia *et al.* (1990) reported median concentrations of cobalt in blood and serum of 0.39 µg/L and 0.21 µg/L, respectively. Alimonti *et al.* (2000) recently reported cobalt concentrations in the range of 0.20–0.43 µg/L in the serum of newborns from an urban area of Rome, suggesting that there is no age dependence in serum cobalt concentrations.

1.2   Production and use

1.2.1   *Production*

(*a*)   *Cobalt*

World production of refined cobalt has increased steadily over the last decade, due partly to new operations and partly to a net increase in production by established producers. World demand for cobalt is strongly influenced by general economic conditions and by demand from industries that consume it in large quantities, such as superalloy melters and manufacturers of rechargeable batteries (Shedd, 2003).

World cobalt resources identified to date are estimated at about 15 million tonnes. The vast majority of these resources are in nickel-bearing laterite deposits or, to a much smaller extent, in nickel–copper sulfide deposits in Australia, Canada and the Russian Federation and in the sedimentary copper deposits of the Democratic Republic of Congo and Zambia. In addition, it is postulated that millions of tonnes of cobalt exist in manganese nodules and crusts on the ocean floor (Shedd, 2003).

Cobalt is extracted from several mineral ores, including arsenide, sulfoarsenide (cobaltite), sulfide (chalcocite, carrollite), arsenic-free cobalt–copper (heterogenite), lateritic and oxide ores. Cobalt is recovered from concentrates and occasionally directly from the ore itself by hydrometallurgical, pyrometallurgical and electrometallurgical processes. Cobalt powder can be produced by a number of methods, but those of industrial importance involve the reduction of oxides, the pyrolysis of carboxylates, and the reduction of cobalt ions in aqueous solution with hydrogen under pressure. Very pure cobalt powder is prepared by the decomposition of cobalt carbonyls. Grey cobalt(II) oxide (CoO) or black cobalt(II)/cobalt(III) oxide (Co_3O_4) is reduced to the metal powder by carbon monoxide or hydrogen. The purity of the powder obtained is 99.5% with a particle size of approximately 4 µm, although the density and particle size of the final product depend on the reduction conditions and on the particle size of the parent oxide. The thermal decomposition of cobalt carboxylates such as formate and oxalate in a controlled reducing or neutral atmosphere produces a high-purity (about 99.9%), light, malleable cobalt powder, with a particle size of approximately 1 µm which is particularly suitable for the manufacture of hard metals (see below). The particle size, form and porosity of the powder grains can be changed by altering the pyrolysis conditions (Hodge, 1993; Donaldson, 2003).

World mine and refinery production figures for cobalt from 1997 to 2001 are presented in Tables 3 and 4, respectively (Shedd, 2001). Available information indicates that cobalt is manufactured by five companies in China, four companies each in India and the United States of America (USA), three companies in Japan, and two companies each in Belgium, Brazil, Canada, the Netherlands, the Russian Federation and the United Kingdom. Argentina, France, Germany, Italy, Mexico, Norway, the Philippines, Poland and Turkey each have one manufacturing company (Chemical Information Services, 2003). Other important cobalt-manufacturing countries include Australia, the Democratic Republic of Congo, Finland, Morocco and Zambia (Shedd, 2001).

Table 3. World cobalt mine production by country (in tonnes of cobalt)[a]

Country[b]	1997	1998	1999	2000	2001
Australia	3 000	3 300	4 100	5 600	6 200
Botswana	334	335	331	308	325
Brazil	400	400	700	900	1 100
Canada	5 709	5 861	5 323	5 298	5 334
China	200	40	250	90	150
Cuba	2 358	2 665	2 537	2 943	3 411
Democratic Republic of the Congo	3 500	5 000	6 000	7 000	4 700
France (New Caledonia)	1 000	1 000	1 100	1 200	1 400
Kazakhstan	300	300	300	300	300
Morocco	714	287	863	1 305	1 300
Russian Federation	3 300	3 200	3 300	3 600	3 800
South Africa	465	435	450	580	550
Zambia	6 037	11 900	5 640	4 600	8 000
Zimbabwe	126	138	121	79	95
Total	27 400	34 900	31 000	33 800	36 700

From Shedd (2001)
[a] Figures represent recoverable cobalt content of ores, concentrates or intermediate products from copper, nickel, platinum or zinc operations.
[b] In addition to the countries listed, Bulgaria, Indonesia, the Philippines and Poland are known to produce ores that contain cobalt, but information is inadequate for reliable estimates of output levels.

(b) Metallic carbides

Carbon reacts with most elements of the periodic table to form a diverse group of compounds known as carbides, some of which have extremely important technological applications.

Table 4. World cobalt refinery production by country (in tonnes of cobalt)[a]

Country[b]	Product	1997	1998	1999	2000	2001
Australia	Metal (including metal powder), oxide, hydroxide	617	1 395	1 700	2 610	3 470
Belgium	Metal powder, oxide, hydroxide	1 200	1 200	950	1 110	1 090
Brazil	Metal	266	364	651	792	889
Canada	Metal (including metal powder), oxide	3 792	4 415	4 196	4 364	4 378
China	Metal	470	410	300	410	450
Democratic Republic of the Congo	Metal	2 808	4 490	5 180	4 320	4 071
Finland	Metal, powder, salts	5 000	5 250	6 200	7 700	8 100
France	Chloride	159	172	181	204	199
India	Metal, salts	110	120	120	206	250
Japan	Metal	264	329	247	311	350
Morocco	Metal	225	242	472	1 200	1 200
Norway	Metal	3 417	3 851	4 009	3 433	3 314
Russian Federation	Unspecified	4 100	3 500	3 600	4 400	5 000
South Africa	Metal, powder, sulfate	316	296	306	397	371
Uganda	Metal	0	0	77	420	634
Zambia	Metal	4 403	4 837	4 236	3 342	4 657
Total		27 100	30 900	32 400	35 200	38 400

From Shedd (2001)

[a] Figures represent cobalt refined from ores, concentrates or intermediate products and do not include production of downstream products from refined cobalt.

[b] In addition to the countries listed, Germany and Slovakia may produce cobalt, but available information is inadequate to make reliable estimates of production.

Metallic carbides (industrial hard carbides) comprise the carbides of metals of groups IVB–VIB. Metallic carbides combine the physical properties of ceramics with the electronic nature of metals; they are hard and strong, but at the same time good conductors of heat and electricity (Oyama & Kieffer, 1992). Tungsten carbide, titanium carbide and tantalum carbide are used as structural materials in extremely high temperatures or in corrosive atmospheres. Carbides are generally stable at high temperatures and metallic carbides are prepared by the direct reaction between carbon and metals at high temperatures. For example, fine tungsten powders blended with carbon and heated in a hydrogen atmosphere at 1400–1500 °C produce tungsten carbide (WC) particles varying in size from 0.5 to 30 μm. Each particle is composed of numerous tungsten carbide crystals. Small amounts of vanadium, chromium or tantalum are sometimes added to tungsten and carbon powders before carburization to produce very fine (< 1 μm) tungsten carbide powders (Stoll & Santhanam, 1992) (Figure 1).

Figure 1. Steps in the manufacture of hard-metal tools

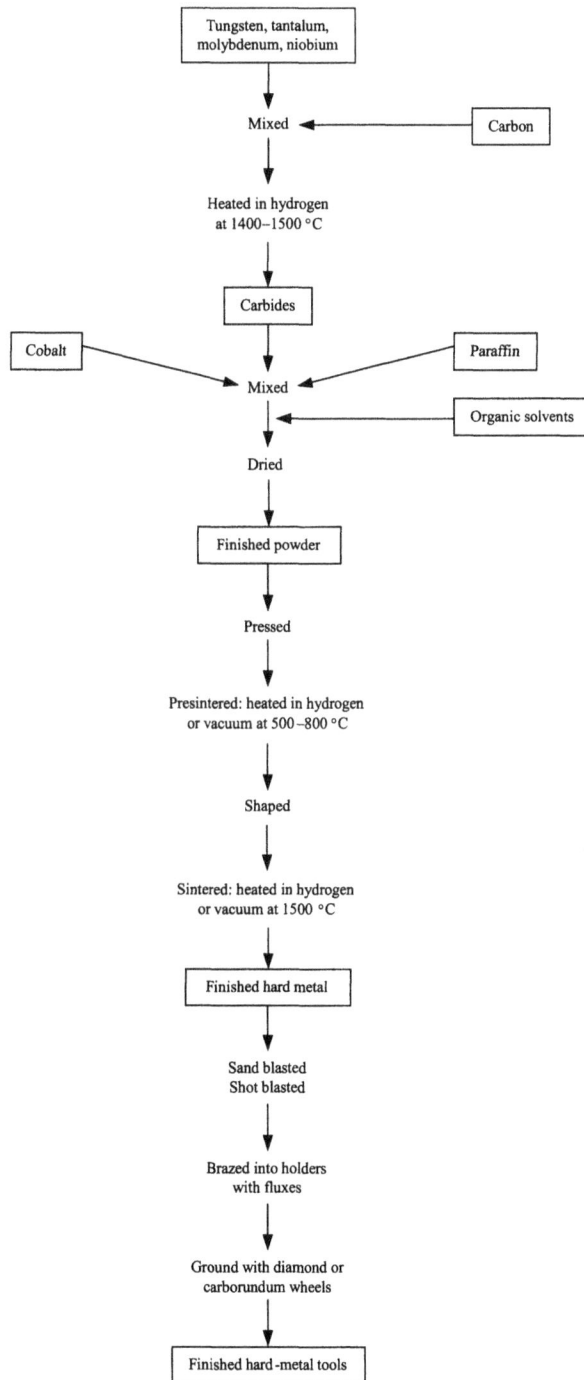

```
            Tungsten, tantalum,
            molybdenum, niobium
                    │
                    ▼
                  Mixed  ◄──────────  Carbon
                    │
                    ▼
            Heated in hydrogen
            at 1400–1500 °C
                    │
                    ▼
                 Carbides
                    │
    Cobalt ──────┐  ▼  ┌────── Paraffin
                  Mixed
                    │  ◄──────  Organic solvents
                    ▼
                  Dried
                    │
                    ▼
             Finished powder
                    │
                    ▼
                 Pressed
                    │
                    ▼
        Presintered: heated in hydrogen
        or vacuum at 500–800 °C
                    │
                    ▼
                 Shaped
                    │
                    ▼
         Sintered: heated in hydrogen
         or vacuum at 1500 °C
                    │
                    ▼
            Finished hard metal
                    │
                    ▼
                Sand blasted
                Shot blasted
                    │
                    ▼
              Brazed into holders
              with fluxes
                    │
                    ▼
          Ground with diamond or
          carborundum wheels
                    │
                    ▼
          Finished hard-metal tools
```

From Kusaka *et al.* (1986)

Available information indicates that tungsten carbide is manufactured by five companies in the USA, four companies in Japan, three companies in Germany and two companies each in Brazil and France. Argentina, Austria, Canada, India, Israel, Portugal and the Republic of Korea each have one manufacturing company (Chemical Information Services, 2003).

(c)   Hard metals

Hard metals are materials in which metallic carbides are bound together or cemented by a soft and ductile metal binder, usually cobalt or nickel. Although the term 'cemented carbide' is widely used in the USA, these materials are better known internationally as 'hard metals' (Santhanam, 1992). Hard metals are manufactured by a powder metallurgy process consisting of a sequence of carefully-controlled steps designed to obtain a final product with specific properties, microstructure and performance (Santhanam, 1992).

Figure 1 illustrates the steps involved in the preparation of hard metals and the manufacture of hard-metal tools. The carbides or carbide solid solution powders are prepared, blended, compacted, presintered and shaped, and subjected to sintering and postsintering operations. The sintered product (finished hard metal) may be either put to use directly, or ground, polished and coated (Santhanam, 1992). [Sintering is the agglomeration of metal powders at temperatures below their melting-point, as in powder metallurgy; while heat and pressure are essential, decrease in surface area is the critical factor; sintering increases strength, conductivity and density (Lewis, 2001).]

The binder metal (cobalt or nickel) is obtained as a very fine powder and is blended with carbide powders in ball mills, vibratory mills or attritors [grinding machines] using carbide balls. The mills are lined with carbide, low-carbon steel or stainless-steel sleeves. Intensive milling is necessary to break up the initial carbide crystallites and disperse the cobalt among the carbide particles to enhance wetting by cobalt during sintering. Milling is performed under an organic liquid such as alcohol, hexane, heptane or acetone; in the process, a solid lubricant such as paraffin wax or poly(ethylene glycol) is added to the powder blend to strengthen the pressed or consolidated powder mix. After milling, the organic liquid is removed by drying. In a spray-drying process, commonly used in the hard-metal industry, a hot inert gas such as nitrogen impinges on a stream of carbide particles to produce free-flowing spherical powder aggregates (Santhanam, 1992).

The milled and dried grade powders are pressed to desired shapes in hydraulic or mechanical presses. Special shapes may require a presintering operation followed by machining or grinding to the final form. Cold isostatic pressing, followed by green forming [forming the powder into the desired shape], is also common in the manufacture of wear-resistant components and metal-forming tools. Rods and wires are formed by an extrusion process (Santhanam, 1992).

For sintering, the pressed compacts are set on graphite trays and are heated initially to approximately 500 °C in an atmosphere of hydrogen or in a vacuum to remove the lubricant. Subsequently, the compacts are heated under vacuum to a final sintering temperature ranging from 1350 to 1550 °C, depending on the amount of metal binder and the microstructure desired. During final sintering, the binder melts and draws the carbide particles together, shrinking the compact by 17–25% (on a linear scale) and yielding a virtually pore-free, fully dense product (Santhanam, 1992).

In the 1970s, the hard-metal industry adapted hot isostatic pressing (HIP) technology to remove any residual internal porosity, pits or flaws from the sintered product. The HIP process involves reheating vacuum-sintered material to a temperature 25–50 °C less than the sintering temperature under a gaseous (argon) pressure of 100–150 MPa (14 500–21 750 psi). An alternative method developed in the early 1980s, the sinter-HIP process, uses low-pressure HIP, up to 7 MPa (1015 psi), combined with vacuum sintering. The pressure is applied at the sintering temperature when the metallic binder is still molten, resulting in void-free products.

After sintering, hard-metal products that require shaping to meet surface finish, tolerance or geometric requirements undergo grinding with metal-bonded diamond wheels or lapping with diamond-containing slurries (Santhanam, 1992).

Recycling of hard-metal scrap is of growing importance and several methods are available. In one method, the scrap is heated to 1700–1800 °C in a vacuum furnace to vaporize some of the cobalt and embrittle the material. After removal from the furnace, the material is crushed and screened. [Screening is the separation of an aggregate mixture into two or more portions according to particle size, by passing the mixture through one or more standard screens.] In chemical recycling, the cobalt is removed by leaching, leaving carbide particles intact. In the zinc reclaim process, commercialized in the late 1970s, the cleaned scrap is heated with molten zinc in an electric furnace at approximately 800 °C under an inert gas. The zinc reacts with the cobalt binder and the carbide pieces swell to more than twice their original volume. The zinc is distilled off under vacuum and reclaimed. The carbide pieces are pulverized and screened to produce a fine powder. The cobalt is still present in the particles and there is no change in grain size from the original sintered scrap. The coldstream reclaim method uses a high velocity airstream to accelerate hard-metal particles with sufficient energy to cause them to fracture against a target surface. This process, so called because the air cools as it expands from the nozzles, is used in combination with the zinc reclaim process (Santhanam, 1992).

(d)   Cobalt alloys

Multimetallic complexes, which include cobalt alloys, are the components in tool steels and Stellite-type alloys that are responsible for hardness, wear resistance and excellent cutting performance (Oyama & Kieffer, 1992). [Stellite is an alloy containing cobalt and chromium, and sometimes other metals.]

1.2.2   Use

Cobalt compounds have been used as blue colouring agents in ceramic and glass for thousands of years, although most of the blue colour of ancient glasses and glazes has been found to be due to copper. Cobalt has been found in Egyptian pottery dating from about 2600 BC, in Persian glass beads dating from 2250 BC, in Greek vases and in pottery from Persia and Syria from the early Christian era, in Chinese pottery from the Tang (600–900 AD) and Ming (1350–1650 AD) dynasties and in Venetian glass from the early fifteenth century. Leonardo Da Vinci was one of the first artists to use cobalt as a brilliant blue pigment in oil paints. The pigment was probably produced by fusing an ore containing cobalt oxide with potash and silica to produce a glass-like material (a smalt), which was then reduced to the powdered pigment. In the sixteenth century, a blue pigment called zaffre was produced from silver–cobalt–bismuth–nickel–arsenate ores in Saxony (IARC, 1991; Donaldson, 2003).

It was not until the twentieth century, however, that cobalt was used for industrial purposes. In 1907, a scientist in the USA, E. Haynes, patented Stellite-type alloys that were very resistant to corrosion and wear at high temperatures (Kirk, 1985). Cobalt was first added to tungsten carbide in 1923 to produce hard metals (Anon., 1989) and permanent magnetic alloys known as Alnicos (cobalt added to alloys of aluminum, nickel and iron) were first described in 1933 (Johnston, 1988; IARC, 1991).

Cobalt is an important metal with many diverse industrial and military applications. Its largest use is in superalloys, which are used primarily to make parts for aircraft gas turbine engines. Cobalt is also an important component of steel when high strength is required, as it increases the tempering resistance of steel; high-strength steels (maraging steels) are used in the aerospace, machine tool and marine equipment industry. Cobalt is also used to make magnets, corrosion- and wear-resistant alloys, high-speed steels, hard-metal and cobalt–diamond tools, cobalt discs and other cutting and grinding tools, catalysts for the petroleum and chemical industries, drying agents for paints, varnishes and inks, ground coats for porcelain enamels, pigments, battery electrodes, steel-belted radial tyres, airbags in automobiles and magnetic recording media (IARC, 1991; Shedd, 2001; Donaldson, 2003).

The major uses of cobalt worldwide in 2003 included: superalloys, 20%; other alloys, 10%; hard metals, 13%; wear-resistant materials, 6%; magnets, 7%; recording materials, 5%; ceramics/enamels/pigments, 12%; batteries, 8%; tyres, paint driers, soaps, 9%; and catalysts, 10% (Hawkins, 2004). According to data from 2002, in the USA, approximately 51% of cobalt was used in superalloys; 8% in cemented carbides (hard metals); 19% in various other metallic uses; and the remaining 22% in a variety of chemical applications (Shedd, 2003).

Cobalt-metal powder (100 mesh or chemical grade) is a common raw material for metal carboxylate production and catalyst manufacture. Fine cobalt powders (400 mesh) are used in hard metals, diamond tools, batteries, magnets, cobalt-containing powdered metal alloys and specialty chemicals. High-purity (99.8%) cobalt briquettes [small lumps or blocks of compressed granular material] are used as raw materials for the production of inorganic cobalt salts and cobalt alloys. Battery-grade cobalt briquettes are used to prepare mixed nitrate solutions for the production of sintered-type nickel hydroxide electrodes. These electrodes are used in nickel–cadmium and nickel–metal hydride batteries. Battery-grade cobalt powders, oxidized as is or after being dissolved in an acid solution, are used as raw materials to produce cobalt oxide precursors for lithium ion and polymer batteries (OM Group, 2003).

The four most important carbides for the production of hard metals are tungsten carbide (WC), titanium carbide (TiC), tantalum carbide (TaC) and niobium carbide (NbC). Traditionally, cemented carbide (hard-metal) inserts and tools for metal-cutting and metal-working have accounted for the largest percentage of carbide industry sales. However, hard-metal tool consumption in non-metal-working fields, notably in the construction and transportation industries, has grown rapidly. In contrast, the demand for primary materials has been somewhat reduced by the use of recycled hard-metal scrap (Santhanam, 1992; Stoll & Santhanam, 1992).

Cobalt sulfate is the usual source of water-soluble cobalt since it is the most economical salt and shows less tendency to deliquesce or dehydrate than the chloride or nitrate salts. It is used in storage batteries, in cobalt electroplating baths, as a drier for lithographic inks and varnishes, in ceramics, enamels and glazes to prevent discolouring and in cobalt pigments for decorating porcelain (O'Neil, 2001).

Uses of other cobalt compounds are described in detail by IARC (1991).

1.3 Occurrence and exposure

1.3.1 *Natural occurrence*

Cobalt occurs in nature in a widespread but dispersed form in many rocks and soils. The cobalt concentration in the earth's crust is about 20 mg/kg. The largest concentrations of cobalt are found in mafic (igneous rocks rich in magnesium and iron and comparatively low in silica) and ultramafic rocks; the average cobalt content in ultramafic rocks is 270 mg/kg, with a nickel:cobalt ratio of 7. Sedimentary rocks contain varying amounts of cobalt, averaging 4 mg/kg in sandstone, 6 mg/kg in carbonate rocks and 40 mg/kg in clays and shales. Concentrations of cobalt in metamorphic rock depend on the amount of the element in the original igneous or sedimentary source. Cobalt has also been found in meteorites (Donaldson *et al.*, 1986; O'Neil, 2001; Donaldson, 2003).

Cobalt salts occur in nature as a small percentage of other metal deposits, particularly copper; cobalt sulfides, oxides and arsenides are the largest mineral sources of cobalt (Schrauzer, 1989; IARC, 1991; Donaldson, 2003).

1.3.2 *Occupational exposure*

Occupational exposure to aerosols containing cobalt metal or solubilized cobalt compounds may occur during the refining of cobalt, the production of alloys, at various stages in the manufacture of hard metals, the maintenance and resharpening of hard-metal tools and blades and during the manufacture and use of diamond tools containing cobalt (see below). However, only about 15% of cobalt produced is used in cemented carbides (hard metals) and diamond tooling and there are many other potential sources of occupational exposure to cobalt (see Section 4, Table 15).

Several studies have reported occupational exposure to cobalt by measuring concentrations in ambient air in industrial sites where hard-metal and diamond grinding wheels were produced. In addition, analytical methods have been recently standardized for the determination of cobalt concentrations in urine and blood (Kristiansen *et al.*, 1997; White, 1999). It should be noted that many workers inhaling different chemical species of cobalt may also be exposed to nickel, tungsten, chromium, arsenic, molybdenum, beryllium, silica and silicates, asbestos, nitrosamines, diamond powders and iron. Exposure to other substances co-occurring with cobalt have also been reported.

(a)   Hard-metal production and use

Exposure to hard-metal dust takes place at all stages of the production of hard metals, but the highest levels of exposure to cobalt have been reported to occur during the weighing, grinding and finishing phases (Reber & Burckhardt, 1970; McDermott, 1971; National Institute for Occupational Safety and Health, 1981; Sprince *et al.*, 1984; Hartung, 1986; Kusaka *et al.*, 1986; Balmes, 1987; Meyer-Bisch *et al.*, 1989; Auchincloss *et al.*, 1992; Stebbins *et al.*, 1992). For example, in two factories in the USA producing hard metals, peak cobalt concentrations in air taken during weighing, mixing and milling exceeded 500 $\mu g/m^3$ in more than half of all samples (Sprince *et al.*, 1984), and in powder rooms with poorly-regulated control of cobalt dusts, concentrations of cobalt in air ranged between 10 $\mu g/m^3$ and 160 $\mu g/m^3$ (Auchincloss *et al.*, 1992).

Table 5 shows the cobalt concentrations in air determined for all stages in the manufacturing process in a study of exposure to hard metals among hard-metal workers in Japan (Kusaka *et al.*, 1986; Kumagai *et al.*, 1996). The concentrations of cobalt and nickel in air were shown to be distributed lognormally (Kusaka *et al.*, 1992; Kumagai *et al.*, 1997). The workers were further studied with respect to prevalence of asthma in association with exposure to cobalt (Kusaka *et al.*, 1996a,b).

Table 6 summarizes data on cobalt concentrations in workplace air and urine of workers in hard-metal production up to 1986 (presented in the previous monograph on cobalt; IARC, 1991), together with more recent studies.

In a factory producing hard metal in Italy, the mean concentration of cobalt in workplace air on Thursday afternoons was 31.7 ± 33.4 $\mu g/m^3$, thus exceeding the current ACGIH threshold limit value (TLV) for occupational exposure of 20 $\mu g/m^3$ (Scansetti *et al.*, 1998; ACGIH Worldwide®, 2003a). Among hard-metal workers in several small factories in northern Italy, cobalt concentrations in the urine of six operators on machines without aspirators were up to 13 times higher than those in the reference population (Cereda *et al.*, 1994).

A British study reported median concentrations of cobalt in urine of 19 nmol/mmol creatinine in workers in the hard-metal industry and 93 nmol/mmol creatinine in workers manufacturing and handling cobalt powders, salts and pigments in the chemical industry (White & Dyne, 1994).

Table 5. Cobalt concentrations in air in different workshops in the hard-metal industry

Workshop	No. of workers	No. of samples of work-place air	Cobalt concentration ($\mu g/m^3$)					
			AM[a]	GM[b]	Min.	Max.	GSD$_W$[c]	GSD$_B$[d]
Powder preparation								
Rotation	15	60	459	211	7	6390	NA	NA
Full-time	2	12	147	107	26	378	1.88	2.27[e]
Press								
Rubber	8	26	339	233	48	2910	2.77	1.00
Steel	23	34	47	31	6	248	2.43	NA
Shaping	67	179	97	57	4	1160	2.56	1.79
Sintering	37	82	24	13	1	145	1.99	1.99
Blasting	3	7	2	2	1	4	1.88	1.00[e]
Electron discharging	10	18	3	2	1	12	2.69	1.00
Wet grinding	191	517	45	21	1	482	2.30	2.31
Dry grinding without ventilation	1	2	1292	NA	1113	1471	NA	NA

From Kusaka *et al.* (1986); Kumagai *et al.* (1996)

NA, not applicable or not available

[a] AM, arithmetic mean

[b] GM, geometric mean

[c] GSD$_w$, geometric standard deviation within-worker variation

[d] GSD$_B$, geometric standard deviation between-worker variation

[e] Because number of workers in this job group was small, the GSD$_B$ value is not reliable.

Concentrations of different tungsten species (W, WC, WO, WO_4^{2-}), cobalt and nickel were studied in air and in urine samples from workers in different areas in a hard-metal factory in Germany. The results are summarized in Tables 7–9 (Kraus *et al.*, 2001).

In addition, the process of depositing carbide coatings, by flame or plasma guns, on to softer substrates to harden their surfaces, may also expose workers to hard metals (Rochat *et al.*, 1987; Figueroa *et al.*, 1992).

Hard metals have applications in tools for machining metals, drawing wires, rods and tubes, rolling or pressing, cutting various materials, drilling rocks, cement, brick, road surfaces and glass, and many other uses in which resistance to wear and corrosion are needed, such as high-speed dental drills, ballpoint pens and tyre studs. During the use of hard-metal tools (e.g. in drilling, cutting, sawing), the levels of exposure to cobalt or hard-metal dust are much lower than those found during their manufacture. However, the grinding of stone and wood with hard-metal tools and the maintenance and sharpening of these tools may release cobalt into the air at concentrations of several hundred micrograms per cubic metre (Mosconi *et al.*, 1994; Sala *et al.*, 1994; Sesana *et al.*, 1994).

Table 6. Biomonitoring of occupational exposure to cobalt in the hard-metal industry

Industry/activity	No. of samples	Sex	Concentration of cobalt in ambient air (mg/m³)[a]	Concentration of cobalt in blood and urine	Comments	Reference
Hard-metal production (two subgroups)	10	M	a. Mean, 0.09 b. Mean, 0.01 (personal samples)	Blood: a. Mean, 10.5 µg/L b. Mean, 0.7 µg/L Urine: a. Mean, [106] µg/L b. Mean, [~3] µg/L Sampling on Friday pm	Significant correlations: air:urine, $r = 0.79$; air:blood, $r = 0.87$; blood:urine, $r = 0.82$	Alexandersson & Lidums (1979); Alexandersson (1988)
Hard-metal production	7	–	Range, 0.180–0.193	Urine: sampling on Sunday (24 h), mean: 11.7 µg/L	Time of sampling: Monday am for basic exposure level; Friday evening for cumulative exposure level	Pellet et al. (1984)
Hard-metal grinding (seven subgroups)	153	–	Up to 61 µg/m³ (stationary samples)	Median values for all subgroups: serum, 2.1 µg/L; urine, 18 µg/L	Significant correlation: serum (x)/urine (y) $y = 2.69x + 14.68$	Hartung & Schaller (1985)
Hard-metal tool production (11 subgroups)	170 5	M F	Mean, 28–367 µg/m³ (personal samples)	Mean: blood, 3.3–18.7 µg/L; urine, 10–235 µg/L Sampling on Wednesday or Thursday at end of shift	Significant correlations (based on mean values): air (x)/urine (y): $y = 0.67x + 0.9$; air (x)/blood (y): $y = 0.004x + 0.23$; urine (x)/blood (y): $y = 0.0065x + 0.23$	Ichikawa et al. (1985)

Table 6 (contd)

Industry/activity	No. of samples	Sex	Concentration of cobalt in ambient air (mg/m^3)[a]	Concentration of cobalt in blood and urine	Comments	Reference
Hard-metal production (six subgroups)	27	–	Breathable dust: range, 0.3–15 with 4–17% cobalt	Mean: serum, 2.0–18.3 µg/L; urine, 6.4–64.3 µg/g creatinine	Significant correlation: serum:urine, $r = 0.93$	Posma & Dijstelberger (1985)
Hard-metal production	26	M	Range, approx. 0.002–0.1; median, approx. 0.01 (personal samples)	Urine: (a) Monday at end of shift, up to 36 µg/L; (b) Friday at end of shift, up to 63 µg/L	Significant correlations: air (x)/urine (y): (a) $y = 0.29x + 0.83$; (b) $y = 0.70x + 0.80$	Scansetti et al. (1985)
Machines with aspirators	6–8	–	Mean ± SD: SS: 3.47 ± 2.15 PS: 4.43 ± 2.70	Urine: GM ± GSD[b], 2.66 ± 1.69 µg/L	SS: stationary sample PS: personal sample	Cereda et al. (1994)
Machines without aspirators	6–16	–	Mean ± SD: SS: 6.68 ± 2.27 PS: 47.75 ± 3.53	Urine: GM ± GSD[b], 28.50 ± 3.97 µg/L	SS: stationary sample PS: personal sample	Cereda et al. (1994)
Hard-metal workers	6	M + F	Mean ± SD (range): Mon: 21.16 ± 17.18 (11–56) Thu: 31.66 ± 33.37 (7–92)	Urine: mean ± SD (range), 13.23 ± 9.92 (2.58–29.8) 30.87 ± 21.94 (8.17–62.6)	Mon: Monday morning Thu: Thursday afternoon	Scansetti et al. (1998)

Updated from Angerer & Heinrich (1988); IARC (1991)

–, not stated

[a] Unless stated otherwise

[b] GM, geometric mean; GSD, geometric standard deviation

Table 7. Concentration of cobalt, nickel and tungsten in air in different workshops in the hard-metal industry

Workshop	Sampling method[a]	No. of samples	Concentration in air ($\mu g/m^3$)		
			Cobalt	Tungsten	Nickel
Forming	P	5	0.61–2.82	7.8–97.4	0.23–0.76
	S	1	1.32	6.2	0.30
Pressing	P	3	0.87–116.0	5.3–211.0	0.32–3.0
Powder processing	P	4	7.9–64.3	177.0–254.0	0.76–1.65
Production of tungsten carbide	P	1	0.39	19.1	0.40
Sintering	P	1	343.0	12.1	29.6
	S	1	1.3	5.9	0.07
Grinding (wet)	P	1	0.20	3.3	0.13
Grinding (dry)	P	1	0.48	81.3	0.31
Heavy alloy production	P	2	0.85–1.84	125.0–417.0	0.48–2.17
	S	3	0.63–8.50	50.0–163.0	0.72–1.70

From Kraus et al. (2001)
[a] P, personal sampling; S, stationary sampling

Coolants are used in the hard-metal industry during the process of grinding of hard-metal tools after sintering and in their maintenance and resharpening. During such operations, the continuous recycling of coolants has been shown to result in increased concentrations of dissolved cobalt in the metal-working liquid and, hence, a greater potential for exposure to (ionic) cobalt in aerosols released from these fluids (Einarsson et al., 1979; Sjögren et al., 1980; Hahtola et al., 2000; Tan et al., 2000). It has been shown that approximately 60% of cobalt trapped in the coolant was in the dissolved form, the remainder being in the form of suspended carbide particles (Stebbins et al., 1992; Linnainmaa et al., 1996). Mists of the coolants in the wet process of grinding hard-metal tools were found to disturb local ventilation systems (Lichtenstein et al., 1975) and, as a result, cobalt concentrations in the air were higher than those from the dry grinding process (Imbrogno & Alborghetti, 1994). Used coolants may contain nitrosamines (Hartung & Spiegelhalder, 1982).

(b)   Cobalt-containing diamond tooling

Diamond tools are used increasingly to cut stone, marble, glass, wood and other materials and to grind or polish various materials, including diamonds. Although these tools are not composed of hard metal, as they do not contain tungsten carbide, they are often considered in the same category. They are also produced by powder metallurgy, whereby microdiamonds are impregnated in a matrix of compacted, extrafine cobalt powder. Consequently, the proportion of cobalt in bonded diamond tools is higher (up to 90%) than in hard metal.

Table 8. Concentration of cobalt, nickel and tungsten in urine of workers in different workshops in the hard-metal industry

Workshop	No. of workers	Metal[a]	Concentration in urine		
			Mean (95% CI) μg/g creatinine	Median μg/g creatinine	Range μg/g creatinine
Forming	23	Co	13.5 (3.7–23.3)	4.2	0.75–106.4
		W	10.7 (6.7–14.6)	9.5	0.33–33.1
		Ni	0.40 (0.19–0.62)	0.3	< DL[b]–2.2
Pressing	30	Co	5.5 (2.9–8.1)	2.8	0.36–35.9
		W	8.6 (4.1–13.1)	6.5	1.5–71.0
		Ni	0.42 (0.28–0.56)	0.4	< DL–1.6
Heavy alloy production	3	Co	1.6 (0.15–3.0)	1.4	1.1–2.2
		W	24.9 (−34.9–84.8)	21.6	2.6–50.5
		Ni	2.9 (−4.8–10.6)	2.2	0.21–6.3
Powder processing	14	Co	28.5 (−5.6–62.7)	11.2	0.75–227.8
		W	12.2 (8.0–16.5)	11.6	2.6–25.1
		Ni	0.53 (0.04–1.0)	0.1	< DL–3.1
Production of tungsten carbide	4	Co	2.1 (−1.9–6.0)	1.1	0.31–5.7
		W	42.1 (4.3–79.9)	48.9	10.0–60.6
		Ni	0.91 (0.13–1.7)	0.8	0.51–1.5
Sintering	6	Co	4.1 (0.12–6.0)	2.6	0.31–9.6
		W	12.5 (−5.7–30.7)	5.5	2.1–46.8
		Ni	0.47 (0.11–0.84)	0.4	< DL–1.0
Grinding	5	Co	2.2 (−0.57–5.0)	1.4	0.19–6.0
		W	94.4 (11.2–177.5)	70.9	10.6–168.6
		Ni	0.25 (0.02–0.48)	0.2	< DL–0.5
Maintenance	2	Co	3.0 (−18.9–24.9)	3.0	1.3–4.7
		W	3.4 (−21.1–27.8)	3.4	1.5–5.3
		Ni	0.63 (−3.5–4.7)	0.6	0.31–1.0

From Kraus *et al.* (2001)
[a] Co, cobalt; W, tungsten; Ni, nickel
[b] DL, detection limit

Exposures to cobalt have been described during the manufacture and use of cobalt–diamond tools. Diamond polishers have been reported to inhale metallic cobalt, iron and silica from so-called cobalt discs during the polishing of diamond jewels (Demedts *et al.*, 1984; Gheysens *et al.*, 1985; Van Cutsem *et al.*, 1987; Van den Eeckhout *et al.*, 1988; Nemery *et al.*, 1990; Van den Oever *et al.*, 1990; Nemery *et al.*, 1992).

Table 9. Monitoring of workplace air and workers' urine for different tungsten species in the hard-metal industry

Workshop	No. of samples[a]	Tungsten species[b]	Air concentration ($\mu g/m^3$) mean (range)	Urine concentration ($\mu g/g$ creatinine) mean (range)
Powder processing	4	W	203.5 (177.0–254.0)	13.8 (2.6–21.1)
Forming, pressing, sintering	8	WC	53.5 (5.3–211.0)	9.5 (2.2–33.1)
Production of tungsten carbide	1	WC, WO, W	19.1	59.6
Grinding (wet)	1	WO_4^{3-}	3.3	70.9
Grinding (dry)	1	WO, WC	81.3	10.6

From Kraus et al. (2001)

[a] Same number of samples for air and for urine

[b] W, tungsten metal; WC, tungsten carbide; WO, tungsten oxide; WO_4^{3-}, tungstenate

Concentrations of cobalt in the workplace air in one study were below 50 $\mu g/m^3$ (range, 0.1–45 $\mu g/m^3$) (Van den Oever et al., 1990). In an Italian factory using diamond wheels to cut wood and stone, mean cobalt concentrations in air were found to be 690 $\mu g/m^3$ and dropped to 115 $\mu g/m^3$ after proper ventilation systems were installed (Ferdenzi et al., 1994). Elevated concentrations of cobalt were also reported in the urine of these workers (Van den Oever et al., 1990; Suardi et al., 1994).

(c)   Alloys containing cobalt

Production and use of cobalt alloys gives rise to occupational exposure to cobalt during the welding, grinding and sharpening processes; the welding process with Stellite alloy (cobalt–chromium) was found to generate average concentrations of cobalt in air of 160 $\mu g/m^3$ (Ferri et al., 1994). A factory producing Stellite tools was reported to have concentrations of cobalt in the air of several hundred micrograms per cubic metre (Simcox et al., 2000), whereas concentrations averaging 9 $\mu g/m^3$ were noted in another Stellite-producing factory (Kennedy et al., 1995).

(d)   Cobalt pigments

Porcelain plate painters in Denmark have been exposed for many decades to cobalt (insoluble cobalt–aluminate spinel or soluble cobalt–zinc silicate) at concentrations which exceeded the hygiene standard by 1.3–172-fold (Tüchsen et al., 1996). During the period 1982–92, the Danish surveillance programme showed a reduction in exposure to cobalt both in terms of concentrations in air and urine; the concentration of cobalt in air decreased from 1356 nmol/m^3 [80 $\mu g/m^3$] to 454 nmol/m^3 [26 $\mu g/m^3$], and that in urine of workers from 100-fold to 10-fold above the median concentration of unexposed control subjects (Christensen & Poulsen, 1994; Christensen, 1995; Poulsen et al., 1995).

A group of workers producing cloisonné [enamel ware] in Japan and exposed to lead, chromium, cadmium, manganese, antimony, copper and cobalt compounds showed peak cobalt concentrations in blood that were twofold higher compared with the referent group, although cobalt concentrations in urine were similar (Arai *et al.*, 1994).

(e)   *Production of cobalt metal and cobalt salts*

In a factory in Belgium engaged in hydrometallurgical purification, workers were exposed to cobalt metal, cobalt oxide and cobalt salts without being exposed to tungsten, titanium, iron or silica, or their carbides, or to diamond. The mean concentration of cobalt in the workplace air was 127.5 $\mu g/m^3$ (median, 84.5 $\mu g/m^3$; range, 2–7700 $\mu g/m^3$). Cobalt concentrations in urine samples from workers taken after the workshift on Fridays averaged 69.8 $\mu g/g$ creatinine (median, 72.4 $\mu g/g$; range, 1.6–2038 $\mu g/g$ creatinine) (Swennen *et al.*, 1993). Cobalt concentrations in urine at the end of the workshift correlated well with workers' exposure on an individual basis to cobalt metal and cobalt salts, but not with exposure to cobalt oxide. Cobalt concentrations of 20 and 50 $\mu g/m^3$ in air would be expected to lead to cobalt concentrations in urine of 18.2 and 32.4 $\mu g/g$ creatinine, respectively (Lison *et al.*, 1994).

Recycling of batteries for the purpose of recovering cobalt, nickel, chromium and cadmium was found to result in cobalt concentrations in workplace air of up to 10 $\mu g/m^3$ (Hengstler *et al.*, 2003).

Workers in a factory in the Russian Federation producing cobalt acetate, chloride, nitrate and sulfates were reported to be exposed to cobalt in dust at concentrations of 0.05–50 mg/m^3 (Talakin *et al.*, 1991). In a nickel refinery also in the Russian Federation, exposures to airborne cobalt of up to 4 mg/m^3 were reported; nickel and cobalt concentrations were strongly correlated, although inhaled concentrations of nickel were far greater than those of cobalt (Thomassen *et al.*, 1999).

In a cobalt plant in Kokkola, Finland, workers were potentially exposed to metallic cobalt and cobalt sulfates, carbonates, oxides and hydroxides (Linna *et al.*, 2003). The highest concentration of cobalt in urine was recorded in a worker in the reduction department (16 000 nmol/L [943 $\mu g/L$]). Among workers in the solution, purification and chemical departments, cobalt concentrations in urine ranging from 300 to 2000 nmol/L [18 to 118 $\mu g/L$] were reported, while mean concentrations of cobalt in the air of all work areas were below 100 $\mu g/m^3$.

In a plant in South Africa converting cobalt metal to cobalt oxide, the highest concentrations of cobalt in ambient air and in urine samples of workers were 9.9 mg/m^3 and 712 $\mu g/g$ creatinine, respectively (Coombs, 1996).

High concentrations of cobalt, as well as antimony, arsenic, cadmium, chromium, lanthanum, lead and selenium, were reported in the lungs of a group of smelter workers in Sweden (Gerhardsson & Nordberg, 1993). Workers from a smelter, a petroleum refinery and a chemical plant in the USA were found to have significantly lower concentrations of cobalt in the seminal plasma, while concentrations of zinc, copper and nickel were high compared with a referent group of hospital workers (Dawson *et al.*, 2000).

(f)   Other exposures

In the United Kingdom, workers in metal thermal spraying were found to inhale cobalt, chromium and nickel. Monitoring of the workplace air and the urine of workers showed concentrations of cobalt in air of 20–30 µg/m^3 and in urine of 10–20 µmol/mol creatinine, a range 10- to 20-fold higher than in unexposed controls (Chadwick *et al.*, 1997).

Non-occupational exposure to cobalt arises from surgical implants and dental prostheses, and from contact with metallic objects, e.g. jewellery. A slight increase in mean cobalt concentrations was reported in the urine of patients with cobalt-alloy knee and hip prostheses (Sunderman *et al.*, 1989).

1.3.3   Environmental exposure

(a)   Air

Cobalt is released into the air from volcanoes and burning fuels (coal, oil). Bertine and Goldberg (1971) estimated a concentration of cobalt of 5 mg/kg in coal and 0.2 mg/kg in oil. The active volcano Mt. Erebus in Antarctica releases considerable amounts of trace elements into the environment, including cobalt (Kyle & Meeker, 1990; Hamilton, 1994). In Mumbai, India, Sadasivan and Negi (1990) found mean concentrations of cobalt in atmospheric aerosols of 1.1 ± 1.5 ng/m^3 (range, 0.3–2.3 ng/m^3), originating from iron debris in the soil. Between 1962 and 1974, average cobalt concentrations in the air in the United Kingdom declined significantly in all but one of seven sampling sites (Hamilton, 1994). Atmospheric concentrations of cobalt in rural areas of developed countries are usually below 1 ng/m^3 (Hamilton, 1994).

(b)   Water and sediments

Cobalt concentrations in sea water range from 0.01–4 µg/L and in fresh and ground waters from 0.1–10 µg/L (Nilsson *et al.*, 1985). Of 720 river water samples examined in the USA, 37% contained traces of cobalt, in the range of 1–5 µg/L, 5 µg/L being the limit of solubility. Because cobalt is present only in low concentrations, no maximal level has been set for drinking-water (Calabrese *et al.*, 1985).

Cobalt concentrations in sediments may vary from < 6 ppm (low) to > 125 ppm (very high) (Hamilton, 1994).

(c)   Soils and plants

Cobalt is omnipresent in soil, but is far from being distributed evenly. Apparently there exists a correlation between the content of cobalt in soil and in the parent rock; as a consequence, soils that are geochemically rich or poor in cobalt can be recognized. Cobalt concentrations in most soils range from 0.1–50 ppm and the amount of cobalt taken up by plants from 0.1 to 2 ppm (Nilsson *et al.*, 1985; Hamilton, 1994). However, industrial pollution may lead to much higher concentrations; close to a hard-metal (tool grinding) factory in the USA, soil was contaminated with cobalt at concentrations up to 12 700 mg/kg (Abraham & Hunt, 1995).

Lack of cobalt in soils results in vitamin B_{12} deficiency in ruminants (Domingo, 1989; Hamilton, 1994).

(d)   *Foods and beverages*

Individual intake of cobalt from food is somewhat variable, but typically in the range 10–100 µg/day. Higher intake may result from taking some vitamin preparations (IARC, 1991).

1.4   Regulations and guidelines

Regulations and guidelines for occupational exposure to cobalt in some countries are presented in Table 10. ACGIH Worldwide® (2003b) recommends a semi-quantitative biological exposure index (BEI) of 15 µg/L in urine and 1 µg/L in blood, and recommends monitoring cobalt in urine or blood of individuals at the end of their last shift of the working week as an indicator of recent exposure.

2.   Studies of Cancer in Humans

2.1   Hard-metal industry

Four mortality studies have been carried out in two cohorts of workers from the hard metal industry in Sweden and France. The key findings are summarized in Table 11.

Hogstedt and Alexandersson (1990) reported on 3163 male workers, each with at least 1 year of occupational exposure to hard-metal dust at one of three hard-metal manufacturing plants in Sweden in 1940–82 and who were followed during the period 1951–82. There were four categories of exposure (with estimated concentrations of cobalt in ambient air prior to 1970 given in parentheses for each category): occasionally present in rooms where hard metal was handled (< 2 µg/m³ cobalt); continuously present in rooms where hard metal was handled, but personal work not involving hard metal (1–5 µg/m³ cobalt); manufacturing hard-metal objects (10–30 µg/m³ cobalt); and exposed to cobalt in powder form when manufacturing hard-metal objects (60–11 000 µg/m³ cobalt). The workers were also exposed to a number of other substances used in the production of hard metal, such as tungsten carbide. There were 292 deaths among persons under 80 years of age during the study period (standardized mortality ratio [SMR], 0.96; 95% confidence interval [CI], 0.85–1.07) and 73 cancer deaths (SMR, 1.05; 95% CI, 0.82–1.32). Seventeen deaths from lung cancer were observed (SMR, 1.34; 95% CI, 0.77–2.13). Comparing the high versus low categories of exposure intensity, SMRs were similar. With regard to latency (time since first exposure), the excess was higher in the subcohort with more than 20 years since first exposure. Among workers with more than 10 years of employment and more than 20 years

Table 10. Occupational exposure limit values and guidelines for cobalt

Country or region	Concentration (mg/m³)[a]	Interpretation[b]	Carcinogen category[c]
Australia	0.05	TWA	Sen
Belgium	0.02	TWA	
Canada			
Alberta	0.05	TWA	
	0.1	STEL	
Ontario	0.02	TWA	
Quebec	0.02	TWA	A3
China	0.05	TWA	
	0.1	STEL	
Finland	0.05	TWA	
Germany	0.5[d]	TWA (TRK)	2; Sah
Ireland	0.1	TWA	
Japan	0.05	TWA	2B; Aw1S1
	0.2	STEL	
Malaysia	0.02	TWA	
Mexico	0.1	TWA	A3
Netherlands	0.02	TWA	
New Zealand	0.05	TWA	A3
Norway	0.02	TWA	Sen
Poland	0.05	TWA	
	0.2	STEL	
South Africa	0.1	TWA	
Spain	0.02	TWA	
Sweden	0.05	TWA	Sen
Switzerland	0.1	TWA	Sen; K
United Kingdom	0.1	TWA (MEL)	
USA[e]			
ACGIH	0.02	TWA (TLV)	A3
NIOSH	0.05	TWA (REL)	
OSHA	0.1	TWA (PEL)	

From Deutsche Forschungsgemeinschaft (2002); Health and Safety Executive (2002); ACGIH Worldwide® (2003a,b,c); Suva (2003)

[a] Most countries specify that the exposure limit applies to cobalt 'as Co'.

[b] TWA, 8-h time-weighted average; STEL, 10–15-min short-term exposure limit; TRK, technical correct concentration; MEL, maximum exposure level; TLV, threshold limit value; REL, recommended exposure level; PEL, permissible exposure level

[c] Sen, sensitizer; A3, confirmed animal carcinogen with unknown relevance to humans; 2, considered to be carcinogenic to humans; Sah, danger of sensitization of the airways and the skin; 2B, possibly carcinogenic to humans: substance with less evidence; Aw1S1, airway sensitizer; K, carcinogenic

[d] Cobalt metal used in the production of cobalt powder and catalysts, hard metal (tungsten carbide) and magnet production (processing of powder, machine pressing and mechanical processing of unsintered articles); all other uses have a TRK of 0.1 mg/m³.

[e] ACGIH, American Conference of Governmental Industrial Hygienists; NIOSH, National Institute for Occupational Safety and Health; OSHA, Occupational Health and Safety Administration

Table 11. Cohort studies of lung cancer in workers in the hard-metal and cobalt industry

Reference, plants	Cohort characteristics	No. of deaths	Exposure categories	Observed/expected or cases/controls	Relative risk (95% CI)	Comments
Hard-metal industry						
Hogstedt & Alexandersson (1990) 3 factories in Sweden	3163 male workers; follow-up, 1951–82	17 deaths		Obs/Exp	**SMR**	No information on smoking
			Whole cohort	17	1.34 [0.77–2.13]	
			Low exposure	11/8.4	1.31 [0.65–2.34]	
			High exposure	6/4.3	1.39 [0.51–3.04]	
			≥ 10 years of exposure and > 20 years since first exposure	7/2.5	2.78 [1.11–5.72]	
			High exposure			
			< 20 years latency	2/2.6	0.77 [0.09–2.78]	
			≥ 20 years latency	4/1.7	2.35 [0.64–6.02]	
Lasfargues et al. (1994) 1 factory in France	709 male workers employed > 1 year; follow-up, 1956–89; vital status, 89.4%; cause of death, 90.7%	10 deaths	Whole cohort	10/4.69	2.13 [1.02–3.93]	National reference. Proportion of smokers comparable with a sample of the French male population
			Duration of employment (years)			
			1–9	7/2.07	3.39 [1.36–6.98]	
			10–19	1/0.81	1.23 [0.03–6.84]	
			≥ 20	1/0.40	2.52 [0.06–14.02]	
			Time since first employment (years)			
			1–9	1/0.54	1.86 [0.05–10.39]	
			10–19	5/1.37	3.65 [1.19–8.53]	
			≥ 20	3/1.38	2.17 [0.45–6.34]	
			Degree of exposure			
			Non-exposed	1/0.66	1.52 [0.04–8.48]	
			Low	0/0.71	0.00 [0.00–5.18]	
			Medium	3/2.08	1.44 [0.30–4.21]	
			High	6/1.19	5.03 [1.85–10.95]	

Table 11 (contd)

Reference, plants	Cohort characteristics	No. of deaths	Exposure categories	Observed/ expected or cases/ controls	Relative risk (95% CI)	Comments
Moulin et al. (1998) 10 factories in France	7459 workers (5777 men, 1682 women); follow-up, 1968–91; vital status, 90.8%; cause of death, 96.8%	63 deaths	Whole cohort	Obs/Exp 63/48.59	1.30 [1.00–1.66]	Information on smoking for 80% of participants but no adjustment for smoking. Includes the factory studied by Lasfargues et al. (1994)
	Nested case–control study; 61 cases (59 men, 2 women) and 180 controls (174 men, 6 women) followed-up at the time the case died and employed > 3 months, matched by gender and age		**Cobalt with tungsten carbide** Levels 2–9/levels 0–1	**Cases/ controls** 35/81	**Odds ratio** 1.9 (1.03–3.6)	
			Levels 0–1	26/99	1.0	
			2–3	8/12	3.4 (1.2–9.6)	
			4–5	19/55	1.5 (0.8–3.1)	
			6–9	8/14	2.8 (0.96–8.1)	
			p for trend		0.08	
			Duration of exposure (levels ≥ 2)			
			Non-exposed	26/99	1.0	
			≤ 10 years	19/52	1.6 (0.8–3.3)	
			10–20 years	12/20	2.8 (1.1–6.8)	
			> 20 years	4/9	2.0 (0.5–8.5)	
			p for trend		0.03	
			Unweighted cumulative dose[a]			
			< 32	6/46	1.0	
			32–142	16/43	2.6 (0.9–7.5)	
			143–299	16/45	2.6 (1.5–11.5)	
			> 299	23/46	4.1 (1.5–11.5)	
			p for trend		0.01	

Table 11 (contd)

Reference, plants	Cohort characteristics	No. of deaths	Exposure categories	Observed/ expected or cases/ controls	Relative risk (95% CI)	Comments
Moulin et al. (1998) (contd)			*Frequency-weighted cumulative dose[a]*	Cases/ controls		
			< 4	8/45	1.0	
			4–27	20/45	2.3 (0.9–6.1)	
			27–164	14/45	1.9 (0.7–5.2)	
			> 164	19/45	2.7 (1.0–7.3)	
			p for trend		0.08	
			Other exposure to cobalt (duration of exposure to levels ≥ 2)	15/30	2.2 (0.99–4.9)	Cobalt alone or simultaneously with agents other than tungsten carbide
Wild et al. (2000) 1 factory in France	2860 workers (2216 men, 644 women); follow-up, 1968–92; cause of death, 96%	46 deaths		Obs/Exp	**SMR** Men	Not adjusted for smoking
			Whole cohort	46/27.11	1.70 (1.24–2.26)	
			Hard-metal dust intensity score ≥ 2	26/12.89	2.02 (1.32–2.96)	
			Before sintering	9/3.72	2.42 (1.10–4.59)	
			After sintering	5/3.91	1.28 (0.41–2.98)	
			Per 10 years of exposure to unsintered hard-metal dust		1.43 (1.03–1.98)	Poisson regression adjusted for smoking and asbestos, PAH, silica, nickel and chromium compounds
			Sintered hard metal dust (yes/no)		0.75 (0.37–1.53)	

Table 11 (contd)

Reference, plants	Cohort characteristics	No. of deaths or cases	Exposure categories	Observed/ expected or cases/ controls	Relative risk (95% CI)	Comments
Cobalt production industry						
Moulin et al. (1993) 1 electro-chemical plant in France	1148 male workers employed 1950–80; follow-up until 1988; vital status, 99%	8 deaths	Exclusively employed in cobalt production Ever employed in cobalt production	3/2.58 4/3.38	1.16 (0.24–3.40) 1.18 (0.32–3.03)	Not adjusted for smoking
Other cobalt compounds						
Tüchsen et al. (1996) 2 porcelain plants in Denmark	1394 female workers (874 exposed; 520 not exposed) employed in the plate underglazing departments 1943–92	15 cases (8 exposed; 7 not exposed)	Exposed to cobalt Not exposed to cobalt	8/3.41 7/3.51	**SIR** 2.35 [1.01–4.62] 1.99 [0.80–4.11]	No information on smoking

PAH, polycyclic aromatic hydrocarbon
[a] Cumulative doses expressed in months × levels

since first exposure, a significant excess of mortality from lung cancer was found (seven cases observed; SMR, 2.78; 95% CI, 1.11–5.72). In addition, there were four deaths from pulmonary fibrosis in this cohort (1.4% of all deaths, which the authors noted to be higher than the national proportion of 0.2%). A survey carried out at the end of the 1970s among hard-metal workers in Sweden showed that their smoking habits were not different from those of the male Swedish population in general (Alexandersson, 1979). [The Working Group noted the small number of exposed lung cancer cases, the lack of adjustment for other carcinogenic exposures and the absence of a positive relationship between intensity of exposure and lung cancer risk.]

A cohort mortality study was carried out among workers at a plant producing hard metals in France (Lasfargues et al., 1994). Seven hundred and nine male workers with at least 1 year of employment were included in the cohort and were followed from 1956 to 1989. Job histories were obtained from company records; however, before 1970 these histories were often missing. Using concentrations of cobalt measured in dust and in urine of workers in 1983, and taking into account improvements in working conditions over time, four categories of exposure were defined: not exposed directly to hard-metal dust; low exposure (cobalt in dust, < 10 $\mu g/m^3$; cobalt in urine, 0.01–0.02 $\mu mol/L$); medium exposure (cobalt in dust, 15–40 $\mu g/m^3$; cobalt in urine, 0.01–0.10 $\mu mol/L$); high exposure (atmospheric mean concentrations of cobalt, > 50 $\mu g/m^3$; cobalt in urine, 0.02–0.28 $\mu mol/L$). Workers who had been employed in jobs with different degrees of exposure were categorized according to their highest exposure and possible previous exposure at other plants was also considered. Of the 709 cohort members, 634 (89.4%) were alive and 295 were still employed at the end of follow-up. Smoking was ascertained for 81% of the workers and 69% of the deceased. The overall mortality did not differ from that expected (75 deaths; SMR, 1.05; 95% CI, 0.82–1.31) whereas mortality due to lung cancer was in excess (10 deaths; SMR, 2.13; 95% CI, 1.02–3.93). This excess was highest among workers employed in the areas with the highest exposures to cobalt (six deaths; SMR, 5.03; 95% CI, 1.85–10.95).

Following the report by Lasfargues et al. (1994) described above, an industry-wide mortality study on the association between lung cancer and occupational exposure to cobalt and tungsten carbide was carried out in the hard-metal industry in France (Moulin et al., 1998). The cohort comprised 7459 workers (5777 men, 1682 women) from 10 factories, including the one previously studied by Lasfargues et al. (1994), from the time each factory opened (between 1945 and 1965) until 31 December 1991. The minimum time of employment was 3 months in nine factories and 1 year in the factory previously studied (Lasfargues et al., 1994). The mortality follow-up period was 1968–91. A total of 1131 workers were considered to be lost to follow-up; of these, 875 were born outside France. The causes of the 684 registered deaths were ascertained from death certificates (633 subjects) and from medical records (29 subjects), but were unknown for 22 subjects (3.2%). The SMR for all causes of mortality was 0.93 (684 deaths; 95% CI, 0.87–1.01), and mortality for lung cancer was increased (63 deaths; SMR, 1.30; 95% CI, 1.00–1.66) when compared with national death rates. [The loss to follow-up will underestimate the SMRs, although analyses from the nested case–control study will probably be less affected by this bias.]

Sixty-one cases (i.e. deaths from lung cancer) and 180 controls were included in a nested case–control study (Moulin *et al.*, 1998). Three controls per case were sampled among cohort participants: (a) under follow-up on the date that the case died, having completed 3 months of employment and known to be alive on that date; and (b) of the same gender and with the same date of birth ± 6 months. Job histories were drawn from administrative records and information on job histories was complemented by interviews with colleagues who were not aware of the case or control status of the subjects. Occupational exposure of cases and controls was obtained using a job–exposure matrix involving 320 job periods and semi-quantitative exposure intensity scores from 0 to 9. Exposure was assessed as (i) simultaneous exposure to cobalt and tungsten carbide specific to hard-metal manufacture and (ii) other exposure to cobalt resulting from other production activities. Exposure to cobalt with tungsten carbide was analysed using the maximum intensity score coded at any period of the job history, the duration of exposure at an intensity of ≥ 2 and the estimated cumulative exposure. Cumulative exposure was expressed as either an unweighted (intensity × duration) or a frequency-weighted (intensity × duration × frequency) score. The cumulative exposure scores were divided into quartiles of the exposure distribution among controls after exposure to cobalt had been classified as exposed versus unexposed. Exposure scores for each risk were based on information up to 10 years prior to the death of the case. Information on smoking habits (defined as never, former or current smokers) was obtained by interviewing colleagues, relatives and the subjects themselves. For analysis, each subject was classified as an ever versus never smoker. Information on smoking habits was available for 80% of the study population. The effect of possible confounders, including potential carcinogens listed in the job–exposure matrix (assessed as 'yes' or 'no'), socioeconomic level and smoking, was assessed using a multiple logistic model.

The odds ratio for workers exposed to cobalt and tungsten carbide was 1.93 (95% CI, 1.03–3.62) for exposure levels 2–9 versus levels 0–1. The odds ratio for cobalt with tungsten carbide increased with duration of exposure and unweighted cumulative dose, but less clearly with level of exposure or frequency-weighted cumulative dose. Exposure to cobalt and tungsten before sintering was associated with an elevated risk (odds ratio, 1.69; 95% CI, 0.88–3.27), which increased significantly with frequency-weighted cumulative exposure ($p = 0.03$). The odds ratio for exposure to cobalt and tungsten after sintering was lower (1.26; 95% CI, 0.66–2.40) and no significant trend was observed for cumulative exposure. Adjustment for exposure to known or suspected carcinogens did not change the results. Adjustment for smoking in the 80% subset with complete smoking data resulted in a slightly higher odds ratio (2.6; 95% CI, 1.16–5.82; versus 2.29; 95% CI, 1.08–4.88). The odds ratio for cobalt alone or with exposures other than to tungsten carbide was 2.21 (95% CI, 0.99–4.90) in a model with only indicators of duration of exposure to cobalt with tungsten carbide.

A study in addition to that of Moulin *et al.* (1998) was conducted in the largest plant already included in the multicentre cohort and used the same job–exposure matrix but made use of the more detailed job histories available (Wild *et al.*, 2000). In this study, which included follow-up from 1968 to 1992, mortality from all causes among 2860 subjects was close to the expected number (399 deaths; SMR for men and women combined, 1.02; 95% CI, 0.92–1.13). Mortality from lung cancer was increased among men (46 deaths; SMR, 1.70; 95% CI, 1.24–2.26). The SMR for exposure to hard-metal dust at an intensity score ≥ 2 was increased (26 deaths; SMR, 2.02; 95% CI, 1.32–2.96). Lung cancer mortality was higher than expected in those working in hard-metal production before sintering (nine deaths; SMR, 2.42; 95% CI, 1.10–4.59); after sintering, the SMR was 1.28 (five deaths; 95% CI, 0.41–2.98). In a Poisson regression model (Table 11) including terms for smoking and other occupational carcinogens, the risk for lung cancer increased with duration of exposure to cobalt with tungsten carbide before sintering (1.43 per 10-year period); there was no evidence of risk from exposure to sintered hard-metal dust.

2.2 Cobalt production industry

Moulin *et al.* (1993) studied the mortality of a cohort of 1148 workers in a cobalt electrochemical plant in France which produced cobalt and sodium by electrochemistry, extending the follow-up of an earlier study (Mur *et al.*, 1987; reported in IARC, 1991). The cohort included all the men who had worked in this plant for a minimum of 1 year between 1950 and 1980. The vital status of the members of the cohort was ascertained up to the end of 1988, and was obtained for 99% of French-born workers using information provided by the registry office of their place of birth. Due to difficulties in tracing workers born outside France, results are presented here only for French-born workers (*n* = 870).

The SMR for all causes of death was 0.95 (247 deaths; 95% CI, 0.83–1.08) and that for all cancer deaths was 1.00 (72 deaths; 95% CI, 0.78–1.26). The SMR for lung cancer mortality was 1.16 (three deaths; 95% CI, 0.24–3.40) among workers employed exclusively in cobalt production and 1.18 (four deaths; 95% CI, 0.32–3.03) for workers ever employed in cobalt production. For workers who worked exclusively as maintenance workers, the SMR for lung cancer was 2.41 (two deaths; 95% CI, 0.97–4.97) and, for those ever employed as maintenance workers, it was 2.58 (eight deaths; 95% CI, 1.12–5.09). There was evidence for an increased risk in this group of workers for those employed more than 10 years in cobalt production and for 30 years or more since first employment in cobalt production. [The Working Group noted that this might be explained by other carcinogenic exposures such as smoking or other occupational exposures such as asbestos.]

2.3 Other cobalt compounds

A study was conducted among 874 women occupationally exposed to poorly soluble cobalt–aluminate spinel and 520 women not exposed to cobalt in two porcelain factories in Denmark (Tüchsen *et al.*, 1996). The period of follow-up was from 1943 (time of first employment) to 1992. Vital status was assessed through the national population register and incident cancer cases were traced through the national cancer register. The observed deaths and incident cancer cases were compared with the expected numbers based on national rates for all Danish women. Cobalt concentrations in air in this plant were high (often > 1000 µg/m³). During the follow-up period, 127 cancer cases were diagnosed in the cohort. The overall cancer incidence was slightly elevated among the exposed women (67 observed; standardized incidence ratio [SIR], 1.20; 95% CI, 0.93–1.52) and close to unity in the reference group (60 observed; SIR, 0.99 [95% CI, 0.76–1.27]). Compared with the national reference rate, both exposed women (eight observed; SIR, 2.35; 95% CI, 1.01–4.62) and the reference group (seven observed; SIR, 1.99; 95% CI, 0.80–4.11) had an increased risk for lung cancer. However, the exposed group had a relative risk ratio of 1.2 (95% CI, 0.4–3.8) when compared with the reference group.

No relation with duration or intensity of exposure was found. The influence of smoking could not be taken into account in this study. Among the eight cases of lung cancer identified in the exposed cohort, three had been exposed to cobalt spinel for less than 3 months. [This study did not provide evidence of an increased risk of lung cancer associated with exposure to cobalt spinel.]

3. Studies of Cancer in Experimental Animals

3.1   Inhalation exposure

There are no data relative to carcinogenicity by inhalation of cobalt metal, cobalt-metal powder or cobalt alloys.

3.1.1   *Mouse*

In a study undertaken by the National Toxicology Program (1998), groups of 50 male and 50 female B6C3F$_1$ mice, 6 weeks of age, were exposed to aqueous aerosols of 0, 0.3, 1 or 3 mg/m³ cobalt sulfate heptahydrate (purity, ≈ 99%; mass median aerodynamic diameter (MMAD), 1.4–1.6 µm; geometric standard deviation (GSD), 2.1–2.2 µm) for 6 h per day on 5 days per week for 105 weeks. No adverse effects on survival were observed in treated males or females compared with chamber controls (survival rates: 22/50 (control), 31/50 (low dose), 24/50 (mid dose) or 20/50 (high dose) in males and 34/50, 37/50, 32/50 or 28/50 in females, respectively; survival times: 662, 695, 670 or 643 days in males and 694, 713, 685 or 680 days in females, respectively). Mean body weights increased in all treated females from week 20 to 105 and decreased in males exposed to the high dose from week 96 to 105 when compared with chamber controls. The incidence of neoplasms and non-neoplastic lesions of the lung is reported in Table 12. Exposure to cobalt sulfate heptahydrate caused a concentration-related increase in benign and malignant alveolar/bronchiolar neoplasms in male and female mice. All the alveolar/bronchiolar proliferative lesions observed within the lungs of exposed mice were typical of those arising spontaneously. However, exposure to cobalt did not cause an increased incidence of neoplasms in other tissues (National Toxicology Program, 1998; Bucher *et al.*, 1999).

Table 12. Incidence of neoplasms and non-neoplastic lesions of the lung in mice in a 2-year inhalation study of cobalt sulfate heptahydrate

Lesions observed	No. of mice exposed to cobalt sulfate heptahydrate at concentrations (mg/m^3) of			
	0 (chamber control)	0.3	1.0	3.0
Males				
Total no. examined microscopically	50	50	50	50
Infiltration cellular, diffuse, histiocyte	1 (3.0)[a]	2 (3.0)	4 (2.3)	10[b] (1.5)
Infiltration cellular, focal, histiocyte	10 (2.7)	5 (2.6)	8 (3.0)	17 (2.7)
Bronchus, cytoplasmic vacuolization	0	18[b] (1.0)	34[b] (1.0)	38[b] (1.0)
Alveolar epithelium hyperplasia	0	4 (2.3)	4 (1.8)	4 (2.3)
Alveolar/bronchiolar adenoma	9	12	13	18[c]
Alveolar/bronchiolar carcinoma	4	5	7	11[c]
Alveolar/bronchiolar adenoma or carcinoma	11	14	19	28[b]
Females				
Total no. examined microscopically	50	50	50	50
Infiltration cellular, diffuse, histiocyte	0	0	0	4 (3.3)
Infiltration cellular, focal, histiocyte	2 (2.0)	5 (1.8)	7 (2.9)	10[c] (2.4)
Bronchus, cytoplasmic vacuolization	0	6[c] (1.0)	31[b] (1.0)	43[b] (1.0)
Alveolar epithelium hyperplasia	2 (1.5)	3 (1.3)	0	5 (2.0)
Alveolar/bronchiolar adenoma	3	6	9	10[c]
Alveolar/bronchiolar carcinoma	1	1	4	9[b]
Alveolar/bronchiolar adenoma or carcinoma	4	7	13[c]	18[b]

From National Toxicology Program (1998)
[a] Average severity grade of lesions in affected animals: 1, minimal; 2, mild; 3, moderate; 4, marked
[b] Significantly different ($p \leq 0.01$) from the chamber control group by the logistic regression test
[c] Significantly different ($p \leq 0.05$) from the chamber control group by the logistic regression test

3.1.2 *Rat*

In a study undertaken by the National Toxicology Program (1998), groups of 50 male and 50 female Fischer 344/N rats, 6 weeks of age, were exposed to aqueous aerosols of 0, 0.3, 1 or 3 mg/m^3 cobalt sulfate heptahydrate (purity, $\approx$ 99%; MMAD, 1.4–1.6 μm; GSD, 2.1–2.2 μm) for 6 h per day on 5 days per week for 105 weeks. No adverse effects on mean body weights nor on survival were observed in treated males or females compared with chamber controls (survival rates: 17/50 (control), 15/50 (low dose), 21/50 (mid dose) or 15/50 (high dose) in males and 28/50, 25/49, 26/50 or 30/50 in females, respectively; survival times: 648, 655, 663 or 643 days in males and 699, 677, 691 or 684 days in females, respectively). Exposure to cobalt sulfate heptahydrate caused a concentration-related increase in the incidence of benign and malignant alveolar bronchiolar neoplasms in male and female rats and benign and malignant pheochromocytomas in female rats. However, exposure to cobalt sulfate did not cause an increased incidence of neoplasms in other tissues. The incidence of neoplasms and non-neoplastic lesions is reported in Table 13. In rats exposed to cobalt sulfate heptahydrate by inhalation, a broad spectrum of inflammatory and proliferative pulmonary lesions was observed. While many of these tumours were highly cellular and morphologically similar to those arising spontaneously, others, in contrast to those seen in mice, were predominantly fibrotic, squamous or mixtures of alveolar/bronchiolar epithelium and squamous or fibrous components. Benign neoplasms typical of those arising spontaneously were generally distinct masses that often compressed surrounding tissue. Malignant alveolar/bronchiolar neoplasms had similar cellular patterns but were generally larger and had one or more of the following histological features; heterogeneous growth pattern, cellular pleomorphism and/or atypia, and local invasion or metastasis. In addition to these more typical proliferative lesions, there were 'fibroproliferative' lesions ranging from less than 1 mm to greater than 1 cm in diameter. Small lesions with modest amounts of peripheral epithelial proliferation were diagnosed as atypical hyperplasia, while larger lesions with florid epithelial proliferation, marked cellular pleomorphism, and/or local invasion were diagnosed as alveolar/bronchiolar carcinomas. While squamous epithelium is not normally observed within the lung, squamous metaplasia of alveolar/bronchiolar epithelium is a relatively common response to pulmonary injury and occurred in a number of rats in this study. Squamous metaplasia consisted of small clusters of alveoli in which the normal epithelium was replaced by multiple layers of flattened squamous epithelial cells that occasionally formed keratin. One male and one female each had a large cystic squamous lesion rimmed by a variably thick band of friable squamous epithelium with a large central core of keratin. These lesions were diagnosed as cysts. In two exposed females, proliferative squamous lesions had cystic areas but also more solid areas of pleomorphic cells and invasion into the adjacent lung; these lesions were considered to be squamous-cell carcinomas. In all groups of male and female rats exposed to cobalt sulfate heptahydrate, the incidence of alveolar proteinosis, alveolar epithelial metaplasia, granulomatous alveolar inflammation and interstitial fibrosis was significantly greater than in the chamber controls. Exposure to cobalt sulfate heptahydrate caused a concentration-related increased incidence of benign and malignant pheochromocytomas in female rats. Although a very common spontaneous neoplasm in male Fischer 344/N rats, pheochromocytomas have a lower spontaneous occurrence in females. The marginally-increased incidence of pheochromocytomas in males was considered an uncertain finding because it occurred only in the group exposed to 1.0 mg/m^3 and was not supported by increased incidence or severity of hyperplasia (National Toxicology Program, 1998; Bucher et al., 1999).

Table 13. Incidence of neoplasms and non-neoplastic lesions of the lung in rats and of the adrenal medulla in female rats in a 2-year inhalation study of cobalt sulfate heptahydrate

Lesions observed	No. of rats exposed to cobalt sulfate heptahydrate at concentrations (mg/m^3) of			
	0 (chamber control)	0.3	1.0	3.0
Males				
Lung				
No. examined microscopically	50	50	48	50
Alveolar epithelium, hyperplasia	9 (1.8)[a]	20[b] (2.0)	20[b] (2.1)	23[c] (2.0)
Alveolar epithelium, hyperplasia, atypical	0	2 (3.0)	2 (3.0)	2 (4.0)
Metaplasia, squamous	0	1 (1.0)	4 (2.0)	2 (3.0)
Alveolar epithelium, metaplasia	0	50[c] (1.9)	48[c] (3.1)	49[c] (3.7)
Inflammation, granulomatous	2 (1.0)	50[c] (1.9)	48[c] (3.1)	50[c] (3.7)
Interstitium, fibrosis	1 (1.0)	50[c] (1.9)	48[c] (3.1)	49[c] (3.7)
Proteinosis	0	16[c] (1.4)	40[c] (2.3)	47[c] (3.4)
Cyst	0	0	0	1 (4.0)
Alveolar/bronchiolar adenoma	1	4	1	6
Alveolar/bronchiolar carcinoma	0	0	3	1
Alveolar/bronchiolar adenoma or carcinoma	1	4	4	7/50[b]
Females				
Lung				
No. examined microscopically	50	49	50	50
Alveolar epithelium, hyperplasia	15 (1.4)	7 (1.6)	20 (1.8)	33[c] (2.0)
Alveolar epithelium, hyperplasia, atypical	0	0	3 (3.7)	5[b] (3.2)
Metaplasia, squamous	0	1 (2.0)	8[c] (2.3)	3 (1.7)
Alveolar epithelium, metaplasia	2 (1.0)	47[c] (2.0)	50[c] (3.6)	49[c] (3.9)
Inflammation, granulomatous	9 (1.0)	47[c] (2.0)	50[c] (3.6)	49[c] (3.9)
Interstitium, fibrosis	7 (1.0)	47[c] (2.0)	50[c] (3.6)	49[c] (3.9)
Proteinosis	0	36[c] (1.2)	49[c] (2.8)	49[c] (3.9)
Cyst	0	0	1 (4.0)	0
Alveolar/bronchiolar adenoma	0	1	10[c]	9[c]
Alveolar/bronchiolar carcinoma	0	2	6[b]	6[b]
Alveolar/bronchiolar adenoma or carcinoma	0	3	15[c]	15[c]
Squamous-cell carcinoma	0	0	1	1
Alveolar/bronchiolar adenoma, alveolar/ bronchiolar carcinoma or squamous-cell carcinoma	0	3	16[c]	16[c]

Table 13 (contd)

Lesions observed	No. of rats exposed to cobalt sulfate heptahydrate at concentrations (mg/m³) of			
	0 (chamber control)	0.3	1.0	3.0
Adrenal medulla				
No. examined microscopically	48	49	50	48
Hyperplasia	8 (1.6)	7 (2.3)	11 (2.1)	13 (2.0)
Benign pheochromocytoma	2	1	3	8[b]
Benign, complex, or malignant pheochromo-cytoma	2	1	4	10[b]

From National Toxicology Program (1998)
[a] Average severity grade of lesions in affected animals: 1, minimal; 2, mild; 3, moderate; 4, marked
[b] Significantly different ($p \leq 0.05$) from the chamber control group by the logistic regression test
[c] Significantly different ($p \leq 0.01$) from the chamber control group by the logistic regression test

3.2   Intratracheal instillation

Rat

Steinhoff and Mohr (1991) reported on the exposure of rats to a cobalt–aluminium–chromium spinel (a blue powder [purity unspecified], with the empirical formula Co[II] 0.66, Al 0.7, Cr[III] 0.3, O 3.66, made of a mixture of CoO, Al(OH)$_3$ and Cr$_2$O$_3$ ignited at 1250 °C; 80% of particles < 1.5 µm). Groups of 50 male and 50 female Sprague-Dawley rats, 10 weeks of age, received intratracheal instillations of 10 mg/kg bw of the spinel in saline every 2 weeks for 18 treatments (then every 4 weeks from the 19th to the 30th treatment) for 2 years. Control groups of 50 males and 50 females received instillations of saline only and other control groups of 50 males and 50 females remained untreated. Animals were allowed to live until natural death or were killed when moribund. No appreciable difference in body weights or survival times was observed between the treated and control groups [exact survival data not given]. Alveolar/bronchiolar proliferation was observed in 0/100 untreated controls, 0/100 saline controls, and in 61/100 rats treated with the spinel. [The Working Group noted that the nature of the bronchoalveolar proliferation or possible association with inflammation was not described.] No pulmonary tumours were observed in 100 untreated or 100 saline controls. In the group that received the spinel, squamous-cell carcinoma was observed in one male rat and two females (Steinhoff & Mohr, 1991).

3.3 Intramuscular injection

Rat

In studies undertaken by Heath (1954, 1956), groups of 10 male and 10 female hooded rats, 2–3 months old, received a single intramuscular injection of 28 mg cobalt-metal powder (spectrographically pure, 400 mesh; 3.5 µm × 3.5 µm to 17 µm × 12 µm with large numbers of long narrow particles of the order of 10 µm × 4 µm) in 0.4 mL fowl serum into the thigh; a control group of 10 males and 10 females received fowl serum only. Average survival times were 71 weeks in treated males and 61 weeks in treated females; survival of controls was not specified. During the observation period of 122 weeks, 4/10 male and 5/10 female treated rats developed sarcomas (mostly rhabdomyosarcomas) at the injection site compared with 0/20 controls. A further group of 10 female rats received a single intramuscular injection of 28 mg cobalt-metal powder in 0.4 mL fowl serum; others received injections of 28 mg zinc powder (five rats) or 28 mg tungsten powder (five rats). Average survival time for cobalt-treated rats was 43 weeks. During the observation period of 105 weeks, sarcomas (mostly rhabdomyosarcomas) developed in 8/10 cobalt powder-treated rats; none occurred in the zinc powder- or tungsten powder-treated rats. No other tumours occurred in any of the cobalt-treated or other rats, except for one malignant lymphoma in a zinc-treated rat (Heath, 1954, 1956). [The Working Group noted the small number of animals and questioned the relevance of the route of administration.]

In a supplementary study, a group of 30 male hooded rats, 2–3 months of age, received a single intramuscular injection of 28 mg cobalt-metal powder (spectrographically pure [particle size unspecified]) in 0.4 mL fowl serum into the right thigh; a control group of 15 males received a single injection of fowl serum only. The rats were killed at daily intervals 1 to 28 days after injection. An extensive and continuing breakdown of the differentiated muscle fibres into free myoblasts, and the transformation of some of these myoblasts were described (Heath, 1960). [The Working Group questioned the relevance of the route of administration.]

In a series of three experiments, each of 80 female hooded rats, 7–9 weeks of age, received intramuscular injections of 28 mg of 'wear' particles obtained by grinding continuously artificial hip or knee prostheses in Ringer's solution or synovial fluid in conditions simulating those occurring in the body. Prostheses were made of cobalt–chromium–molybdenum alloy (66.5% cobalt, 26.0% chromium, 6.65% molybdenum, 1.12% manganese). Particles (diameter, down to 0.1 µm [mostly 0.1–1 µm]) were injected in 0.4 mL horse serum and the rats were observed for up to 29 months [survival not specified]. No control group was reported. Sarcomas developed at the injection site in 3/16, 4/14 and 16/50 rats in the three series, respectively. Approximately half of the tumours were rhabdomyosarcomas; the remainder were mostly fibrosarcomas (Heath *et al.*, 1971; Swanson *et al.*, 1973).

3.4   Intramuscular implantation

3.4.1   *Rat*

As a follow-up to the studies by Heath *et al.* (1971) and Swanson *et al.* (1973) (see above), groups of female Wistar and hooded rats, weighing 190–310 and 175–220 g, respectively, received intramuscular implants of 28 mg coarse (100–250 μm diameter; 51 Wistar rats) or fine (0.5–50 μm diameter, 85% 0.5–5 μm; 61 Wistar and 53 hooded rats) particles of a dry powder, obtained by grinding a cobalt–chromium–molybdenum alloy (68% cobalt, 28% chromium, 4% molybdenum). The animals were observed for life. A sham-operated control group of 50 female Wistar rats was included. Survival at 2 years was: 11/51 rats receiving the coarse particles, 7/61 Wistar rats receiving the fine particles, 0/53 hooded rats receiving the fine particles and 5/50 Wistar controls. No tumour was noted at the implantation site of rats treated with either coarse or fine alloy particles nor in sham-operated control animals (Meachim *et al.*, 1982).

3.4.2   *Guinea-pig*

In a similar study in guinea-pigs (Meachim *et al.*, 1982), a group of 46 female Dunkin-Hartley guinea-pigs, weighing 550–930 g, received intramuscular implants of 28 mg powdered cobalt–chromium–molybdenum alloy (68% cobalt, 28% chromium, 4% molybdenum; particle diameter, 0.5–50 μm) and were observed for life; 12/46 animals were alive at 3 years. No control group was reported. No tumours were observed at the implantation site; nodular fibroblastic hyperplasia was observed at the implantation site in eight animals (Meachim *et al.*, 1982).

3.5   Subcutaneous implantation

Rat

Groups of five male and five female Wistar rats, 4–6 weeks of age, received subcutaneous implants of one pellet (approximately 2 mm in diameter) of a cobalt–chromium–molybdenum (Vitallium) alloy. The percentage composition of the metal constituents of the Vitallium alloy was not given. Animals were observed for up to 27 months [survival of animals receiving cobalt–chromium–molybdenum alloy not given]. No sarcomas developed in rats that received the pellets (Mitchell *et al.*, 1960).

3.6   Intra-osseous implantation

3.6.1   *Rat*

Groups of 10–17 male and 8–15 female Sprague-Dawley rats, 30–43 days of age, received implants of one of seven test materials containing cobalt alloyed with chromium and nickel, molybdenum, tungsten and/or zirconium, with traces of other elements (as small rods, 1.6 mm diameter and 4 mm length, powders or porous compacted wire), in the femoral bone and were observed for up to 30 months. Groups of 13 male and 13 female untreated and sham-operated controls were available. Average survival time was more than 21 months. Sarcomas at the implant site were observed in 1/18 rats given cobalt-based alloy powder containing 41% cobalt, 3/26 rats (given a nickel–cobalt-based powder containing 33% cobalt and 3/32 rats given porous compacted wire containing 51% cobalt. No tumours were observed in two groups of 25 rats given rods containing 69 or 47% cobalt, in two groups of 26 rats given rods containing 0.11 or 33% cobalt, in two groups of 25 and 26 untreated rats nor in a group of 26 sham-treated control rats (Memoli *et al.*, 1986).

3.6.2   *Rabbit*

Two groups of 15–20 rabbits [strain, sex and age unspecified] received an implantation in the femoral cavity of metallic chromium dust or metallic cobalt dust [purity and particle size unspecified]. Physical examination by palpation and X-ray examination 3 years after implantation revealed no implantation-site tumour in survivors of the chromium-treated group nor in the six survivors of the cobalt-treated group (Vollman, 1938). In a follow-up study of survivors [number unspecified] at intervals up to 6 years after implantation, sarcomas were observed at the implantation site in three chromium-treated and two cobalt-treated rabbits (Schinz & Uehlinger, 1942). [The Working Group noted the limited reporting.]

3.7   Intraperitoneal injection

Rat

Groups of 10 male and 10 female Sprague-Dawley rats, 10 weeks of age, received three intraperitoneal injections at 2-month intervals of saline or cobalt–aluminium–chromium spinel powder (see Section 3.2) in saline (total dose, 600 mg/kg bw). Animals were allowed to live their natural lifespan or were sacrificed when moribund [survival not given]. Malignant peritoneal tumours occurred in 1/20 controls (histiocytoma) and 2/20 spinel-treated animals (one histiocytoma, one sarcoma) (Steinhoff & Mohr, 1991).

3.8   Intrarenal administration

Rat

Groups of female Sprague-Dawley rats, weighing 120–140 g, received a single injection of 5 mg metallic cobalt powder (20 rats) or cobalt sulfide powder (18 rats) [purity and particle size unspecified] suspended in 0.05 mL glycerine into each pole of the right kidney. A control group of 16 female rats received injections of 0.05 mL glycerine into each pole of the kidney. After 12 months, necropsies were performed on all rats; no tumours were observed in the kidneys of treated or control rats (Jasmin & Riopelle, 1976). [The Working

Group noted the short duration and inadequate reporting of the experiment and the unusual site of administration.]

3.9   Intrathoracic injection

Rat

Two groups of 10 female hooded rats, 2–3 months of age, received intrathoracic injections of 28 mg cobalt-metal powder (spectrographically pure; particle size, < 400 mesh; 3.5 μm × 3.5 μm to 17 μm × 12 μm, with many long narrow particles of the order of 10 μm × 4 μm) in serum [species unspecified] through the right dome of the diaphragm (first group) or through the fourth left intercostal space (second group) and were observed for up to 28 months. Death occurred within 3 days of the treatment in 6/10 rats injected through the diaphragm and in 2/10 rats injected through the intercostal space. The remaining rats in the first group (diaphragm) survived 11–28 months and in the second group (intercostal space), 7.5–17.5 months. Of the 12 rats that survived the injection, four developed intra-thoracic sarcomas (three of mixed origin, including rhabdomyosarcomatous elements; one rhabdomyosarcoma arising in the intercostal muscles) (Heath & Daniel, 1962). [The Working Group noted the small numbers of animals and the questionable route of administration.]

4.   Other Data Relevant to an Evaluation of Carcinogenicity and its Mechanisms

4.1   Deposition, retention, clearance and metabolism

Several reviews of the toxicology of cobalt, including toxicokinetic aspects, are available (IARC, 1991; Midtgård & Binderup, 1994; Lauwerys & Lison, 1994; Lison, 1996; Barceloux, 1999). This section will focus on the toxicokinetic data published since the previous IARC evaluation (1991) and potentially relevant for cancer. Particular emphasis will be put on studies that examined the fate of inhaled hard-metal particles and related components, when available.

Solubilization of cobalt from tungsten carbide–cobalt powder

It has been shown that tungsten carbide–cobalt powder (WC–Co) is more toxic to murine macrophages *in vitro* than pure cobalt-metal particles, and that the cellular uptake of cobalt is enhanced when the metal is present in the form of WC–Co (Lison & Lauwerys, 1990). In a further study by the same authors, the solubilization of cobalt in the extracellular milieu was shown to increase in the presence of WC. This phenomenon, however, does not explain the greater toxicity of the WC–Co mixture, because increasing the amount of solubilized cobalt in the extracellular medium in the absence of WC did not result in increased toxicity. Moreover, the amount of cobalt solubilized from a toxic dose of WC–Co was insufficient alone to affect macrophage viability. A toxic effect was only observed when the WC–Co mixture came into direct contact with the cells. These results indicate that the toxicity of the WC–Co mixture does not result simply from an enhanced bioavailability of its cobalt component and suggest that hard-metal dust behaves as a specific toxic entity (Lison & Lauwerys, 1992).

4.1.1   *Humans*

(*a*)   *Deposition and retention*

Since the previous IARC (1991) evaluation, no additional relevant data concerning the deposition and/or retention of inhaled cobalt-containing particles in humans have been located.

In several studies conducted on lung tissue or bronchoalveolar lavage fluid (BALF) from patients with lung disease induced by hard-metal particles (hard-metal disease), the presence of tungsten, tantalum or titanium particles was detected, but no or insignificant amounts of cobalt were found (Lison, 1996).

Citizens in Catalonia, Spain, were found to have cobalt in their lungs at the limit of detection (Garcia *et al.*, 2001). In contrast, citizens of Mexico City showed remarkably high concentrations of cobalt in their lungs over three decades, which was attributed to air pollution (Fortoul *et al.*, 1996). In an autopsy study carried out in Japan, cobalt concentrations in the lung were reported to be related mainly to blood concentrations and were found to be lower in patients who had died from lung cancer than from other causes (Adachi *et al.*, 1991; Takemoto *et al.*, 1991). A study of uranium miners in Germany demonstrated by NAA that cobalt, associated with uranium, arsenic, chromium and antimony was present at high concentrations in the lungs, with or without concurrent lung tumours, even 20 years after cessation of mining (Wiethege *et al.*, 1999).

(*b*)   *Intake and absorption*

There are few data on the respiratory absorption of inhaled cobalt-containing materials in humans. The absorption rate is probably dependent on the solubility in biological fluids and in alveolar macrophages of the cobalt compounds under consideration. Increased excretion of the element in post-shift urine of workers exposed to soluble cobalt-containing particles (cobalt metal and salts, hard-metal particles) has been interpreted as an indirect indication of rapid absorption in the lung; in contrast, when workers were exposed to the less soluble cobalt oxide particles, the pattern of urinary excretion indicated a lower absorption rate and probably a longer retention time in the lung (Lison & Lauwerys, 1994; Lison *et al.*, 1994). The importance of speciation and solubility for respiratory absorption has also been highlighted by Christensen and Mikkelsen (1986). These authors found that cobalt concentrations in blood and urine increased (0.2–24 µg/L and 0.4–848 µg/L, respectively) in pottery plate painters using a soluble cobalt paint compared to the control group of painters without cobalt exposure (0.05–0.6 µg/L and 0.05–7.7 µg/L, respectively). The pottery painters exposed to slightly soluble cobalt paint had only slightly increased cobalt concentrations compared to controls (see Section 1.3.2(d)).

The absorption of cobalt compounds has been estimated to vary from 5 to 45% of an orally-administered dose (Valberg *et al.*, 1969; Smith *et al.*, 1972; Elinder & Friberg, 1986). The mean urinary excretion within 24 h of radioactive cobalt (from cobalt chloride) given orally at 20 µmoles to 17 volunteers was estimated to be about 18% (Sorbie *et al.*, 1971). In a short-term cross-over study in volunteers, the gastrointestinal uptake of soluble cobalt chloride measured as cobalt concentrations in urine was found to be considerably higher than that of insoluble cobalt oxide (urine ranges, < 0.17–4373 and < 0.17–14.6 nmol/mmol creatinine, respectively). It was also shown that ingestion of controlled amounts of soluble cobalt compounds resulted in significantly higher cobalt concentrations in urine ($p < 0.01$) in women (median, 109.7 nmol/mmol creatinine) than in men (median, 38.4 nmol/mmol creatinine), suggesting that the gastrointestinal uptake of cobalt is higher in women than men (Christensen *et al.*, 1993).

Cobalt has been detected in pubic hair, toe nails and sperm of some but not all workers diagnosed with hard-metal disease (Rizzato *et al.*, 1992, 1994; Sabbioni *et al.*, 1994a).

It was found that absorption of cobalt through the skin and gastrointestinal tract also contributed to concentrations of cobalt in urine in occupationally-exposed individuals (Christensen *et al.*, 1993; Scansetti *et al.*, 1994; Christensen, 1995; Linnainmaa & Kiilunen, 1997). Concentrations of cobalt in urine of smokers at a hard-metal factory were higher than those in nonsmokers (10.2 nmol/L [0.6 µg/L] versus 5.1 nmol/L [0.3 µg/L] on average), while no difference in concentrations of cobalt in blood was detected (Alexandersson, 1988). However, the cobalt excreted in urine was found not to be derived from cobalt contained in cigarettes nor from daily intake of vitamin B_{12}, but through eating and smoking with cobalt-contaminated hands at work (Linnainmaa & Kiilunen, 1997).

After absorption, cobalt is distributed systemically but does not accumulate in any specific organ, except the lung in the case of inhalation of insoluble particles. Normal cobalt concentrations in human lung have been reported to be 0.27 ± 0.40 (mean ± SD) µg/g dried lung based on tissue samples taken from 2274 autopsies in Japan (Takemoto *et al.*, 1991). A majority of the autopsies were carried out on subjects with malignant neoplasms. There was no increase in cobalt concentration with age, no gender difference and no association with degree of emphysema nor degree of contamination (the grade of particle deposition in the lung).

The normal concentration of cobalt in blood is in the range of 0.1–0.5 µg/L and that in urine is below 2 µg/L in non-occupationally exposed persons. The concentrations of cobalt in blood, and particularly in urine, increase in proportion to the degree of occupational (inhalation) exposure and may be used for biological monitoring in order to assess individual exposure (Elinder & Friberg, 1986). As well as the high concentrations found in workers exposed to cobalt, increased concentrations of cobalt have been found in blood (serum) of uraemic patients (Curtis *et al.*, 1976; Lins & Pehrsson, 1976) and in urine of individuals taking multivitamin preparations (as cyanocobalamin, a source of cobalt) (Reynolds, 1989) (see also IARC, 1991 and Section 1.1.5).

(c)   Excretion

The major proportion of systemically-distributed cobalt is cleared rapidly (within days) from the body, mainly via urine, but a certain proportion (10%) has a longer biological half-life, in the range of 2–15 years (Newton & Rundo, 1970; Elinder & Friberg 1986). Of an oral dose of cobaltous chloride, 6–8% was eliminated within 1 week in normal healthy persons (Curtis *et al.*, 1976). The elimination of cobalt is considerably slower in patients undergoing haemodialysis, which supports the importance of renal clearance (Curtis *et al.*, 1976). In workers in the hard-metal industry, it has been shown that concentrations of cobalt in urine increase rapidly in the hours that follow cessation of exposure, with a peak of elimination about 2–4 h after exposure, and a subsequent decrease (more rapid in the first 24 h) in the following days (Apostoli *et al.*, 1994).

4.1.2   *Experimental systems*

Following subcutaneous administration of cobalt chloride (250 µmol/kg bw) to rats, cobalt was found predominantly (> 95%) in plasma, from which it was rapidly eliminated (half-life ($t_{1/2}$), approximately 25 h) (Rosenberg, 1993). In-vitro studies (Merritt *et al.*, 1984) have shown that cobalt ions bind strongly to circulating proteins, mainly albumin. Edel *et al.* (1990) reported the in-vitro interaction of hard metals with human lung and plasma components and identified three biochemical pools of cobalt with different molecular weights in the lung cytosol. It has been suggested that cobalt binding to proteins may be of significance for immunological reactions involving cobalt as a hapten (Sjögren *et al.*, 1980). Wetterhahn (1981) showed that oxyanions of chromium, vanadium, arsenic and tungsten enter cells using the normal active transport system for phosphate and sulfate and may inhibit enzymes involved in phosphoryl or sulfuryl transfert reactions. Similarly, the divalent ions of cobalt may complex small molecules such as enzymes and alter their normal activity.

While cobalt-metal particles are practically insoluble in water, the solubilization of these particles is greatly enhanced in biological fluids due to extensive binding to proteins (0.003 mg/L in physiological saline, but 152.5 mg/L in human plasma at 37 °C) (Harding, 1950) and is increased up to sevenfold in the presence of WC particles (in oxygenated phosphate buffer at 37 °C) (Lison *et al.*, 1995).

(a)   In-vivo studies

Gastrointestinal absorption of cobalt in rats is dependent on the dose, the ratio of iron to cobalt and the status of body iron stores (Schade *et al.*, 1970). It has been shown that following oral administration of cobalt chloride, 75% is eliminated in faeces and the highest accumulation of cobalt is found in liver, kidney, heart and spleen (Domingo *et al.*, 1984; Domingo, 1989; Ayala-Fierro *et al.*, 1999).

Following intravenous administration of cobalt chloride to rats, 10% of the dose was found to be excreted in faeces, indicating that cobalt can be secreted in the bile. Elimination was triphasic. During the first 4 h, cobalt was rapidly cleared from blood with a half-life of 1.3 h. The second phase from 4 h to 12 h demonstrated a slower clearance rate with a half-life of 4.3 h. The final phase from 12 h to 36 h had a half-life of 19 h (Ayalu-Fierro *et al.*, 1999).

Kyono *et al.* (1992) exposed rats to ultrafine metallic cobalt particles (mean primary diameter, 20 nm) using a nebulizer producing droplets (MMAD, 0.76 μm; GSD, 2.1; concentration, 2.12 ± 0.55 mg/m^3) for 5 h per day for 4 days and induced reversible lung lesions. Clearance from the lung followed two phases: 75% of the cobalt was cleared within 3 days with a biological half-life of 53 h; the second phase from 3 days to 28 days had a slower clearance rate with a half-life of 156 h.

Kreyling *et al.* (1993) performed clearance studies using inhalation of monodisperse, porous cobalt oxide particles (MMAD, 1.4 and 2.7 μm) in Long-Evans rats. Of the small and large particles, 37% and 38%, respectively, were eliminated in the faeces within 3 days. The half-life for long-term thoracic retention was 25 and 53 days, respectively. After 6 months, large and small cobalt particles were still distributed in the bodies of the rats, mainly in the lung (91 and 52%), skeleton (6 and 22%) and in soft tissue (1.4 and 17%), respectively.

Lison and Lauwerys (1994) found that when non-toxic doses of cobalt metal were administered intratracheally to rats either alone (0.03 mg/100 g body weight) or mixed with tungsten carbide (0.5 mg/100 g body weight; WC–Co containing 6% of cobalt-metal particles), the retention time of the metal in the lung was longer in cobalt- than in WC–Co-treated animals. After 1 day, the lungs of animals instilled with cobalt alone contained twice as much cobalt as in those administered the same amount of cobalt as WC–Co (12 versus 5 μg cobalt/g lung after 24 h).

Slauson *et al.* (1989) induced patchy alveolitis, bronchiolitis and inflammation in the lungs of calves using parainfluenza-3 virus followed by a single inhalation exposure to an aerosol of submicronic cobalt oxide (total dose, about 80 mg). The virus-exposed calves retained 90% of initial cobalt lung burden at day 7 compared with 51% retention in controls. This difference was still present at day 21. Pneumonic calves also exhibited decreased translocation of particles to regional lymph nodes. The authors suggested impaired particulate clearance from acutely-inflamed lungs, which implicated decreased mucociliary clearance and interstitial sequestration within pulmonary alveolar macrophages as the major contributing factors.

(b)   In-vitro studies

In-vitro dissolution of monodisperse, 2.7-μm cobalt oxide particles in baboon alveolar macrophage cell cultures was found to be three times higher than in a cell-free system; the daily dissolution rate was 0.25% versus 0.07% and 0.09% for beads containing particles only and particles combined with alveolar macrophages, respectively (Lirsac *et al.*, 1989). Kreyling *et al.* (1990) studied in-vitro dissolution of cobalt oxide particles in human and canine alveolar macrophages and found that smaller particles had faster dissolution rates. In-vitro dissolution rates were found to be similar to in-vivo translocation rates previously found for human and canine lung. Dissolution of ultrafine cobalt powder in artificial lung fluid was six times higher than that of standard cobalt powder (Kyono et al., 1992).

Collier et al. (1992) studied factors influencing in-vitro dissolution rates in a simple non-cellular system using 1.7 μm count median diameter (CMD) porous cobalt oxide particles and cobalt-labelled fused aluminosilicate (Co-FAP). Less than 0.5% of cobalt oxide and 1.8% of Co-FAP dissolved over 3 months. The difference in dissolution was much greater in the first week than in the following weeks, with Co-FAP being 20 times more soluble. The dissolution rate for cobalt oxide was higher at lower pH. Lundborg et al. (1992) measured and changed phagolysosomal pH within rabbit alveolar macrophages. No clear effect on cobalt dissolution rate was detected for 0.6-μm cobalt oxide particles at pH values ranging between 5.1 and 5.6. Lundborg et al. (1995) also studied the effect of phagolysosomal size on cobalt dissolution in rabbit alveolar macrophages incubated with sucrose and in human alveolar macrophages from smokers and non-smokers. The authors found no difference in cobalt dissolution in either rabbit or human cells in spite of large differences in morphological appearance of the macrophages.

Lison and Lauwerys (1994) found that cellular uptake of cobalt was greater when the metal was presented to mouse macrophages as WC–Co. This increased bioavailability of cobalt from hard-metal particles has been interpreted as the result of a physicochemical interaction between cobalt metal and tungsten carbide particles (Lison et al., 1995).

In-vitro exposure of HeLa (tumour) cells to cobalt chloride has been shown to result in the intracellular accumulation of cobalt (Hartwig et al., 1990).

4.2   Toxic effects

4.2.1   *Humans*

The health effects resulting from exposure to metallic cobalt-containing particles may be subdivided into local and systemic effects. Local effects are those that occur at the points of contact or deposition of the particles, the skin and the respiratory tract; these effects may be due to the particles themselves (as a result of surface interactions between the particles and biological targets) and/or to cobalt ions solubilized from the particles. Toxic effects outside the respiratory tract are unlikely to be caused by the metallic particles themselves, but result from the release of cobalt ions from the particles and their subsequent absorption into the circulation. (Systemic effects may also be indirect consequences of the damage caused in the lungs).

(*a*)   *Dermal effects*

The skin sensitizing properties of cobalt are well known, both from human experience and from animal testing (Veien & Svejgaard, 1978; Wahlberg & Boman, 1978; Fischer & Rystedt, 1983). Exposure to cobalt may lead to allergic contact dermatitis, sometimes having features of an airborne dermatitis, particularly in hard-metal workers (Dooms-Goossens et al., 1986). Urticarial reactions have also been described. Cross-reaction with nickel (as well as co-sensitization) is frequent (Shirakawa *et al.*, 1990). The dermal effects of cobalt may occur with all forms of cobalt, i.e. cobalt metal and other cobalt compounds, such as salts.

(b)   Respiratory effects

The various respiratory disorders caused by the inhalation of metallic cobalt-containing particles have been extensively reviewed (Balmes, 1987; Cugell *et al.*, 1990; Seghizzi *et al.*, 1994; Lison, 1996; Barceloux, 1999; Nemery *et al.*, 2001a). These particles may cause non-specific mucosal irritation of the upper and lower airways leading to rhinitis, sinusitis, pharyngitis, tracheitis or bronchitis, but the main diseases of concern are bronchial asthma and a fibrosing alveolitis known as hard-metal lung disease.

(i)   Bronchial asthma

Bronchial asthma, which like contact dermatitis is presumably based on immunological sensitization to cobalt, has been described in workers exposed to various forms of cobalt, i.e. not only in workers exposed to hard-metal dust, but also in those exposed to 'pure' cobalt particles (Swennen *et al.*, 1993; Linna *et al.*, 2003), as well as in subjects exposed to other cobalt compounds, such as cobalt salts. Occupational asthma is more frequent than fibrosing alveolitis in hard-metal workers or workers exposed to cobalt dust, but occasionally the two conditions co-exist (Davison *et al.*, 1983; Van Cutsem *et al.*, 1987; Cugell *et al.*, 1990). Chronic bronchitis is reported to be quite prevalent in hard-metal workers, particularly in older studies when dust exposure was considerable and smoking status was not well ascertained (Tolot *et al.*, 1970). It is not clear whether those patients with airway changes (asthma or chronic obstructive lung disease) represent 'airway variants' of the same respiratory disease, or whether the pathogenesis of these airway changes is altogether different from that of parenchymal changes. Earlier autopsy studies frequently indicated the presence of emphysema in patients with hard-metal lung disease.

(ii)   Hard-metal lung disease

Interstitial (or parenchymal) lung disease caused by metallic cobalt-containing particles is a rare occupational lung disease. Several reviews are available on this fibrosing alveolitis which is generally called hard-metal lung disease (Bech *et al.*, 1962; Anthoine *et al.*, 1982; Hartung, 1986; Balmes, 1987; Van Den Eeckhout *et al.*, 1988; Cugell, 1992; Seghizzi *et al.*, 1994; Lison, 1996; Newman *et al.*, 1998; Nemery *et al.*, 2001a,b). A discussion of the occurrence and features of interstitial lung disease caused by metallic cobalt-containing compounds is not only relevant in itself, but it may also have a bearing on the risk of lung cancer, because fibrosing alveolitis and lung cancer may be related mechanistically with regard to both oxidative damage and inflammatory events. Moreover, there is some evidence from observations in humans that lung fibrosis represents a risk for lung cancer, although this evidence is not unequivocal (Bouros *et al.*, 2002).

Terminology of (interstitial) hard-metal lung disease

The terminology used to label this disease is complex and confusing. Especially in the earlier literature, hard-metal disease was mostly referred to as a pneumoconiosis (e.g. hard-metal pneumoconiosis or tungsten carbide pneumoconiosis). This term is justified inasmuch as pneumoconiosis is defined as "the non-neoplastic reaction of the lungs to inhaled mineral or organic dust and the resultant alteration in their structure, but excluding asthma, bronchitis and emphysema" (Parkes, 1994). However, it can be argued that the term pneumoconiosis is not entirely appropriate, because it suggests that the disease results from the accumulation of high quantities of dust in the lungs and this is not always the case in hard-metal workers. Indeed, like hypersensitivity pneumonitis and chronic beryllium disease, hard-metal lung disease differs from the common mineral pneumoconioses in that the occurrence of the disease is not clearly related to the cumulative dust burden, but is more probably due to individual susceptibility. Thus the term hard-metal pneumoconiosis has tended to be abandoned in favour of 'hard-metal lung disease'. An advantage, but also a drawback, of the latter term is that the respiratory effects of exposure to hard-metal dust include not only interstitial lung disease (pneumonitis, fibrosis), but also (and probably more frequently) airway disorders, such as bronchitis and occupational asthma. Therefore, the phrases 'hard-metal lung', 'hard-metal disease' or 'hard-metal lung disease' usually encompass more than just the parenchymal form of the disease (Nemery et al., 2001a).

In its most typical pathological presentation, this interstitial lung disease consists of a giant-cell interstitial pneumonia (GIP), one of the five types of interstitial pneumonias originally described by Liebow (1975). GIP is now accepted as being pathognomonic of hard-metal lung disease. Ohori et al. (1989) reviewed the published literature and concluded that GIP is indeed highly specific for hard-metal lung, since they found only three published cases of GIP that had not had exposure to cobalt or hard metal. However, while there is no doubt that GIP should be considered to be hard-metal lung disease unless proven otherwise, not all patients with hard-metal lung disease have a 'textbook presentation' of GIP. Indeed, the lung pathology in hard-metal lung disease is variable depending on, among other factors, the stage of the disease and probably also on its pathogenesis in individual patients. The pathology in some patients may be more reminiscent of mixed dust pneumoconiosis (Bech, 1974). Moreover, a pathological diagnosis is not always available.

The most compelling argument against the term hard-metal lung disease is that the disease may also occur without exposure to hard-metal dust. This was established when GIP was found in diamond polishers in Belgium shortly after the introduction of a new technology to facet diamonds. These workers were exposed not to hard-metal dust, but to cobalt-containing dust that originated from the use of high-speed cobalt–diamond polishing discs (Demedts et al., 1984). This observation confirmed an earlier hypothesis that cobalt, rather than tungsten carbide, is responsible for hard-metal lung, and it led to the proposal that the interstitial lung disease should be called 'cobalt pneumopathy' or 'cobalt lung', rather than hard-metal lung (Lahaye et al., 1984). Nevertheless, the term cobalt lung is not entirely appropriate either, because not all types of exposure to cobalt appear to lead to interstitial lung disease.

Table 14 summarizes the various terms which have been used for the interstitial lung disease caused by hard-metal and cobalt dust. The most correct term is probably 'cobalt-related interstitial lung disease', but this would add more confusion and thus the term hard-metal lung disease will be used here.

Table 14. Terminology of hard-metal lung disease

Name of disease (term)	Features supporting use of term	Features not supporting use of term	Reference
Hard-metal pneumoconiosis	Lung disease is caused by exposure to hard-metal dust	Pathogenesis (hypersensitivity) differs from that of mineral pneumoconiosis.	Parkes (1994)
Tungsten carbide pneumoconiosis	Tungsten carbide is main component of hard metal	Tungsten carbide is not the actual causative agent.	
Hard-metal disease Hard-metal lung Hard-metal lung disease	Lung disease is caused by exposure to hard metal	Terms encompass interstitial lung disease as well as airway disease, such as asthma; similar disease may be caused by exposure to materials other than hard metal.	Nemery et al. (2001a)
Cobalt lung or cobalt pneumopathy	Cobalt is the most critical toxic agent	Not all types of exposure to cobalt lead to interstitial lung disease.	Lahaye et al. (1984)
Giant cell interstitial pneumonia (GIP)	Pathognomonic pathological feature	Pathology not always available and not always present in individual cases	Ohori et al. (1989)

Adapted from Nemery et al. (2001a)

Pathogenesis of hard-metal lung disease

There is little doubt that cobalt plays a critical role in the pathogenesis of hard-metal lung disease. Studies in experimental systems have demonstrated that WC–Co particles exhibit a unique pulmonary toxicity compared with cobalt particles (see Section 4.2.2). The toxicity is probably due, at least in part, to the production of toxic oxygen species. However, the hard-metal lung disease that occurs in humans has never been reproduced in experimental animals; neither the typical pattern of inflammation (GIP), nor the progressive nature of the fibrosis.

The basis of individual susceptibility to develop hard-metal lung disease is not known. Cobalt is known to elicit allergic reactions in the skin, probably via cell-mediated pathways (Veien & Svejgaard, 1978), but the relationship, if any, between this cell-mediated allergy and GIP is unknown. Occasionally, patients have been found to have both cobalt dermatitis and interstitial lung disease (Sjögren et al., 1980; Cassina et al., 1987; Demedts & Ceuppens, 1989). Immunological studies (Shirakawa et al., 1988; Kusaka et al., 1989;

Shirakawa *et al.*, 1989, 1990; Kusaka *et al.*, 1991; Shirakawa *et al.*, 1992) have found both specific antibodies and positive lymphocyte transformation tests against cobalt (as well as nickel) in some patients with hard-metal asthma. However, to date, the immunopatho-genesis of GIP is unknown. In one case of GIP, expression of intracellular transforming growth factor β1 (TGF-β1) was shown in alveolar macrophages, including multinucleate forms, and in hyperplastic alveolar epithelium (Corrin *et al.*, 1994). In another case of GIP, immunolocalization of tumour necrosis factor α (TNF-α) was found to be highly asso-ciated with the infiltrating mononuclear cells within the interstitium and with cannibalistic multinucleated giant cells in the alveolar spaces (Rolfe *et al.*, 1992). The involvement of autoimmune processes is suggested by the report of the recurrence of GIP in a transplanted lung (Frost *et al.*, 1993) and in one case of hard-metal alveolitis accompanied by rheuma-toid arthritis (Hahtola *et al.*, 2000).

Recent evidence also indicates that the susceptibility to develop cobalt-related interstitial (hard-metal) lung disease is associated with the HLA-DPB1*02 allele, i.e. with the presence of glutamate at position 69 in the HLA-DPB chain (Potolicchio *et al.*, 1997), probably because of a high affinity of the HLA-DP molecule for cobalt (Potolicchio *et al.*, 1999). It should be noted that the HLA-DPB1*02 allele is the same as that associated with susceptibility to chronic beryllium disease (Richeldi *et al.*, 1993).

Clinical presentation

The clinical presentation of hard-metal lung disease is variable: some patients present with subacute alveolitis and others with chronic interstitial fibrosis (Balmes, 1987; Cugell *et al.*, 1990; Cugell, 1992). In this respect, hard-metal lung disease is somewhat similar to hypersensitivity pneumonitis (extrinsic allergic alveolitis). Thus, the patient may expe-rience work-related bouts of acute illness, which may lead progressively to pronounced disease with more persistent shortness of breath; but in other instances, the course of the disease is more insidious and the work-relatedness of the condition is not clearly apparent.

Most studies have found no relation between disease occurrence and length of occu-pational exposure. Subacute presentations may be found in young workers after only a few years exposure, but may also occur in older workers with very long careers. Chronic presentations are more likely in older subjects. The role of smoking in the susceptibility to hard-metal disease has not been evaluated thoroughly, but it is possible that non-smokers are slightly over-represented (Nemery *et al.*, 2001a).

Epidemiology

Descriptions of the epidemiology of hard-metal lung disease can be found in Lison (1996) and Newman *et al.* (1998). Precise incidence figures are not available. Clinical surveys and cross-sectional studies in the hard-metal industry have shown that typical hard-metal lung disease is a relatively rare occurrence, affecting a small percentage of the workforce at most (Miller *et al.*, 1953; Bech *et al.*, 1962; Dorsit *et al.*, 1970; Coates & Watson, 1971; Sprince *et al.*, 1984; Kusaka *et al.*, 1986; Sprince *et al.*, 1988; Meyer-Bisch *et al.*, 1989; Tan *et al.*, 2000), unless conditions of hygiene are very poor (Auchincloss *et al.*, 1992; Fischbein *et al.*, 1992). In diamond polishers in Belgium, the prevalence of cobalt-related occupational respiratory disease, including both airway and interstitial lung disease, has been estimated at about 1% of the total workforce (Van den Eeckhout *et al.*, 1988). A cross-sectional survey of 10 workshops, involving a total of 194 polishers, found no cases of overt lung disease, but there was a significant inverse relationship between spirometric indices of pulmonary function and mean levels of exposure to cobalt as assessed by ambient air or biological monitoring (Nemery *et al.*, 1992).

Lung disease has been associated not only with the manufacture of cobalt–diamond tools (Migliori *et al.*, 1994), but also with their use, at least in the case of high-speed cobalt–diamond discs used for diamond polishing (Demedts *et al.*, 1984; Lahaye *et al.*, 1984; Wilk-Rivard & Szeinuk, 2001; Harding, 2003). This could be explained by the fact that the projection of cobalt in bonded diamond tools is higher (up to 90%) than in hard metal.

Carbide coatings can now also be deposited by flame or plasma guns onto softer substrates to harden their surfaces, and this process also exposes workers to a risk of hard-metal lung disease (Rochat *et al.*, 1987; Figueroa *et al.*, 1992).

A detailed and comprehensive cross-sectional survey of 82 workers exposed to cobalt compounds in a plant in Belgium involved in cobalt refining and 82 sex- and age-matched controls from the same plant found no radiological or functional evidence of interstitial lung disease in spite of substantial exposure to cobalt (mean duration of exposure, 8 years; range, 0.3–39.4 years; mean cobalt concentration in air, 125 $\mu g/m^3$ with about a quarter of the workers having had exposures above 500 $\mu g/m^3$) and (subclinical) evidence for other effects of cobalt (thyroid metabolism and haematological parameters) (Swennen *et al.*, 1993). The absence of interstitial lung disease in workers exposed to cobalt-metal particles in the absence of other compounds such as tungsten carbide has recently been confirmed in another cross-sectional survey of 110 current and former cobalt refinery workers and 140 control workers in Finland (Linna *et al.*, 2003). These cross-sectional studies suggest (but do not prove) that exposure to even relatively high levels of cobalt-metal particles does not lead to interstitial lung disease (although such exposure does lead to asthma).

There is no published evidence for the occurrence of typical 'hard-metal lung disease' in workers exposed to cobalt-containing alloys, although adverse respiratory effects may be associated with the manufacture or maintenance of some cobalt-containing alloys (Deng *et al.*, 1991; Kennedy *et al.*, 1995). Dental technicians (who are exposed to a variety of agents, including cobalt) may also develop interstitial lung disease (Lob & Hugonnaud, 1977; De Vuyst *et al.*, 1986; Sherson *et al.*, 1988; Selden *et al.*, 1995).

Interstitial lung disease has not been described in workers exposed to cobalt salts, except for a study describing four cases of pulmonary fibrosis in a cobalt carbonate factory that operated before the Second World War (Reinl *et al.*, 1979).

It is conceivable that full-blown hard-metal lung represents a 'tip of the iceberg pheno-menon' and that there is other less specific pulmonary damage in many more subjects. The relationship of overt or latent disease with exposure levels remains unknown. This is due, in part, to the role of individual susceptibility factors, but also to the nature of the hard-metal industry, which is often composed of relatively small tool manufacturing plants or repair workshops, thus making large and comprehensive surveys of the industry rather difficult. In addition, epidemiological studies of a rare and specific condition, such as hard-metal lung, are also difficult because of the poor sensitivity of conventional epidemiological techniques such as questionnaire studies, pulmonary function testing and chest X-ray. Moreover, cross-sectional studies are not the best method to detect clinical cases of hard-metal lung disease, because of the healthy worker effect, and possibly also because of a 'healthy workshop effect' (Nemery et al., 1992). The latter refers to the frequently-experienced fact that the factories with the poorest occupational hygiene practice, and therefore probably those with the highest attack rates, are also the least likely to participate in health surveys (Auchincloss et al., 1992).

(c)   Extrapulmonary effects

Cobalt exerts a number of toxic effects outside the respiratory system (IARC, 1991; Lison, 1996), which are not specific for metallic cobalt-containing particles. Cobalt stimulates erythropoiesis, thus possibly causing polycythaemia, and has been used in the past for the treatment of anaemia (Alexander, 1972; Curtis et al., 1976). Cobalt is toxic to the thyroid (Kriss et al., 1955; Little & Sunico, 1958) and it is cardiotoxic (see IARC, 1991). The occurrence of cardiomyopathy in occupationally-exposed workers has been investigated and there is some evidence that it may occur, although this is still debated (Horowitz et al., 1988; Jarvis et al,. 1992; Seghizzi et al., 1994).

Possible neuropsychological sequelae, consisting of deficits in encoding or slowed memory consolidation, have been reported in patients with hard-metal disease (Jordan et al., 1990, 1997).

4.2.2   Experimental systems

Cobalt and its various compounds and/or alloys have been shown in experimental systems to produce non-neoplastic toxicity in different organs including the respiratory tract, the thyroid gland, erythropoietic tissue, myocardium and reproductive organs (Lison, 1996; National Toxicology Program, 1998; Barceloux, 1999). This section focuses on effects that may contribute to the evaluation of the carcinogenicity of inhaled hard-metal dusts and their components and is therefore limited mainly to studies examining effects on the respiratory tract.

(a)   Cobalt metal, hard metals and other alloys

(i)   Inflammation and fibrosis: in-vivo studies

A series of early experimental studies, initiated in the 1950s, explored the potential mechanisms of the respiratory diseases observed in workers in plants producing hard metal in Germany, the United Kingdom and the USA (see IARC, 1991). These studies were essentially designed to compare the effects of cobalt metal or oxide, tungsten, tungsten carbide and hard-metal mixtures.

Rats

Harding (1950) was probably the first to describe severe and fatal pulmonary oedema and haemorrhage in piebald rats administered cobalt-metal powder by intratracheal instillation (500 µg/rat), and suggested that this acute pulmonary toxicity might be related to the high solubility of cobalt metal in protein-containing fluids, presumably through some attachment of cobalt metal to protein.

Kaplun and Mezencewa (1960) found that the lung toxicity induced in rats by a single intratracheal instillation of cobalt-metal particles (5 or 10 mg/animal) was exacerbated by the simultaneous addition of tungsten or titanium (10 mg of a mixture containing 8–15% cobalt). Examination of the lungs after 4, 6 and 8 months revealed that pathological changes induced by the mixtures were identical to those produced by cobalt alone but more marked. The authors described a 'thickening' of the lung parenchyma with accumulation of lymphocytes, histiocytes and fibroblasts, hyperplasia of the walls of airways and blood vessels, and the presence of adenomas occurring several months after a single dose. The enhanced toxicity of the tungsten carbide–cobalt mixture was explained by the higher solubility of cobalt in the presence of tungsten (4–5-fold increase in 0.3% HCl during 24 h) (Kaplun & Mezencewa, 1960).

Kitamura et al. (1980) examined the pulmonary response of male Sprague-Dawley rats to a single administration of cemented tungsten carbide powder obtained after grinding pre-sintered alloy with diamond wheels. The powder was administered intratracheally at a dose of 23 mg/100 g bw. About 20% of the animals died during the first 3 days after exposure; histological examination of the lungs revealed marked haemorrhagic oedema with intense alveolar congestion. Among survivors, a transient reduction in body weight gain was also observed during the first week post-exposure. Six months after exposure, all sacrificed animals showed pulmonary lesions of patchy fibrosis in the vicinity of deposited dust (peri-bronchiolar and perivascular regions), occasionally associated with traction emphysema. There was no definitive inflammatory reaction nor interstitial pneumonitis (alveolitis). The lesions were suggested to result from condensation of collapsed alveoli without noticeable dense collagenization. In rats sacrificed at 12 months, the lesions had apparently regressed and two-thirds of the animals had neither fibrosis nor dust retention; the remaining animals showed changes similar to those observed at 6 months. The toxic effect on the lung was attributed, without experimental evidence, to the cytotoxic action of cobalt released from the particles. Neither cobalt metal nor tungsten carbide alone were tested.

Tozawa et al. (1981) examined the lung response to pre-sintered cemented carbides (WC:Co, 98:2 or WC:Co:TiC:TaC, 64:16:6:14) in male Sprague-Dawley rats, 6 and 12 months after a single intratracheal administration. They observed marked fibrotic foci after 6 months that were to some extent reversed 6 months later. They also noted that cobalt was eliminated more rapidly than tungsten from the lung. Neither cobalt metal nor tungsten carbide alone were tested.

Lasfargues *et al.* (1992) carried out studies in female Sprague-Dawley rats to compare the acute toxicity of hard-metal particles (WC–Co mixture containing 6% of cobalt-metal particles; d_{50}, 2 µm) with tungsten carbide particles (WC; cobalt content, 0.002%) and with an equivalent dose of cobalt-metal particles alone. After intratracheal instillation of a high dose of cobalt-metal particles (1 mg/100 g bw; median particle size d_{50}, 4 µm), a significantly increased lung weight was noted at 48 h. The lung weights of the animals exposed to WC (15.67 mg/100 g bw) were no different from those of control rats, but significant increases were noted in animals exposed to the hard metal (16.67 mg/100 g bw). These increases were much more substantial in the WC–Co group than in those animals instilled with an equivalent dose of cobalt particles alone. Increased mortality was observed in the group of animals exposed to WC–Co but not in those instilled with cobalt metal or WC alone. A second series of experiments with non-lethal doses (cobalt metal, 0.06 mg/ 100 g bw; tungsten carbide particles, 1 mg/100 g bw; hard-metal mixture, 1 mg/100 mg bw) was performed in order to analyse the cellular fraction of BALF and lung histology 24 h after dosing. While histological lung sections from rats instilled with cobalt alone or tungsten carbide particles were almost normal, an intense alveolitis was observed in rats exposed to the hard-metal mixture. In rats exposed to cobalt metal alone, no significant biochemical or cellular modifications in BALF were observed. Analysis of the cellular fraction of BALF from animals exposed to hard-metal particles showed a marked increase in the total cell number, similar to that induced by the same dose of crystalline silica; the increase in the neutrophil fraction was even more pronounced than that in the silica-treated group. Similarly, biochemical analyses of the cell-free fraction of BALF showed an increase in lactate dehydrogenase (LDH) activity, total protein and albumin concentration in the group instilled with hard metal, while exposure to the individual components of the mixture, i.e. Co or WC, did not produce any significant modification of these parameters (Lasfargues *et al.*, 1992). No change in the ex-vivo production of the inflammatory mediators interleukin-1 (IL-1) and TNF-α, a growth factor fibronectin or a proteinase inhibitor cystatin-c by lung phagocytes was found 24 h after administration of cobalt metal (0.06 mg/100 g bw), WC (1 mg/100 g bw) and WC–Co (1 mg/100 g bw) (Huaux *et al.*, 1995).

Lasfargues *et al.* (1995) also examined the delayed responses after single intratracheal administrations of tungsten carbide or hard-metal particles (WC or WC–Co, 1, 5 or 10 mg/ 100 g bw) or cobalt-metal particles (0.06, 0.3 or 0.6 mg/100 g bw) alone. The lung response to the hard-metal mixture was characterized by an immediate toxic response (increased cellularity and LDH, *N*-acetylglucosaminidase, total protein and albumin concentrations) in BALF followed by a subacute response after 28 days. The effects of cobalt or tungsten carbide alone were very modest, occurring at the highest doses only. Four months after instillation, fibrosis could not be identified histologically in the lungs of the animals treated with the hard-metal powder. This reversibility of the lesions was considered reminiscent of the natural history of hard-metal disease in humans. After repeated intratracheal administrations (once a month for 4 months) of the different particles (1 mg/100 g bw WC or WC–Co, or 0.06, 0.3 or 0.6 mg/100 g bw cobalt), no effect on the lung parenchymal architecture was observed in the groups treated with tungsten carbide or cobalt alone. In contrast, clear fibrotic lesions were observed in the group instilled with hard metal. No giant multi-nucleated cells were observed in BALF nor lung tissue of animals treated with WC–Co.

Kyono *et al.* (1992) examined the effect of ultrafine cobalt-metal particles (mean diameter, 20 nm) on the lungs of Sprague-Dawley-Jcl rats exposed by inhalation (2 mg/m³) for 5 h per day for 4 days. The rats were killed at 2 h, or at 3, 8 or 28 days after the end of exposure. Focal hypertrophy and proliferation of the lower airway epithelium, damaged macrophages and type I pneumocytes as well as proliferation of type II cells, fibroblasts and myofibroblasts were observed early after exposure. Morphological transformation of damaged type I cells to the 'juvenile' form (large nucleolus, abundant smooth endoplasmic reticulum, prominent Golgi apparatus and cytoplasm) was also reported, and interpreted as a sign of active biosynthesis and a capability of self-repair of this cell type. Cobalt was shown to be removed from the lung in two phases with estimated half-lives of 53 and 156 h, respectively. The morphological lesions caused by ultrafine cobalt under the presented conditions were reversible after 1 month: severe fibrosis was not detected in the lungs examined at 28 days. In a companion study, a single intratracheal instillation of ultra-fine cobalt metal (0.5 or 2 mg) into rats caused alveolar septal fibrosis detectable 15 months after treatment. Therefore, the authors noted that the possibility that fibrosis can develop after prolonged exposure to ultrafine cobalt metal must be considered.

Adamis *et al.* (1997) examined the lung response in male Sprague-Dawley rats exposed to respirable dust samples collected at various stages of hard-metal production in a plant in Hungary. Samples included finished powder for pressing (8% cobalt content), heat-treated, pre-sintered material (8% cobalt) and wet grinding of sintered hard metal (3% cobalt). The animals were administered 1 and/or 3 mg of dust suspended in saline and were killed after 1, 4, 7 or 30 days. Analyses of BALF (LDH, acid phosphatase protein and phospholipids) indicated the occurrence of an inflammatory reaction, a damage of the cell membrane and an increase of capillary permeability which varied with the type of powder used, with the pre-sintered sample showing the greatest toxicity. Histological studies showed that the pathological changes induced by the three powders were essentially the same, consisting of oedema, neutrophil and lymphocyte infiltration, together with an accumulation of argyro-philic fibres in the interalveolar septa and in the lumina of alveoli and bronchioli.

Zhang *et al.* (1998) compared ultrafine cobalt-metal particles (mean diameter, 20 nm; 47.9 m²/g surface area) with ultrafine nickel and titanium dioxide powders for their capa-city to produce inflammation after a single intratracheal instillation into male Wistar rats (1 mg/animal). All indices measured in BALF indicated that ultrafine nickel was the most toxic material. In the group of animals treated with cobalt particles, the lung:body weight ratio was significantly increased at days 1, 3, 7 and 15 after exposure and returned to normal after 30 days; LDH activity, total protein, lipid peroxide concentrations and inflam-matory cells in BALF were significantly increased for up to 30 days.

Guinea-pigs

A single intratracheal instillation of cobalt metal (10–50 mg) into guinea-pigs [strain not specified] was shown to result in the development of acute pneumonia with diffuse eosinophilic infiltration and bronchiolitis obliterans. The subchronic response assessed 8–12 months after the dose was characterized by the presence of multinucleated cells and a lack of cellular reaction within the alveolar walls. It was concluded that cobalt metal is not fibrogenic and does not provoke a chronic lesion in the regional lymph nodes (Delahant, 1955; Schepers, 1955a).

In the comparative studies (Schepers, 1955b), instillation of cobalt metal mixed with tungsten carbide (150 mg in a 9:91 ratio, i.e. a dose of 13.5 mg cobalt metal) into guinea-pigs [strain not specified] induced a transient inflammatory reaction with residual papillary hypertrophy of bronchial mucosa and peribronchial and periarterial fibrosis in the vicinity of retained particles. In inhalation experiments, a similar mixture (ratio 1:3) caused severe inflammation with focal pneumonia and bronchial hyperplasia and metaplasia, but fibrosing alveolitis was not observed after treatment with the cobalt metal–tungsten carbide mixture. Multinucleated giant cells were also found in animals treated with a combination of tungsten carbide and carbon (without cobalt); tungsten carbide alone was not tested (Schepers, 1955c).

Rabbits

Exposure of rabbits by inhalation to cobalt metal (0.2 and 1.3 mg/m^3) for 4 weeks did not produce any inflammatory reaction; the particles were not taken up by macrophages and the phagocytic capacity of these cells was not impaired (Johansson et al., 1980).

Mini-pigs

Kerfoot et al. (1975) exposed mini-pigs for 3 months (6 h per day, 5 days per week) to aerosols of cobalt metal at concentrations of 0.1 and 1.0 mg/m^3. The animals were first submitted to a sensitization period of 5 days' exposure to cobalt followed by a 10-day removal from exposure before the 3 months of exposure. Post-exposure lung function studies demonstrated a dose-dependent and reversible reduction in lung compliance but no radiographic or histological signs of fibrosis, except some increased collagen deposition which could only be detected electron-microscopically. The authors interpreted these changes as demonstrating functional impairment. Functional alterations were no longer detectable 2 months after the end of cobalt exposure. [The Working Group noted that the collagen increase was not assessed quantitatively.]

(ii)   *Cytotoxicity: in-vitro studies*

Mouse macrophage cells

Cytotoxic effects on mouse peritoneal macrophages of a range of metallic particles of orthopaedic interest have been examined. High doses of cobalt metal and cobalt–chromium alloy (0.5 mg metal/2 mL/3 × 10^6 cells) caused membrane damage as indicated by the increased release of cytoplasmic LDH. Decreased glucose-6-phosphate dehydrogenase (G6PD) activity was also observed after 10 h incubation with the same dose of metal (Rae, 1975).

In mouse peritoneal and alveolar macrophages, Lison and Lauwerys (1990) showed that the cytotoxicity of cobalt-metal particles, assessed by LDH release and morphological examination, was significantly increased in the presence of tungsten carbide particles either as industrial hard-metal powders or a WC–Co mixture reconstituted in the laboratory (6% cobalt in weight). Both particles (Co and WC) needed to be present simultaneously in order to exert their increased cytotoxic action. The interaction between tungsten carbide and cobalt particles was associated with an increased solubilization of cobalt in the culture medium. However, in the test system used, the toxicity of hard-metal particles could not be ascribed to solubilized cobalt ions, because the effects could not be reproduced with cobalt chloride or with cobalt ions solubilized from the WC–Co mixture (Lison & Lauwerys, 1992). In further studies, the cytotoxic potential of cobalt-metal particles, based on the measurement of glucose uptake, G6PD activity and superoxide anion production, was assessed *in vitro* in mouse peritoneal and alveolar macrophages incubated in a culture medium supplemented with 0.1% lactalbumin hydrolysate. Glucose uptake and superoxide anion production were significantly more depressed by a WC–Co mixture than by cobalt alone, while G6PD activity was decreased by both WC–Co and cobalt-metal particles alone (Lison & Lauwerys, 1991). Cobalt-metal particles (d_{50}, 4 µm and 12 µm; 10–100 µg/ 10^6 cells) affected cell integrity only marginally (Lison & Lauwerys, 1991).

Using the LDH release assay in mouse peritoneal macrophages, a similar toxic interaction between cobalt-metal particles and other metallic carbides (titanium, niobium and chromium carbides) but not latex beads (Lison & Lauwerys, 1992), crystalline silica, iron or diamond particles, was found. It was noted that the interaction between the carbides and the cobalt particles was dependent to some extent on the specific surface area of the particles suggesting the involvement of a surface chemistry (physicochemical) phenomenon (Lison & Lauwerys, 1995).

Lison *et al.* (1995) found that butylated hydroxytoluene protected macrophage cultures from the toxicity of a WC–Co (94:6) mixture, suggesting the involvement of lipid peroxidation in the cytotoxic activity of these particles. Lipid peroxidation was also demonstrated by the formation of thiobarbituric acid-reactive substances when arachidonic acid was incubated with WC–Co particles. Lison and Lauwerys (1993) had shown earlier that other enzymes that detoxify activated oxygen species such as catalase and superoxide dismutase (SOD), and scavengers such as sodium azide, benzoate, mannitol, taurine or methionine, did not protect against the cytotoxicity of WC–Co particles.

Rat fibroblasts, alveolar macrophages and type II pneumocyte cells

Thomas and Evans (1986) observed no effect of a cobalt–chromium–molybdenum alloy (0.5–10 mg/mL) on the proliferation of rat fibroblasts in culture nor on production of collagen.

Roesems *et al.* (1997), using the same experimental model as Lison and Lauwerys (1990; see above) but incubating the preparations in the absence of lactalbumin hydrolysate, which was replaced by foetal calf serum, showed that rat alveolar type II pneumocytes were less sensitive than alveolar macrophages to cobalt-metal particles *in vitro*, and that human type II pneumocytes were even less sensitive than rat type II pneumocytes. In contrast, using the dimethylthiazol diphenyl tetrazolium (MTT) assay, rat type II pneumocytes were found to be more sensitive than alveolar macrophages (25 μg/600 000 cells) and the toxicity of cobalt-metal particles could be reproduced by cobalt ions (Roesems *et al.*, 2000).

In the experimental system used by Roesems *et al.* (2000), an increased cytotoxicity of cobalt-metal particles associated with WC was confirmed in rat alveolar macrophage cell cultures, but not in type II pneumocytes. In contrast to the results presented by Lison and Lauwerys (1992), Roesems *et al.* (2000) found that cobalt ions played a role in the cytotoxic effect of cobalt-metal particles whether associated or not with WC. This difference was probably due to the presence of lactalbumin hydrolysate which was found to quench cobalt ions and may have masked their cytotoxicity in the experiments by Lison and Lauwerys (1992) and Lison (2000). *In vivo*, however, the bioavailability of cobalt ions is relatively limited because these cations precipitate in the presence of physiological concentrations of phosphates ($Co_3(PO_4)_2$; K_{sp}: 2.5×10^{-35} at 25 °C) (Lison *et al.*, 1995; 2001) and bind to proteins such as albumin (Merritt *et al.*, 1984).

In rat type II pneumocytes, cobalt-metal particles (15–1200 μg/3×10^6 cells) were found to stimulate the hexose monophosphate shunt in a dose- and time-dependent manner, indicating that these particles caused oxidative stress (Hoet *et al.*, 2002).

(iii)   *Biochemical effects*

As a transition element, cobalt shares a number of chemical properties with iron and thus it has been suggested that it may catalyse the decomposition of hydrogen peroxide by a Fenton-like mechanism. While several studies have indeed indicated that reactive oxygen species (ROS) are formed in the presence of a mixture of cobalt(II) ions and hydrogen peroxide (Moorhouse *et al.*, 1985), the exact nature of the radicals formed is still a matter of speculation. These free radicals have been proposed to account for several toxic properties of cobalt compounds, including their genotoxic activity.

Lison and Lauwerys (1993) using a deoxyribose degradation assay reported a significant formation of hydroxyl radicals *in vitro* when cobalt-metal particles (d_{50}, 4 μm; 6 μg/mL) were incubated with hydrogen peroxide. However, this effect was less than that seen with an equivalent concentration of cobalt(II) ions. The activity of cobalt-metal particles was increased about threefold when associated with tungsten carbide particles (d_{50}, 2 μm). It was also noted that the latter behaved as a strong oxidizing compound, but the exact role of this activity in the interaction with cobalt metal or cobalt(II) ions could not be elucidated.

Using electron spin resonance (ESR) and 5,5-dimethyl-1-pyrroline *N*-oxide (DMPO) spin trapping in a cell-free system, Lison *et al.* (1995) reported that cobalt-metal particles (d_{50}, 4 μm; 1 mg/mL phosphate buffer) produced small amounts of activated oxygen species, presumed to be hydroxyl radicals. This activity was observed in the absence of hydrogen peroxide and could not be reproduced with cobalt(II) ions, indicating that a Fenton-like mechanism was not involved. The production of activated oxygen species by cobalt-metal particles was markedly increased in the presence of tungsten carbide particles (Co:WC, 6:94). It was proposed that this reaction could be the consequence of a solid-solid interaction between particles whereby molecular oxygen is reduced at the surface of WC particles by electrons migrating from cobalt-metal particles, which are consequently oxidized and solubilized. The resulting Co(II) did not drive the production of ROS (see Figure 2). Further investigations of the surface interaction between cobalt metal and tungsten carbide particles (Zanetti & Fubini, 1997) indicated that the association of the two solids behaves like a new chemical entity, with physico-chemical properties different from those of the individual components, and which provides a lasting source of ROS as long as metallic cobalt is present. Radical generation originates from reactive oxygen formed at the carbide surface. When compared to other metals (iron, nickel), cobalt metal was the most active in the above reaction (Fenoglio *et al.*, 2000). In the presence of hydrogen peroxide, the WC–Co mixture exhibits a peroxidase-like activity (Fenoglio *et al.*, 2000; Prandi, 2002).

Figure 2. Mechanism proposed for release of reactive oxygen species (ROS) from buffered aqueous suspensions of cobalt/tungsten carbide (Co/WC) mixtures (hard metals)

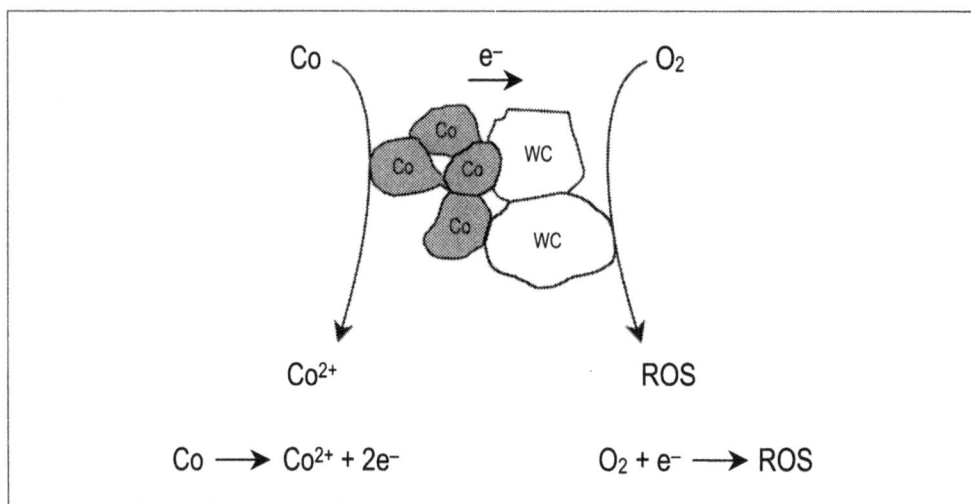

Adapted from Zanetti & Fubini (1997)
Cobalt is progressively oxidized and solubilized; oxygen is activated at the carbide surface.
e⁻, electron

Leonard *et al.* (1998), using ESR, confirmed that cobalt-metal particles in aqueous suspension reduced molecular oxygen. The authors proposed that the species generated is likely to be a cobalt(I)-bound superoxide anion (Co(I)-OO•) which exhibits strong oxidizing properties. This product was further shown in the presence of SOD, to generate hydrogen peroxide which reacts with Co(I) to produce a hydroxyl radical and Co(II) via a Co(I)-

mediated Fenton-like reaction. In the presence of proper chelators, such as glutathione (GSH), Gly-Gly-His and anserine, the cobalt(II) ions formed by the molecular oxygen oxidation of cobalt produce hydroxyl radicals and Co(III) through a Co(II)-mediated Fenton-like mechanism.

Keane *et al.* (2002) confirmed the generation of hydroxyl radicals by hard-metal materials in aqueous suspension by examining the properties of detonation coating, a hard metal-related material made of a WC and cobalt mixture (6.7 and 5.4% of cobalt in pre- and post-detonation powders, respectively). The post-detonation powder was a much stronger generator of hydroxyl radicals than the pre-detonation material.

(b)   Other relevant cobalt compounds

(i)   Inflammation and fibrosis: in-vivo studies

In a dose-finding study for a carcinogenicity assay (Bucher *et al.*, 1990), male and female Fischer 344/N rats and B6C3F$_1$ mice were exposed to cobalt sulfate heptahydrate aerosols of 0, 0.3, 1.0, 3.0, 10 or 30 mg/m^3 for 6 h per day on 5 days per week for 13 weeks. The main histopathological effects in both species were limited to the respiratory tract. Lesions included degeneration of the olfactory epithelium, squamous metaplasia of the respiratory epithelium, inflammation in the nose, and fibrosis, histiocytic infiltrates, bronchiolar epithelial regeneration and epithelial hyperplasia in the alveoli of the lungs. In rats, inflammation, necrosis, squamous metaplasia, ulcers and inflammatory polyps of the larynx were observed; mice developed metaplasia of the trachea. The most sensitive tissue in rats was the larynx: squamous metaplasia was observed with the lowest exposure concentration (0.3 mg/m^3). Degeneration of the olfactory epithelium was noted at the two highest doses tested (10 and 30 mg/m^3). A no-observed-adverse-effect level was not reached in these studies.

In the subsequent carcinogenicity study (0, 0.3, 1.0 or 3.0 mg/m^3 cobalt sulfate heptahydrate, 6 h per day, 5 days per week for 104 weeks) conducted in the same rodent strains (National Toxicology Program, 1998; Bucher *et al.*, 1999) (see also Section 3.1.1), similar non-neoplastic effects were noted. Degeneration of olfactory epithelium was more pronounced than in the 13-week dose-finding study and was observed at the lowest dose tested (0.3 mg/m^3). In rats, proteinosis, alveolar epithelial metaplasia, granulomatous alveolar inflammation and interstitial fibrosis were observed at all dose levels. The non-neoplastic lesions were less severe in mice and mainly consisted of cytoplasmic vacuolization of the bronchi. Diffuse and focal histiocytic cell infiltrations were also observed in lungs of mice with neoplasms and were therefore considered to be a consequence of the neoplasms rather than a primary effect of cobalt sulfate.

Wehner *et al.* (1977) examined the influence of lifetime inhalation of cobalt oxide (10 μg/L, 7 h/day, 5 days/week) alone or in combination with cigarette smoke in male Syrian golden hamsters with the aim of detecting a carcinogenic effect. No tumorigenic action of cobalt oxide was observed and there was no additive effect of exposure to smoke.

Some pulmonary changes consisting of focal interstitial fibrosis, granulomas, hyperplasia of alveolar cells and emphysema were observed in animals exposed to cobalt oxide alone.

Lewis *et al.* (1991) showed that the intratracheal instillation of cobalt chloride (1–1000 μg/kg) into Syrian golden hamster lungs induced biochemical changes compatible with the development of oxidative stress (i.e. decreased concentrations of reduced glutathione, increased concentrations of oxidized glutathione, and stimulation of the pentose phosphate pathway). Similar changes were also observed *in vitro* after incubation of lung slices with cobalt chloride (0.1–10 mM), and preceded the detection of cellular toxicity indicating their possible early involvement in the pulmonary toxicity of cobalt (II) ions. It was later shown in the same in-vitro model that simultaneous treatment with hydrogen peroxide or 1,3-bis(2-chloroethyl)-1-nitrosourea, a glutathione reductase inhibitor, potentiated the oxidative stress induced by cobalt chloride; however, this effect was not associated with an enhancement of cell dysfunction observed with cobalt chloride alone or cobalt chloride and hydrogen peroxide together. Furthermore, on the basis of comparative analysis of the results with the known oxidant tert-butyl hydroperoxide, glutathione oxidation did not appear to be the cause of the cellular dysfunction caused by cobalt chloride (Lewis *et al.*, 1992).

In a series of studies on the effects of various cobalt compounds on rabbit lung morphology (Johansson *et al.*, 1983, 1986), exposure to 0.4, 0.5 and 2.0 mg/m³ soluble cobalt chloride for 1 and 4 months increased the number of alveolar macrophages and their oxidative metabolic activity. Exposure to 0.4 and 2 mg/m³ cobalt chloride for 14–16 weeks (6 h/day, 5 days/week) induced a combination of lesions characterized by nodular aggregation of type II pneumocytes, abnormal accumulation of enlarged, vacuolated alveolar macrophages and interstitial inflammation (Johansson *et al.*, 1987). The effect of cobalt chloride (0.5 mg/m³ for 4 months) on rabbit lung, i.e. formation of noduli of type II cells, was potentiated by simultaneous exposure to nickel chloride administered at the same dose (Johansson *et al.*, 1991). Camner *et al.* (1993) reported that the inflammatory reaction (as indicated by the presence of neutrophils and eosinophils in BALF) induced by the inhalation of cobalt chloride (2.4 mg/m³, 6 h/day for 2 weeks) was more pronounced in guineapigs that had been pre-sensitized to cobalt by repeated application of cobalt chloride.

(ii)   *Cytotoxicity: in-vitro studies*

At relatively high doses (0.1–1 mM), cobalt(II) ions have been shown to inhibit exocytosis and respiratory burst in rabbit neutrophils through an interaction with a calcium-dependent intracellular mechanism (Elferink & Deierkauf, 1989).

In U-937 cells and human alveolar macrophages, cobalt ions (0.5–1 mM as cobalt chloride) induced apoptosis and accumulation of ubiquitinated proteins. It was suggested that cobalt-induced apoptosis contributed to cobalt-induced lung injury (Araya *et al.*, 2002). In neuronal PC12 cells cobalt chloride triggered apoptosis in a dose- and time-dependent manner, presumably via the production of ROS and the increase of the DNA-binding activity of transcriptor factor AP-1 (Zou *et al.*, 2001). A subsequent study showed that caspase-3 and p38 mitogen-activated protein kinase-mediated apoptosis was induced by cobalt chloride in PC12 cells (Zou *et al.*, 2002).

Microtubule disorganization has been reported in 3T3 cells exposed to high concentrations of cobalt sulfate (100 μM) for 16 h (Chou, 1989).

Soluble cobalt compounds (40 μM [5 μg/mL] as cobalt chloride), but not particulate materials, have been reported to induce cytotoxicity and neoplastic transformation in the C3H10T½ assay (Doran *et al.*, 1998) (see Section 4.4).

(iii) *Biochemical effects*

Using two assays to detect hydroxyl radicals (HO•), based either on the degradation of deoxyribose or the hydroxylation of phenol or salicylate, Moorhouse *et al.* (1985) found that, in an acellular system at physiological pH, cobalt(II) ions promoted the formation of hydroxyl-like radicals in the presence of hydrogen peroxide (H_2O_2, 1.44 mM); the formation of the radicals was decreased by catalase, but not by SOD or ascorbic acid. Ethylenediaminetetraacetic acid (EDTA) in excess of Co(II) accelerated the formation of ROS, and hydroxyl radical scavengers such as mannitol, sodium formate, ethanol or urea, blocked deoxyribose degradation by the cobalt(II)–H_2O_2 mixture. Lison and Lauwerys (1993) reported similar findings, i.e. a significant degradation of deoxyribose in the presence of cobalt(II) (0.1 mM) mixed with hydrogen peroxide (1.44 mM), suggesting the formation of hydroxyl radicals.

Using an ESR spin-trapping technique (with DMPO), Kadiiska *et al.* (1989) found that cobalt(II) ions, unlike iron(II) ions, did not react with hydrogen peroxide by the classic Fenton reaction at physiological pH, either in a chemical system or in rat liver microsomes. They suggested that superoxide anions, not hydroxy radicals, were primarily formed. In a subsequent study using the same technique, Hanna *et al.* (1992) used several ligands to complex cobalt(II) ions and further documented the formation of superoxide anions, but not hydroxyl radicals, in the presence of hydrogen peroxide.

Using ESR, Wang *et al.* (1993) detected the ascorbic acid radical *in vivo* in circulating blood after intravenous administration of ascorbic acid (100 mM) and cobalt(II) at two separate sites into male Sprague-Dawley rats. Similar but less intense signals were also observed with nickel(II) and iron(II) ions. The formation of the ascorbic acid radical was interpreted as the in-vivo formation of free radicals in animals overloaded with cobalt(II) ions; the mechanism of this radical formation was, however, not addressed. The authors suggested that their findings might explain the mechanism of the toxicity observed in workers exposed to cobalt-containing materials.

The in-vitro generation of ROS by cobalt(II) from hydrogen peroxide and related DNA damage have also been examined by Mao *et al.* (1996). The formation of hydroxyl radicals and/or singlet oxygen ($^1O_2$) showed that the oxidation potential of cobalt(II) could be modulated by several chelators such as anserine or 1,10-phenanthroline. Shi *et al.* (1993) examined the modulation of ROS production from cobalt(II) ions and hydroperoxides and showed that several chelating agents, including endogenous compounds such as reduced GSH, facilitated the production of these species.

Sarkar (1995) hypothesized that oligopeptides or proteins represent other ligands that can modulate the redox potential of cobalt(II) ions. The presence of such proteins (histones) in the nucleus might allow the production of ROS in close proximity to biologically-relevant targets such as DNA. It has also been suggested that the ability of cobalt(II) to substitute for zinc(II) in the DNA-binding domain of nuclear (transcription factor) proteins might allow the in-situ formation of free radicals that may damage genetic regulatory/response elements and may explain the mutagenic potential of these metals.

(iv)   *Other effects*

Cobalt also interferes with cellular mechanisms that control the degradation of regulatory proteins such as p53, which is involved in the control of the cell cycle, genome maintenance and apoptosis. An *et al.* (1998) reported that, in mammalian cells, cobalt chloride (100 μM) activates hypoxia-inducible factor-1α which in turn induces accumulation of p53 through direct association of the two proteins. Cobalt sulfate (50 μg/mL [178 μM]) has been shown to induce p53 proteins in mouse cells treated *in vitro* (Duerksen-Hughes *et al.*, 1999). Inhibition of proteasome activity by cobalt (1 mM), subsequent accumulation of ubiquitinated proteins and increased apoptosis have been reported in human alveolar macrophages and U-937 cells (Araya *et al.*, 2002) (See Section 4.4.2). Whether these biochemical mechanisms are involved in the carcinogenic responses observed with some cobalt compounds remains, however, to be examined.

4.3   Reproductive and developmental effects

Only a few studies have been conducted with soluble cobalt compounds to explore their potential effects on development.

Wide (1984) reported that a single intravenous injection of cobalt chloride hexahydrate into pregnant NMRI mice (5 mM per animal in the tail vein; [120 μg/animal]) on day 8 of gestation significantly affected fetal development (71% of skeletal malformations versus 30% in controls); in animals injected at day 3 of gestation, no interference with implantation was noted. In the same experiment but replacing cobalt chloride by tungstate (25 mM of W per animal; [460 μg/animal]) a significant increase in the number of resorptions was observed (19% versus 7% in controls), but no skeletal malformations.

In a study undertaken by Pedigo and colleagues (1988), following 13 weeks of chronic exposure to 100 to 400 ppm [100–400 μg/mL] cobalt chloride in drinking water, male CD-1 mice showed marked dose-related decreases in fertility, testicular weight, sperm concentration and motility, and increases in circulating levels of testosterone. Pedigo and Vernon (1993) reported that cobalt chloride (400 ppm in drinking-water for 10 weeks) increased pre-implantation losses per pregnant female in the dominant lethal assay by compromising the fertility of treated male mice.

Paksy *et al.* (1999) found that in-vitro incubation of postblastocyst mouse embryos with cobalt(II) ions (as cobalt sulfate) adversely affected the development stages at a concentration of 100 μM and decreased the trophoblast area (at a concentration of 10 μM).

In pregnant Wistar rats, oral administration of cobalt(II) ions as cobalt chloride (12, 24 or 48 mg/kg bw per day from day 14 of gestation through to day 21 of lactation) significantly affected the late period of gestation as well as postnatal survival and development of the pups. Signs of maternal toxicity were apparently also noted but the details are not reported (Domingo et al., 1985).

A study conducted in pregnant Sprague-Dawley rats (Paternain et al., 1988) concluded that the administration of cobalt chloride (up to a dose of 100 mg/kg by gavage, from day 6–15 of gestation) was not embryotoxic nor teratogenic, despite signs of maternal toxicity.

Sprague-Dawley rats maintained on diets (15 g per day) containing 265 ppm [31.8 mg/kg bw per day] cobalt for up to 98 days showed degenerative changes in the testis from day 70 to the end of the treatment; given that cobalt was not detected in testis, these changes were considered secondary to hypoxia due to blockage of veins and arteries by red blood cells and changes in permeability of the vasculature and seminiferous tubules (Mollenhaur et al., 1985). Decreased sperm motility and/or increased numbers of abnormal sperm were noted in mice, but not in rats, exposed to 3 mg/m^3 or higher concentrations (30 mg/m^3) in 13-week inhalation studies with cobalt sulfate (National Toxicology Program, 1991).

The fetal and postnatal developmental effects of cobalt sulfate have been compared in C57BL mice, Sprague-Dawley rats and/or New Zealand rabbits (Szakmáry et al., 2001). Several developmental alterations (elevated frequency of fetuses with body weight or skeletal retardation, embryolethality, increased anomalies in several organs) were observed in pregnant mice and rats treated with cobalt sulfate by gavage on days 1–20 of gestation (25, 50 or 100 mg/kg bw per day, respectively). In rabbits, cobalt sulfate at 20 mg/kg bw was embryotoxic with inhibition of skeletal development. No teratogenic effects were noted in rabbits treated with up to 200 mg/kg per day during days 6–20 of gestation. Postnatal developmental parameters were transiently altered in the pups of rats treated daily with 25 mg/kg cobalt sulfate. [The Working Group noted that the doses used in these studies were relatively high and produced maternal toxicity. The interpretation of these data should, therefore, be considered with caution].

4.4   Genetic and related effects

4.4.1   *Humans*

(*a*)   *Sister chromatid exchange*

Five studies have been conducted to date on the possible cytogenetic effects induced by cobalt compounds in lymphocytes (or leukocytes) of individuals exposed to metals.

Results of sister chromatid exchange have been obtained in two studies in which exposure was to a mixture of metals. Occupational exposure to metals was studied by Gennart et al. (1993) who determined sister chromatid exchange in 26 male workers (aged 23–59 years) exposed to cobalt, chromium, nickel and iron dust in a factory producing metal powder and in 25 controls (aged 24–59 years), who were clerical workers, matched for age, smoking habits and alcohol consumption. The metal particle sizes ranged from 2 to 100 μm. Slight exposure to nickel or chromium oxides could not be excluded, since, at one stage of the production process, the metals are melted in an oven. The workers had been employed for at least 2 years (range, 2–20 years). The atmospheric concentrations of cobalt were measured at two different work areas in 1986 and in 1989, at the time of the cytogenetic survey. An improvement in the local exhaust ventilation system took place between the two sampling times. At the work area where the ovens were located, the (geometric) mean cobalt concentration in the air (based on 4–8 values) was 92 μg/m^3 in 1986 and 40 μg/m^3 in 1989. At the second work area, the individual values ranged from 110 to 164 μg/m^3 in 1986 and from 10 to 12 μg/m^3 in 1989. The differences in the concentrations of cobalt in the urine in exposed persons (cobalt geometric mean, 23.6 μg/g creatinine; range, 6.4–173.1) and controls (cobalt geometric mean, 1.1 μg/g creatinine; range, 0.2–3.2) were statistically significant. Analysis of variance revealed that both exposure status (exposed versus controls) and smoking habits (smokers and former smokers versus never smokers) had statistically-significant effects on the sister chromatid exchange or high-frequency cell (HFC) rank values. These effects may not be attributable to cobalt alone.

Stea *et al.* (2000) compared sister chromatid exchange in patients who had chrome–cobalt alloy prostheses and in those with other metal alloys. The study population consisted of 30 patients (11 men and 19 women; mean age, 63.8 years; range, 33–78) with joint (28 hip and two knee) prostheses and 17 control subjects (11 men and six women; mean age, 58.65 years; range, 40–71) matched for age, sex, and exposure to occupational and environmental risk factors such as chemicals, antineoplastic drugs and traffic smog. Ten subjects (mean age, 65.1 years; range, 51–76) had prostheses made of titanium–aluminium–vanadium alloys, 14 subjects (mean age, 61.9 years; range, 33–75) had prostheses made of chrome–cobalt alloys and five (mean age, 65 years; range, 57–78) had mixed prostheses. Of the prostheses, 13 were cemented (in some cases only one component was cemented) and 17 were cementless. The average duration of the implant was 7.5 years (range, 0.5–25) for the hip prostheses and 2.5 years for the two knee prostheses. The mean sister chromatid exchange rate in subjects with prostheses (5.2 ± 1.5) was not statistically different from that in subjects without prostheses (4.4 ± 1.3). Subjects with titanium–aluminium–vanadium alloy prostheses had a significantly higher sister chromatid exchange frequency (6.3 ± 2.3) than the controls (4.4 ± 1.3) whereas subjects with prostheses made of chrome–cobalt alloys or mixed prostheses had a higher, but not significantly, sister chromatid exchange frequency (4.7 ± 1.1 and 5.0 ± 2.1, respectively) than the controls. The number of sister chromatid exchanges was not affected by the presence of bone-cement used in prosthesis fixation nor by duration of the implant. There was no difference in the incidence of sister chromatid exchange between the two populations (those with prostheses and controls) considered globally and the considered risk factors, including smoking. The HFC values (> 9 exchanges per cell) were also recorded. Among the cases studied, three patients with implants (one with a prosthesis made of chrome–cobalt alloy and two with

mixed prostheses) showed markedly elevated percentages of HFCs (> 10%). It was concluded that the indication of possible cytogenetic damage in the patient populations should
be considered with caution, since the sample population was small.

(b) Micronuclei, DNA damage

Burgaz *et al.* (2002) applied the micronucleus test to assess the effect of occupational
exposure to metal alloys in both exfoliated nasal cells, and *in vitro* in lymphocytes, with
the cytochalasin-B technology which allows discrimination between lymphocytes that
have divided once during the in-vitro culture periods (binucleates) and those that have not
(mononucleates). The groups studied consisted of 27 male dental laboratory technicians
(mean age, 29.2 ± 10.8 years) exposed to metal alloys (35–65% cobalt, 20–30% chromium,
0–30% nickel) in dental laboratories during the production of skeletal prostheses, and 15
male controls (mean age, 28.4 ± 9.5 years) from the faculty of pharmacy. The differences
in concentrations of cobalt in urine of technicians and controls were statistically significant
(urinary cobalt, 0.12 ± 0.24 µg/g creatinine in controls and 24.8 ± 24.1 µg/g creatinine in
technicians). The mean frequencies of micronucleated binucleates among peripheral lymphocytes were significantly higher (4.00 ± 2.98) in the dental technicians than in controls
(1.40 ± 1.30). A statistically-significant difference was also found between the mean frequencies of micronuclei in nasal cells among the dental technicians (3.5 ± 1.80) and the
controls (1.19 ± 0.53). The correlation between duration of exposure (13.1 ± 9.1 years) and
frequencies of micronuclei was statistically significant in lymphocytes, but not in nasal
cells of technicians. The results of multifactorial variance analysis revealed that occupational exposure was the only factor that significantly influenced the induction of micronuclei. In the exposed group, a significant correlation was found between urinary cobalt
concentrations and frequencies of micronuclei in nasal cells, but not in lymphocytes.

The possible genotoxic effects of occupational exposure to cobalt alone or to hard-
metal dust (WC–Co) was explored in a study using the in-vitro cytochalasin-B micronucleus test in lymphocytes as end-point for mutations (De Boeck *et al.*, 2000). Micronuclei were scored both as binucleates and as mononucleates to discriminate between
micronuclei accumulated during chronic exposure *in vivo* (mononucleates) and additional
micronuclei expressed during the culture period *in vitro* (binucleates). The authors aimed
to assess genotoxic effects in workers from cobalt refineries and hard-metal plants who
were exposed at the time of the study to the TLV/time-weighted average (TWA) of cobalt-
containing dust. The study comprised three groups of male workers: 35 workers (mean
age, 38.5 ± 7.7 years; range, 27.7–55.3) exposed to cobalt dust from three refineries, 29
workers (mean age, 40.7 ± 12.4 years; range, 20.7–63.6) exposed to hard-metal dust
(WC–Co) from two production plants and 27 matched control subjects (mean age, 38.0 ±
8.8 years; range, 23.3–56.4) recruited from the respective plants. In these three groups, the
(geometric) mean concentration of cobalt in urine was 21.5 µg/g creatinine (range,
5.0–82.5) in workers exposed to cobalt, 19.9 µg/g creatinine (range, 4.0–129.9) in workers
exposed to hard-metal dust and 1.7 µg/g creatinine (range, 0.6–5.5) in controls. The study
design integrated additional complementary biomarkers of DNA damage: 8-hydroxy-

deoxyguanosine (8-OHdG) in urine, DNA single-strand breaks and formamido-pyrimidine DNA glycosylase (FPG)-sensitive sites with the alkaline Comet assay in mononuclear leukocytes. No significant increase in genotoxic effects was detected in workers exposed to cobalt-containing dust compared with controls. No difference in any genotoxicity biomarker was found between workers exposed to cobalt and to hard-metal dusts. The only statistically-significant difference observed was a higher frequency of micronucleated binucleate cytokinesis-blocked lymphocytes in workers exposed to cobalt compared to workers exposed to hard-metal dusts, but not in comparison with their concurrent controls. The frequency of micronucleated mononucleates did not vary among the different worker groups. Multiple regression analysis indicated that workers who smoked and were exposed to hard-metal dusts had elevated 8-OHdG and micronucleated mononucleate values. The authors concluded that workers exposed solely to cobalt-containing dust at TLV/TWA (20 μg cobalt/g creatinine in urine, equivalent to TWA exposure to 20 μg/m^3) did not show increased genotoxic effects but that workers who smoked and were exposed to hard-metal dusts form a specific occupational group which needs closer medical surveillance.

Hengstler *et al.* (2003) concluded from a study of workers co-exposed to cadmium, cobalt, lead and other heavy metals, that such mixed exposure may have genotoxic effects. The authors determined DNA single-strand break induction by the alkaline elution method in cryopreserved mononuclear blood cells of 78 individuals co-exposed to cadmium (range of concentrations in air, 0.05–138 μg/m^3), cobalt (range, 0–10 μg/m^3) and lead (range, 0–125 μg/m^3) and of 22 subjects without occupational exposure to heavy metals (control group). Non-parametric correlation analysis showed significant correlations between DNA single-strand breaks and cobalt ($p < 0.001$; $r = 0.401$) and cadmium ($p = 0.001$; $r = 0.371$) concentrations in air, but not lead concentrations. They elaborated a model with a logistic regression analysis and concluded from it that more than multiplicative effects existed for co-exposure to cadmium, cobalt and lead. Some concerns about the study were addressed by Kirsch-Volders and Lison (2003) who concluded that it did not provide convincing evidence to support the alarming conclusion of Hengstler *et al.* (2003).

4.4.2   *Experimental systems* (see Table 15 for references)

(a)   *Metallic cobalt*

The results of tests for genetic and related effects of metallic cobalt, cobalt-metal alloys and cobalt (II) and (III) salts, with references, are given in Table 15.

Cobalt metal is active not only as a solid particle but also as a soluble compound.

The genetic toxicology of cobalt compounds has been reviewed by Domingo (1989), Jensen and Tüchsen (1990), Léonard and Lauwerys (1990), Beyersmann and Hartwig (1992), Hartwig (1995), Lison *et al.* (2001), National Institute of Environmental Health Sciences (2002) and De Boeck *et al.* (2003a). A report of the European Congress on Cobalt and Hard Metal Disease, summarizing the state of the art was published by Sabbioni *et al.* (1994b). The interactions of cobalt compounds with DNA repair processes (Hartwig, 1998;

Table 15. Genetic and related effects of cobalt

Test system	Result[a] Without exogenous metabolic system	Result[a] With exogenous metabolic system	Dose[b] (LED/HID)	Reference
Cobalt				
DNA breaks, alkaline elution, purified DNA (3T3 mouse cells)	+[c]		1 µg/mL (d_{50} = 4 µm)	Anard et al. (1997)
DNA breaks, alkaline elution, purified DNA (3T3 mouse cells)	r[c,d]		1 µg/mL + Na formate (d_{50} = 4 µm)	Anard et al. (1997)
Cell transformation, C3H10T1/2 mouse fibroblast cells, in vitro	–[c]		500 µg/mL (d_{50} ≤ 5 µm)	Doran et al. (1998)
Induction of FPG-sensitive sites, alkaline Comet assay, human mononuclear leukocytes, in vitro	–[c]		6 µg/mL (d_{50} = 4 µm)	De Boeck et al. (1998)
DNA breaks, alkaline elution, human lymphocytes, in vitro	+[c]		3 µg/mL (d_{50} = 4 µm)	Anard et al. (1997)
DNA single-strand breaks and alkali-labile sites, alkaline Comet assay, human mononuclear leukocytes, in vitro	+[c]		4.5 µg/mL (d_{50} = 4 µm)	Anard et al. (1997)
DNA single-strand breaks and alkali-labile sites, alkaline Comet assay, human mononuclear leukocytes, in vitro	+[c]		0.6 µg/mL (d_{50} = 4 µm)	Van Goethem et al. (1997)
DNA single-strand breaks and alkali labile sites, alkaline Comet assay, human mononuclear leukocytes, in vitro	+[c]		0.3 µg/mL (d_{50} = 4 µm)	De Boeck et al. (1998)
DNA single-strand breaks and alkali labile sites, human mononuclear leukocytes, in vitro	+		0.6 µg/mL (d_{50} = 4 µm)	De Boeck et al. (2003b)
DNA repair inhibition, alkaline Comet Assay, human mononuclear leukocytes, in vitro	+		5.5 µg/mL MMS, post-treatment 1.2 µg/mL Co (d_{50} = 4 µm)	De Boeck et al. (1998)
DNA repair inhibition, alkaline Comet Assay, human mononuclear leukocytes, in vitro	+		co-exposure 5.5 µg/mL MMS, 1.2 µg/mL Co (d_{50} = 4 µm)	De Boeck et al. (1998)
Micronucleus formation, binucleates, cytochalasin-B assay, human lymphocytes, in vitro	+[c]		0.6 µg/mL (d_{50} = 4 µm)	Van Goethem et al. (1997)
Micronucleus formation, binucleates, cytochalasin-B assay, human lymphocytes, in vitro	+		3 µg/mL (d_{50} = 4 µm)	De Boeck et al. (2003b)
Cell transformation (foci), human non-tumorigenic osteosarcoma osteoblast-like in vitro	–		3 µg/mL (d_{50} = 1–4 µm)	Miller et al. (2001)

Table 15 (contd)

Test system	Result[a]		Dose[b] (LED/HID)	Reference
	Without exogenous metabolic system	With exogenous metabolic system		
Cobalt alloys				
Co–Cr alloy				
Cell transformation, C3H10T1/2 mouse fibroblast cells, *in vitro*	–[c]		500 µg/mL ($d_5 \leq 5$ µm)	Doran *et al.* (1998)
rW–Ni–Co alloy				
DNA single-strand breaks, alkaline elution, human non-tumorigenic osteosarcoma cells, *in vitro*	+		5 mg/mL (d_{50} = 1–5 µm)	Miller *et al.* (2002)
Sister chromatid exchange, human non-tumorigenic osteosarcoma cells, *in vitro*	+		5 mg/mL (d_{50} = 1–5 µm)	Miller *et al.* (2002)
Micronucleus formation, human non-tumorigenic osteosarcoma osteoblast-like cells, *in vitro*	+		25 µg/mL (d_{50} = 1.5 µm)	Miller *et al.* (2001)
Micronucleus formation, human non-tumorigenic osteosarcoma osteoblast-like cells, *in vitro*	+		5 mg/mL (d_{50} = 1–5 µm)	Miller *et al.* (2002)
Cell transformation (foci), human non-tumorigenic osteosarcoma osteo-blast-like cells, *in vitro*	+		50 µg/mL (d_{50} = 1–5 µm)	Miller *et al.* (2001)
Cell transformation (foci), human non-tumorigenic osteosarcoma osteo-blast-like cells, *in vitro*	+		10 mg/mL (d_{50} = 1–5 µm)	Miller *et al.* (2002)
Cobalt-containing metal carbides				
Cr_3C_2–Co				
DNA single-strand breaks and alkali-labile sites, alkaline Comet assay, human mononuclear leukocytes *in vitro*	?		0.6 µg Co eq./mL	De Boeck *et al.* (2003b)
DNA single-strand breaks and alkali-labile sites, alkaline Comet assay, human mononuclear leukocytes, *in vitro*	e[d]		6 µg Co eq./mL	De Boeck *et al.* (2003b)
Micronucleus formation, binucleates, cytochalasin-B assay, human lymphocytes, *in vitro*	+		3 µg Co eq./mL	De Boeck *et al.* (2003b)
Micronucleus formation, binucleates, cytochalasin-B assay, human lymphocytes, *in vitro*	e[d]		6 µg Co eq./mL	De Boeck *et al.* (2003b)

Table 15 (contd)

Test system	Result[a] Without exogenous metabolic system	With exogenous metabolic system	Dose[b] (LED/HID)	Reference
Mo₂C-Co				
DNA single-strand breaks and alkali-labile sites, alkaline Comet assay, human mononuclear leukocytes, *in vitro*	?		0.6 μg Co eq./mL	De Boeck *et al.* (2003b)
DNA single-strand breaks and alkali-labile sites, alkaline Comet assay, human mononuclear leukocytes, *in vitro*	s[d]		6 μg Co eq./mL	De Boeck *et al.* (2003b)
Micronucleus formation, binucleates, cytochalasin-B assay, human lymphocytes, *in vitro*	−		6 μg Co eq./mL	De Boeck *et al.* (2003b)
Micronucleus formation, binucleates, cytochalasin-B assay, human lymphocytes, *in vitro*	s[d]		6 μg Co eq./mL	De Boeck *et al.* (2003b)
NbC-Co				
DNA single-strand breaks and alkali-labile sites, alkaline Comet assay, human mononuclear leukocytes, *in vitro*	?		0.6 μg Co eq./mL	De Boeck *et al.* (2003b)
DNA single-strand breaks and alkali-labile sites, alkaline Comet assay, human mononuclear leukocytes, *in vitro*	s[d]		6 μg Co eq./mL	De Boeck *et al.* (2003b)
Micronucleus formation, binucleates, cytochalasin-B assay, human lymphocytes, *in vitro*	+		3 μg Co eq./mL	De Boeck *et al.* (2003b)
Micronucleus formation, binucleates, cytochalasin-B assay, human lymphocytes, *in vitro*	e[d]		6 μg Co eq./mL	De Boeck *et al.* (2003b)
WC-Co				
Induction of FPG-sensitive sites, alkaline Comet assay, human mononuclear leukocytes, *in vitro*	−[c]		6 μg Co eq./mL (Co d$_{50}$ = 4 μm) (WC d$_{50}$ < 1 μm)	De Boeck *et al.* (1998)
DNA breaks, alkaline elution, alkaline Comet assay, human lymphocytes, *in vitro*	+[c] s[e]		1.5 μg Co eq/mL (Co d$_{50}$ = 4 μm) (WC d$_{50}$ = 2 μm)	Anard *et al.* (1997)
DNA single-strand breaks and alkali-labile sites, alkaline Comet assay, human lymphocytes, *in vitro*	+[c]		3 μg Co eq./mL (Co d$_{50}$ = 4 μm) (WC d$_{50}$ = 2 μm)	Anard *et al.* (1997)
DNA single-strand breaks and alkali-labile sites, alkaline Comet assay, human mononuclear leukocytes, *in vitro*	+[c]		0.6 μg Co eq./mL (Co d$_{50}$ = < 1 μm) (WC d$_{50}$ = 4 μm)	Van Goethem *et al.* (1997)

Table 15 (contd)

Test system	Result[a] Without exogenous metabolic system	With exogenous metabolic system	Dose[b] (LED/HID)	Reference
DNA single-strand breaks and alkali-labile sites, alkaline Comet assay, human mononuclear leukocytes, in vitro	e[d]		6 μg Co eq./mL (Co d_{50} = 4 μm) (WC d_{50} = <1 μm)	Van Goethem et al. (1997)
DNA single-strand breaks and alkali-labile sites, alkaline Comet assay, human mononuclear leukocytes, in vitro	+[c]		0.3 μg Co eq./mL (Co d_{50} = 4 μm) (WC d_{50} = <1 μm)	De Boeck et al. (1998)
DNA single-strand breaks and alkali-labile sites, alkaline Comet assay, human mononuclear leukocytes, in vitro	e[d]		6 μg Co eq./mL (Co d_{50} = 4 μm) (WC d_{50} = <1 μm)	De Boeck et al. (1998)
DNA single-strand breaks and alkali-labile sites, alkaline Comet assay, human mononuclear leukocytes, in vitro	?		0.6 μg Co eq./mL (Co d_{50} = 4 μm) (WC d_{50} = <1 μm)	De Boeck et al. (2003b)
DNA single-strand breaks and alkali-labile sites, alkaline Comet assay, human mononuclear leukocytes, in vitro	e[d]		6 μg Co eq./mL (Co d_{50} = 4 μm) (WC d_{50} = <1 μm)	De Boeck et al. (2003b)
Micronucleus formation, binucleates, cytochalasin-B assay, human lymphocytes, in vitro	+[c]		0.6 μg Co eq./mL (Co d_{50} = 4 μm) (WC d_{50} = <1 μm)	Van Goethem et al. (1997)
Micronucleus formation, binucleates, cytochalasin-B assay, human lymphocytes, in vitro	e[d]		3 μg Co eq./mL (Co d_{50} = 4 μm) (WC d_{50} = <1 μm)	Van Goethem et al. (1997)
Micronucleus formation, binucleates, cytochalasin-B assay, human lymphocytes, in vitro	+		0.6 μg Co eq./mL (Co d_{50} = 4 μm) (WC d_{50} = <1 μm)	De Boeck et al. (2003b)
Micronucleus formation, binucleates, cytochalasin-B assay, human lymphocytes, in vitro	e[d]		6 μg Co eq./mL(Co d_{50} = 4 μm) (WC d_{50} < 1 μm)	De Boeck et al. (2003b)
DNA single-strand breaks or alkali-labile sites, alkaline Comet assay, male Wistar rat type II pneumocytes, in vivo	+		16.6 mg/kg, i.t. (Co d_{50} = 4 μm) (WC d_{50} < 1 μm)	De Boeck et al. (2003c)
DNA single-strand breaks or alkali-labile sites, alkaline Comet assay, male Wistar rat BALF cells, in vivo	−		16.6 mg/kg i.t. (Co d_{50} = 4 μm) (WC d_{50} < 1 μm)	De Boeck et al. (2003c)
DNA single-strand breaks or alkali-labile sites, alkaline Comet assay, Wistar male rat mononuclear leukocytes, in vivo	−		16.6 mg/kg i.t. (Co d_{50} = 4 μm) (WC d_{50} < 1 μm)	De Boeck et al. (2003c)
Micronucleus formation, male Wistar rat type II pneumocytes, in vivo	+		16.6 mg/kg i.t. (Co d_{50} = 4 μm) (WC d_{50} < 1 μm)	De Boeck et al. (2003c)
Micronucleus formation, cytochalasin-B assay, male Wistar rat lymphocytes, in vivo	−		49.8 mg/kg i.t. (Co d_{50} = 4 μm) (WC d_{50} < 1 μm)	De Boeck et al. (2003c)

Table 15 (contd)

Test system	Result[a] Without exogenous metabolic system	With exogenous metabolic system	Dose[b] (LED/HID)	Reference
rWC–Co particles				
Co(II) salts				
DNA breaks, alkaline elution, purified DNA (3T3 mouse cells)	+[c] e[d]		1 μg Co eq./mL (Co d_{50} = 4 μm) (WC d_{50} = 2 μm)	Anard et al. (1997)
DNA breaks, alkaline elution, purified DNA (3T3 mouse cells)	r[e]		1 μg Co eq./mL + Na formate (Co d_{50} = 4 μm) (WC d_{50} = 2 μm)	Anard et al. (1997)
DNA breaks, alkaline elution, human mononuclear lymphocytes, in vitro	+[c]		1.5 μg Co eq/mL (Co d_{50} = 4 μm) (WC d_{50} = 2 μm)	Anard et al. (1997)
Cobalt compounds				
Co(II) salts				
Cobalt(II) acetate				
Inhibition of repair of UV-induced pyrimidine dimers, nucleoid sedimentation, human HeLa S-3 cells, in vitro	+		100 μM	Snyder et al. (1989)
Enhancement of cell transformation by simian adenovirus SA7, Syrian hamster embryo cells, in vitro	+		0.2 mM	Casto et al. (1979)
DNA base damage (products of hydroxyl radical attack), female and male Fischer 344/NCr rats, in vivo	+ (kidney > liver > lung)		ip, single, 50 μM/kg	Kasprzak et al. (1994)
Cobalt(II) chloride				
Reduction of fidelity of DNA replication by substitution of Mg^{2+} Escherichia coli DNA polymerase, sea-urchin nuclear DNA polymerase, avian myeloblastosis virus DNA polymerase	+		1 mM [130 μg/mL]	Sirover & Loeb (1976)
Prophage induction, Escherichia coli	−		~ 320 μM[f] [415 μg/mL]	Rossman et al. (1984)
Escherichia coli WP2, inhibition of protein synthesis	+		6.25 μg/mL	Leitão et al. (1993)
Escherichia coli AB1886, inhibition of protein synthesis	+		6.25 μg/mL	Leitão et al. (1993)
Salmonella typhimurium TA100, reverse mutation	−		NG	Ogawa et al. (1986)
Salmonella typhimurium TA102, reverse mutation	−	−	40 ppm [40 μg/mL]	Wong (1988)

Table 15 (contd)

Test system	Result[a] Without exogenous metabolic system	Result[a] With exogenous metabolic system	Dose[b] (LED/HID)	Reference
Salmonella typhimurium TA1535, reverse mutation	–		NG	Arlauskas et al. (1985)
Salmonella typhimurium TA1535, reverse mutation	–	–	40 ppm [40 µg/mL]	Wong (1988)
Salmonella typhimurium TA1537, reverse mutation	–		NG	Arlauskas et al. (1985)
Salmonella typhimurium TA1537, reverse mutation	–		1000 µmol/plate [130 000 µg/plate]	Ogawa et al. (1986)
Salmonella typhimurium TA1537, reverse mutation	+	–	40 ppm [40 µg/mL]	Wong (1988)
Salmonella typhimurium TA1538, reverse mutation	–		NG	Arlauskas et al. (1985)
Salmonella typhimurium TA98, reverse mutation	–		NG	Arlauskas et al. (1985)
Salmonella typhimurium TA98, reverse mutation	–		NG	Ogawa et al. (1986)
Salmonella typhimurium TA98, reverse mutation	+	–	40 ppm [40 µg/mL]	Wong (1988)
Salmonella typhimurium TA2637, reverse mutation	–		1000 µmol/plate [130 000 µg/plate]	Ogawa et al. (1986)
Salmonella typhimurium, TA97 preincubation assay	+		100 µM [13 µg/mL]	Pagano & Zeiger (1992)
Salmonella typhimurium, TA97 preincubation assay	r[d]		100 µM [13 µg/mL] + DEDTC 420 µM	Pagano & Zeiger (1992)
Escherichia coli SY1032/pKY241 transfected with pUB₃, *supF* tRNA locus, mutation	+		20 µM [2.6 µg/mL]	Ogawa et al. (1999)
Bacillus subtilis rec strains H17/M45, growth inhibition	–		[325 µg/plate]	Nishioka (1975)
Bacillus subtilis rec strain H17, growth inhibition	+		[325 µg/plate]	Kanematsu et al. (1980)
Saccharomyces cerevisiae SBTD-2B, 'petite' mutation, respiratory deficiency	+		2 mM [260 µg/mL]	Prazmo et al. (1975)
Saccharomyces cerevisiae, strain 197/2d, 'petite' mutation	+		4 mM [520 µg/mL]	Putrament et al. (1977)
Saccharomyces cerevisiae, strain 197/2d, erythromycin-resistant mutation	+		4 mM [520 µg/mL]	Putrament et al. (1977)
Saccharomyces cerevisiae, 'petite' mutation, respiratory deficiency	(+)		640 µg/mL	Egilsson et al. (1979)
Saccharomyces cerevisiae D7, *ilv* gene mutation	–		10 mM [1300 µg/mL]	Fukunaga et al. (1982)
Saccharomyces cerevisiae D7, trp gene conversion	+		10 mM [1300 µg/mL]	Fukunaga et al. (1982)
Saccharomyces cerevisiae D7, *ilv* gene mutation	–		100 mM [13 000 µg/mL]	Singh (1983)
Saccharomyces cerevisiae D7, trp gene conversion	(+)		100 mM [13 000 µg/mL]	Singh (1983)

Table 15 (contd)

Test system	Result[a]		Dose[b] (LED/HID)	Reference
	Without exogenous metabolic system	With exogenous metabolic system		
Saccharomyces cerevisiae D7, trp gene conversion	(+)		1500 µg/mL [11.5 mM]	Kharab & Singh (1985)
Saccharomyces cerevisiae D7, *ilv* gene mutation	(+)		3000 µg/mL [23 mM]	Kharab & Singh (1985)
Saccharomyces cerevisiae D7, 'petite' mutation, respiratory deficiency	+		750 µg/mL [5.76 mM]	Kharab & Singh (1987)
Drosophila melanogaster, gene mutation or mitotic recombination, wing spot test mwh/flr	+		2 mM [260 µg/mL]	Ogawa et al. (1994)
Drosophila melanogaster, gene mutation or reduced mitotic recombination, wing spot test mwh/TM3	–		8 mM [1040 µg/mL]	Ogawa et al. (1994)
DNA strand breaks, alkaline sucrose gradient, Chinese hamster ovary cells, *in vitro*	+		2 mM [260 µg/mL]	Hamilton-Koch et al. (1986)
DNA strand breaks, nucleoid sedimentation assay, Chinese hamster ovary cells, *in vitro*	–		10 mM [1300 µg/mL]	Hamilton-Koch et al. (1986)
DNA strand breaks, nucleoid sedimentation, human HeLa cells, *in vitro*	+		50 µM [65 µg/mL]	Hartwig et al. (1990)
DNA-protein cross links, rat Novikoff ascites hepatoma cells, *in vitro*	+		1 mM [130 µg/mL]	Wedrychowski et al. (1986)
Gene mutation, Chinese hamster V79 cells, *Hprt* locus, *in vitro*	(+)		0.2 mM [26 µg/mL]	Miyaki et al. (1979)
Gene mutation, Chinese hamster V79 cells, *Hprt* locus, *in vitro*	+		100 µM [13 µg/mL]	Hartwig et al. (1990)
Gene mutation, Chinese hamster V79 cell line, *Gpt* locus, *in vitro*	–		100 µM [13 µg/mL]	Kitahara et al. (1996)
Gene mutation, Chinese hamster transgenic cell line G12, *Gpt* locus, *in vitro*	+		50 µM [6.5 µg/mL]	Kitahara et al. (1996)
Sister chromatid exchanges, mouse macrophage-like cells P388D₁, *in vitro*	+		100 µM [13 µg/mL]	Andersen (1983)
Cell transformation, C3H10T1/2 mouse fibroblast cells, *in vitro*	+[c]		38 µM [5 µg/mL]	Doran et al. (1998)
Reduction in colony forming, V79 Chinese hamster cells, *in vitro*	+ (42%)		180 µM [24 µg/mL]	Kasten et al. (1992)
Reduction of cloning efficiency, Chinese hamster ovary cells, *in vitro*	+ (50%)		4 mM [520 µg/mL]	Hamilton-Koch et al. (1986)
Displacement of acridine orange from DNA, calf thymus DNA and *Micrococcus luteus* DNA	+		0.33 mM [43 µg/mL]	Richardson et al. (1981)
Formation of metal–DNA complex, calf thymus B-DNA	+		NG	Aich et al. (1999)

Table 15 (contd)

Test system	Result[a]		Dose[b] (LED/HID)	Reference
	Without exogenous metabolic system	With exogenous metabolic system		
Induction of reporter gene expression under the control of the promoter region of the metallothionein gene, chick embryo liver cells transfected with luciferase or chloramphenicol acetyl transferase, *in vitro*	+		112 μM [15 μg/mL]	Lu *et al.* (1996)
Production of reactive oxygen species (fluorescence from oxidation of DCFH-DA), A549 cells, human lung cells, *in vitro*	+		200 μM [26 μg/ml]	Salnikow *et al.* (2000)
Production of reactive oxygen species (fluorescence from oxidation of DCFH-DA), A549 cells, human lung cells, *in vitro*	r[d]		300 μM [39 μg/mL] + 2-mercapto-ethanol	Salnikow *et al.* (2000)
Production of reactive oxygen species (fluorescence from oxidation of DCFH-DA), A549 cells, human lung cells, *in vitro*	r[d]		300 μM [39 μg/mL] + vitamin E	Salnikow *et al.* (2000)
DNA strand breaks, fluorescence analysis of DNA unwinding, human white blood cells, *in vitro*	+		50 μM [6.5 μg/mL]	McLean *et al.* (1982)
DNA strand breaks, alkaline sucrose gradient, human diploid fibroblasts, *in vitro*	+		5 mM	Hamilton-Koch *et al.* (1986)
DNA strand breaks, nick translation, human diploid fibroblasts, *in vitro*	+		10 mM [1300 μg/mL]	Hamilton-Koch *et al.* (1986)
DNA strand breaks, nucleoid sedimentation, human diploid fibroblasts, *in vitro*	–		10 mM [1300 μg/mL]	Hamilton-Koch *et al.* (1986)
DNA strand breaks and alkali-labile sites, alkaline Comet assay, human mononuclear leukocytes, *in vitro*	+[c]		0.3 μg/mL	De Boeck *et al.* (1998)
Induction of gene expression (*Cap43*), A549 cells, human lung cells, *in vitro*	+		100 μM [13 μg/mL]	Salnikow *et al.* (2000)
Induction of gene expression (*Cap43*), A549 cells, human lung cells, *in vitro*	s[d]		100 μM [13 μg/mL] + 2-mercapto-ethanol	Salnikow *et al.* (2000)
Induction of gene expression (*Cap43*), A549 cells, human lung cells, *in vitro*	s[d]		100 μM [13 μg/mL] + H₂O₂	Salnikow *et al.* (2000)
Sister chromatid exchange, human lymphocytes, *in vitro*	+		10 μM [1.3 μg/mL]	Andersen (1983)
Aneuploidy, human lymphocytes, *in vitro*	+		3.7 μg/mL	Resende de Souza-Nazareth (1976)

Table 15 (contd)

Test system	Result[a] Without exogenous metabolic system	With exogenous metabolic system	Dose[b] (LED/HID)	Reference
Aneuploidy, pseudodiploidy and hyperploidy, bone marrow of male hamsters, in vivo	+		400 mg/kg bw ip[g]	Farah (1983)
Aneuploidy, pseudodiploidy and hyperploidy, testes of hamsters, meiosis 1, in vivo	+		400 mg/kg bw ip[g]	Farah (1983)
Inhibition of binding of p53 protein to p53 consensus sequence on linear DNA fragment	+ (full)		> 100 μM (300 μM)	Palecek et al. (1999)
Inhibition of binding of p53 protein to supercoiled DNA	+		600 μM	Palecek et al. (1999)
Affinity of reconstituted apopolypeptide (Zn finger protein) with estrogen response element consensus oligonucleotide	r		NG (K_D 0.720 μM)[h]	Sarkar (1995)
Inactivation of bacterial Fpg protein (with Zn finger domain), conversion of supercoiled bacteriophage PM2 DNA into open circular form, electrophoresis	–		1000 μM	Asmuss et al. (2000)
Inhibition of XPA (with Zn finger domain) binding to UV-irradiated oligonucleotide, gel mobility shift analysis	+		50 μM [6.5 μg/mL]	Asmuss et al. (2000)
Cobalt(II) chloride hexahydrate				
Lysogenic induction, Escherichia coli WP2$_s$ (λ)	r[j]		(10 μg/mL)[f] + UV	Leitão et al. (1993)
Lysogenic induction, Escherichia coli K12 ABI886 (λ)	r[j]		(10 μg/mL)[k] + UV	Leitão et al. (1993)
Lysogenic induction, Escherichia coli ABI157 (λ)	+		100 μg/mL – Mg	Leitão et al. (1993)
Phage reactivation, Escherichia coli ABI157 (λ)	–		250 μg/mL – UV	Leitão et al. (1993)
Phage reactivation, Escherichia coli ABI157 (λ)	e[i]		62.5 μg/mL + UV	Leitão et al. (1993)
Escherichia coli WP2, reverse mutation	–[l]		20 μg/mL [84 μM]	Kada & Kanematsu (1978)
Escherichia coli WP2 uvrA, reverse mutation	–[l]		NG	Arlauskas et al. (1985)
Escherichia coli WP2$_s$ gene mutation	–[l]		50 μg/mL [210 μM]	Leitão et al. (1993)
Escherichia coli WP2$_s$ gene muation	r[l]		50 μg/mL + UV	Leitão et al. (1993)
Saccharomyces cerevisiae D7, 'petite' mutation, respiratory deficiency	+		[130 μg/mL]	Lindegren et al. (1958)
Salmonella typhimurium TA100, reverse mutation	–		100 mM [23 800 μg/mL]	Tso & Fung (1981)
Salmonella typhimurium TA100, reverse mutation	–		NG	Arlauskas et al. (1985)

Table 15 (contd)

Test system	Result[a] Without exogenous metabolic system	With exogenous metabolic system	Dose[b] (LED/HID)	Reference
Salmonella typhimurium TA1538, reverse mutation	–[2]		20 µg/mL [84 µM]	Mochizuki & Kada (1982)
Salmonella typhimurium TA98, reverse mutation	–[2]		20 µg/mL [84 µM]	Mochizuki & Kada (1982)
Bacillus subtilis strain NIG 1125, reverse mutation	–[3]		30 µg/mL [126 µM]	Inoue et al. (1981)
Gene mutation, mouse lymphoma L5178Y cells, *Tk* locus, *in vitro*	–		57.11 µg/mL	Amacher & Paillet (1980)
Gene mutation, Chinese hamster V79 cells, *8AG* locus, *in vitro*	–		9 µM [2 µg/mL]	Yokoiyama et al. (1990)
Gene mutation, Chinese hamster V79 cells, *8AG* locus, *in vitro*	r[i]		3 µM [0.7 µg/mL] + γ rays	Yokoiyama et al. (1990)
Micronucleus formation, BALB/c mouse bone marrow, *in vitro*	–		50 µg/mL [385 µM]	Suzuki et al. (1993)
DNA strand breaks, alkaline elution, human lymphocytes, *in vitro*	–	–	102 µM [25 µg/mL]	Anard et al. (1997)
Inhibition of nucleotide excision repair (incision and polymerization steps) of UV-induced DNA damage, alkaline unwinding, VH16 human fibroblasts	+		50 µM [12 µg/mL]	Kasten et al. (1997)
Inhibition of nucleotide excision repair (ligation step) of UV-induced DNA damage, alkaline unwinding, VH16 human fibroblasts	–		200 µM [48 µg/mL]	Kasten et al. (1997)
Inhibition of UV-induced cyclobutane pyrimidine dimers (incision step), alkaline unwinding + T4 endonuclease V, VH16 human fibroblasts	+		150 µM [86 µg/mL]	Kasten et al. (1997)
Micronucleus formation, male BALB/c AnNCrj mouse bone marrow, *in vivo*	+		50 mg/kg bw	Suzuki et al. (1993)
Micronucleus formation, male BALB/c AnNCrj mouse bone marrow, *in vivo*	e[i]		50 mg/kg + DMH 20 mg/kg	Suzuki et al. (1993)
Micronucleus formation, male BALB/c AnNCrj mouse bone marrow, *in vivo*	e[i]		50 mg/kg + benzo(a)pyrene 50 mg/kg	Suzuki et al. (1993)
Micronucleus formation, male BALB/c AnNCrj mouse bone marrow, *in vivo*	e[i]		50 mg/kg + 2-naphthylamine 200 mg/kg	Suzuki et al. (1993)

Table 15 (contd)

Test system	Result[a] Without exogenous metabolic system	With exogenous metabolic system	Dose[b] (LED/HID)	Reference
Cobalt(II) molybdenum(VI) oxide				
Enhancement of cell transformation by simian adenovirus SA7, Syrian hamster embryo cells, *in vitro*	+		250 μM [55 μg/mL]	Casto *et al.* (1979)
Cobalt(II) nitrate				
Chromosome aberrations (numerical), human diploid fibroblasts W1.38 and MRC$_5$, *in vitro*	–		0.08 μM[g] [0.015 μg/mL]	Paton & Allison (1972)
Chromosome aberrations (numerical), human mononuclear leucocytes, *in vitro*	–		0.8 μM[g] [0.15 μg/mL]	Paton & Allison (1972)
Cobalt(II) nitrate hexahydrate				
Drosophila melanogaster (flr^3/In(3LR)TM3, r^{p}Sep bx^{34e}e^{s}Ser) × *(mwh).mwh and fbr^3*, gene mutations, chromosomal deletion, non disjunction or mitotic recombination (small single spots and large single spots), SMART test	+		1 mM [291 μg/mL]	Ye°ilada (2001)
Drosophila melanogaster (flr^3/In(ELR)TM3, r^{p}Sep bx^{34e}e^{s}Ser) × *(mwh).mwh and fbr^3*, mitotic recombination (twin spots), SMART test	+		10 mM [2910 μg/mL]	Ye°ilada (2001)
Cobalt(II) sulfate				
Allium cepa, chromosomal aberrations	+		20 μM [3 μg/mL]	Gori & Zucconi (1957)
Allium cepa, aneuploidy	+		100 μM [15 μg/mL] for 5 days + H$_2$0 for 3 days	Gori & Zucconi (1957)
Production of reactive oxygen species (degradation of 2-deoxyribose), malondialdehyde assay	+		1 μM [0.155 μg/mL]	Ball *et al.* (2000)
Production of reactive oxygen species (degradation of 2-deoxyribose), malondialdehyde assay	r[d]		50 μM [7.8 μg/mL] + desferrioxamine 1 mM	Ball *et al.* (2000)
Chemical changes in DNA bases GC/MS-SIM, calf thymus DNA	–		25 μM [4 μg/mL]	Nackerdien *et al.* (1991)
Chemical changes in DNA bases GC/MS-SIM, calf thymus DNA	+[d]		25 μM [4 μg/mL] + H$_2$O$_2$ 208 mM	Nackerdien *et al.* (1991)

Table 15 (contd)

Test system	Result[a] Without exogenous metabolic system	With exogenous metabolic system	Dose[b] (LED/HID)	Reference
Bacillus subtilis rec strain H17, growth inhibition	(+)		388 µg/plate	Kanematsu et al. (1980)
Cytoskeletal perturbation of microtubules and microfilaments, mouse cells SWISS 3T3, *in vitro*	+		100 µM [15.5 µg/mL]	Chou (1989)
Chemical changes in DNA bases GC/MS-SIM, isolated human chromatin, K562 cells	−		25 µM [4 µg/mL]	Nackerdien et al. (1991)
Chemical changes in DNA bases GC/MS-SIM, isolated human chromatin, K562 cells	+[d]		25 µM + H_2O_2 208 mM	Nackerdien et al. (1991)
Chemical changes in DNA bases GC/MS-SIM, isolated human chromatin, K562 cells	r[m]		25 µM [4 µg/mL] + H_2O_2 208 mM + EDTA 120 µM	Nackerdien et al. (1991)
Chemical changes in DNA bases GC/MS-SIM, isolated human chromatin, K562 cells	r[m]		25 µM + H_2O_2 208 mM + mannitol 50 mM	Nackerdien et al. (1991)
Chemical changes in DNA bases GC/MS-SIM, isolated human chromatin, K562 cells	r[m]		25 µM + H_2O_2 208 mM + DMSO 50 mM	Nackerdien et al. (1991)
Chemical changes in DNA bases GC/MS-SIM, isolated human chromatin, K562 cells	r[m], e[m]		25 µM + H_2O_2 208 mM + glutathione 1 mM	Nackerdien et al. (1991)
Chemical changes in DNA bases GC/MS-SIM, isolated human chromatin, K562 cells	e[m]		25 µM + H_2O_2 208 mM + SOD 200 units/mL	Nackerdien et al. (1991)
Induction of human metal-inducible genes (*MT-IIA*, *hsp70*, *c-fos*), HeLa human cervical carcinoma cells, *in vitro*	+[n]		500 µM	Murata et al. (1999)
Metal responsive element (MRE)-DNA binding activity, HeLa human cervical carcinoma cells, *in vitro*	−		500 µM	Murata et al. (1999)
Heat shock element (HSE)-DNA binding activity, HeLa human cervical carcinoma cells, *in vitro*	?		500 µM	Murata et al. (1999)
Cobalt(II) sulfate monohydrate				
Cell transformation, Syrian hamster embryo cells, *in vitro*	+		0.125 µg/mL [0.75 µM]	Kerckaert et al. (1996)

Table 15 (contd)

Test system	Result[a]		Dose[b] (LED/HID)	Reference
	Without exogenous metabolic system	With exogenous metabolic system		
Cobalt(II) sulfate heptahydrate				
Salmonella typhimurium TA100, reverse mutation	+	–°	100 µg/plate	Zeiger et al. (1992)
Salmonella typhimurium TA98, TA1535, reverse mutation	–	–	10 000 µg/plate	Zeiger et al. (1992)
Induction of p53, ELISA assay, NCTC929 mouse fibroblasts, *in vitro*	+		50 µg/mL [178 µM]	Duerksen-Hughes et al. (1999)
CO(II)acetate tetrahydrate				
Chromosomal aberrations, human lymphocytes, *in vitro*	–		0.6 µg/mL [2.4 µM]	Voroshilin et al. (1978)
Co(III)hexaamine ions and Co(III) amine complexes				
Conformational changes of DNA oligonucleotides, circular dichroism and NMR spectroscopy	+		µM range (< 24 µM)	Bauer & Wang (1997)
Co(III) complexes				
Escherichia coli, strains AB1157 (wild type), AB1886 *uvrA6*, GW801 *recA56*, GW802 *rec56 uvrA6*, GW803 *recA56 lexA⁻*, PAM 5717 *lexA* and AB1899 *lon*, DNA repair assay	+ (8/15)[p] (+) (7/15)[p]		NG NG	Schultz et al. (1982)
Salmonella typhimurium, strain TA100, TA98, TA92, reverse mutation	+ (4/15)		0.1–0.5 µmol/plate	Schultz et al. (1982)
Salmonella typhimurium, strain TA1535, 1537, 1538. reverse mutation	–		2 µmol/plate	Schultz et al. (1982)
Co(III) salts				
CoNO₃				
Pisum abyssinicum chlorophyll mutation	+[q]		0.1–1 mM [18.3–183 µg/mL]	von Rosen (1964)
Co(III) complexes with desferal				
Plasmid PBR322, scission of double-stranded DNA	+		≤ 42.5 µM[f] + H_2O_2 4 mM	Joshi & Ganesh (1992)
Co(OH)₃				
Bacillus subtilis rec strain H17, growth inhibition	(+)		[2750 µg/plate]	Kanematsu et al. (1980)

Table 15 (contd)

Test system	Result[a]		Dose[b] (LED/HID)	Reference
	Without exogenous metabolic system	With exogenous metabolic system		
Co(III) Schiff–base complex				
Inhibition of Zn-finger transcription factor, HNMR spectroscopy	+		0.5 mM	Louie & Meade (1998)
Inhibition of Zn-finger transcription factor, Sp1, gel shift, filter binding assay	+		10 μM	Louie & Meade (1998)
Cobalt sulfides (2⁺) and (4⁺)				
CoS particles				
DNA strand breaks, alkaline sucrose gradient, Chinese hamster CHO cells, *in vitro*	+		10 μg/mL	Robison *et al.* (1982)
Gene mutation, Chinese hamster transgenic cell line G10, *Gpt* locus, *in vitro*	–		1 μg/cm²	Kitahara *et al.* (1996)
Gene mutation, Chinese hamster transgenic cell line G10, *Gpt* locus, *in vitro*	s[d]		1 μg/cm² + H_2O_2 10 μM	Kitahara *et al.* (1996)
Gene mutation, Chinese hamster transgenic cell line G12, *Gpt* locus, *in vitro*	+		0.5 μg/cm²	Kitahara *et al.* (1996)
Gene mutation, Chinese hamster transgenic cell line G12, *Gpt* locus, *in vitro*	s[d]		0.5 μg/cm² + H_2O_2 10 μM	Kitahara *et al.* (1996)
CoS (amorphous)				
Cell transformation, Syrian hamster embryo cells, *in vitro*	(+)		10 μg/mL (d_{50} = 2.0 μm)	Abbracchio *et al.* (1982); Costa *et al.* (1982)
CoS₂ (crystalline)				
Cell transformation, Syrian hamster embryo cells, *in vitro*	+[g]		1 μg/mL (d_{50} = 1.25 μm)	Abbracchio *et al.* (1982); Costa *et al.* (1982)

Table 15 (contd)

rW-Ni-Co alloy, reconstituted mixture of W (92%), Ni (5%) and Co (3%) particles; rWC-Co, reconstituted mixture of WC (94%) and Co (6%) particles; HNMR, proton nuclear magnetic resonance

[a] ?, inconclusive; +, positive; (+), weak positive; −, negative; r = reduction; e = enhancement; s = stable

[b] LED, lowest effective dose; HID, highest ineffective dose; ip, intraperitoneally; po, orally; i.t., intratracheal instillation; MMS, methylmethane sulfonate; DMSO, dimethylsulfoxide; EDTA, ethyldiaminetetraacetate; DEDTC, diethyldithiocarbamate; NG, not given; DCFH-D, 2′,7′-dichlorofluorescine diacetate; DMH, 1,1-dimethylhydrazine; 8AG, 8-azaguanine; H_2O_2, hydrogen peroxide; SOD, superoxide dismutase; UV, ultraviolet irradiation

[c] Refer to the same experiment where Co and WC–Co were compared

[d] as compared to CO

[e] as compared to rWC–Co

[f] Estimated from a graph in the paper

[g] Total dose given to each animal over nine days

[h] This value corresponds to the dissociation constant (K_D) for cobalt-reconstituted polypeptide binding with estrogen response element consensus oligonucleotide

[i] as compared to the other mutagen used

[j] toxic dose; highest ineffective subtoxic dose was not given.

[k] Similar effect to strain *E. coli* WP2s(λ), but data not shown in the paper

[l,1,2 or 3] antimutagenic effect; [l,1] inhibition of mutagenesis induced by *N*-methyl-*N*-nitrosoguanidine (MNNG); [l,2] inhibition of mutagenesis induced by 3-amino-1,4-dimethyl-*5H*-pyrido[4,3-*b*]indole (Trp-P-1) or [l,3] inhibition of spontaneous mutability

[m] as compared to Co + H_2O_2

[n] metallothionein (*MT-IIA*) and heat shock protein (*hsp70*) genes were induced but not *c-fos* gene.

[o] Tested at doses up to 10 000 μg/plate

[p] The ratio corresponds to the number of Co(III) complexes positive for DNA repair assay on the total number of Co(III) complexes tested

[q] Co as EDTA chelate (Co-EDTA) was also positive.

[r] Optimal concentration for 100% DNA cleavage; slight increase in concentration over this value lead to extensive degradation.

[s] more than corresponding amorphous salt

Hartwig & Schwerdtle, 2002) and with zinc finger proteins (Hartwig, 2001) and their effect on gene expression (Beyersman, 2002) have been reviewed. An evaluation of carcinogenic risks of cobalt and cobalt compounds was published in 1991 (IARC, 1991).

Metallic cobalt particles (median diameter (d_{50}) = 4 μm) have been shown with alkaline elution technology to induce DNA breakage and/or alkali-labile sites in DNA purified from 3T3 mouse cells. Similar changes have been demonstrated *in vitro* in human mononuclear leukocytes by both the alkaline elution and the Comet assay methods. Oxidative DNA damage was not detected at FPG-sensitive sites with the Comet methodology. In experiments run in parallel, a statistically-significant induction of micronuclei in binucleated human lymphocytes was obtained with the cytochalasin-B method. In-vitro cell transformation was not induced in mouse fibroblast cells by cobalt particles ($d_{50} ≤ 5$ μm) nor in human osteoblast-like cells by approximately same size ($d_{50} = 1$–4 μm) cobalt particles.

Metallic cobalt ($d_{50} = 1$–5 μm) has been tested in combination with tungsten and nickel particles. *In vitro*, the mixture induced DNA single-strand breaks as shown by alkaline elution methodology, micronuclei, and cell transformation in human non-tumorigenic osteosarcoma osteoblast-like cell line (TE85, clone F-5).

(b)   Hard-metal particles

When tested *in vitro* over a range of cobalt equivalent concentrations, a mixture of tungsten carbide and cobalt metal (WC–Co), caused significantly more (on average threefold more) DNA breaks than cobalt particles alone, both in isolated human DNA and in cultured human lymphocytes (alkaline elution and Comet assays); this DNA damage was inhibited by scavenging activated oxygen species. In the same assay run in parallel, cobalt chloride did not cause DNA breaks. Dose-dependency and time-dependency of DNA breakage and of induction of alkali-labile sites were shown for hard-metal particles in the Comet assay (De Boeck *et al.*, 2003b). A similarly greater genotoxic activity of hard metal compared with cobalt-metal particles alone has been found with the cytokinesis-blocked micronucleus test when applied *in vitro* to human lymphocytes. The data demonstrate clearly that interaction of cobalt with tungsten carbide particles leads to enhanced mutagenicity. Recently, this observation has been extended to other carbides. In the in-vitro cytokinesis-blocked micronucleus test, while the metal carbides alone did not increase the micronucleus frequency, cobalt alone and the four tested carbide–cobalt mixtures induced statistically-significant concentration-dependent increases in micronucleated binucleates. As with the tungsten carbide–cobalt metal mixture, nobium carbide and chromium carbide particles were able to interact with cobalt, producing greater mutagenic effects than those produced by the particles of the individual metals. Molybdenum carbide particles did not display interactive mutagenicity with cobalt in the micronucleus test, possibly because of their small specific surface area, compactness and/or spherical shape (De Boeck *et al.*, 2003b). However, with the Comet assay, when also performed directly at the end of the treatment, no firm conclusion could be made.

From a mechanistic point of view, the in-vitro studies comparing the effects of cobalt metal alone and the hard-metal mixture (WC–Co) provide convincing evidence that the mutagenic activity of metallic cobalt is not exclusively mediated by the ionic form dissolved in biological media (Anard *et al.*, 1997). However, the dissolved cations do play an important role through direct or indirect mutagenic effects as reviewed separately for the soluble Co(II) and Co(III) compounds.

In-vivo experimental data on the mutagenicity of cobalt particles alone are lacking. Evidence of the in-vivo mutagenic potential of hard-metal dust was obtained recently in type II pneumocytes of rats (De Boeck, 2003c). DNA breaks/alkali-labile sites (alkaline Comet assay) and chromosome/genome mutations (micronucleus test) were assessed after a single intratracheal instillation of hard metal (WC–Co), and dose–effect and time trend relationships were examined. In addition, the alkaline Comet assay was performed on cells obtained from BALF and on peripheral blood mononucleated cells (PBMC). Protein content, LDH activity, total and differential cell counts of BALF were evaluated in parallel as parameters of pulmonary toxicity. In type II pneumocytes, WC–Co induced a statistically-significant increase in tail DNA (12-h time point) and in micronuclei (72 h) after a single instillation in rats at a dose which produced mild pulmonary toxicity. In PBMC, no increase in DNA damage nor in micronuclei was observed.

Cobalt compounds, like other metallic compounds, are known to be relatively inactive in prokaryotic systems (Rossman, 1981; Swierenga *et al.*, 1987).

(c)   Cobalt(II) chloride

Cobalt(II) chloride was found to be inactive in the λ prophage induction assay, and gave conflicting results in the *Bacillus subtilis rec*[+/−] growth inhibition assay; when a cold preincubation procedure was used, positive results were observed (Kanematsu *et al.*, 1980). Lysogenic induction and phage reactivation was found in *Escherichia coli* in the absence of magnesium. Also in *E. coli*, reduction of fidelity of DNA replication by substitution of magnesium and inhibition of protein synthesis were observed. Cobalt(II) chloride was inactive in all but two bacterial mutagenicity tests. One study gave positive results in the absence, but not in the presence, of an exogenous metabolic system, and in the second study, a preincubation procedure was used.

In bacteria, cobalt(II) chloride has been reported to reduce the incidence of spontaneous mutations and to inhibit mutations induced by *N*-methyl-*N'*-nitrosoguanidine and 3-amino-1,4-dimethyl-5*H*-pyrido[4,3-*b*]indole. It was found to be comutagenic with several heteroaromatic compounds such as benzo(a)pyrene and naphthylamine.

In *Saccharomyces cerevisiae*, cobalt(II) chloride induced gene conversion and petite ρ⁻ mutation in mitochondrial DNA but not other types of mutation.

In *Drosophila melanogaster*, mitotic recombination was found.

In mammalian cells cultured *in vitro*, positive results were obtained for induction of DNA–protein cross-linkage, DNA strand breakage and sister chromatid exchange in most studies. Cobalt(II) chloride induced mutations at the *Hprt* locus in Chinese hamster V79 cells, but not at the *8AG* and the *Gpt* loci. At the same *Gpt* locus in a transgenic Chinese hamster V79 G12 cell line, lower concentrations of cobalt(II) chloride did induce gene mutations. In a single study, at the *Tk* locus in mouse lymphoma L5178Y cells, the results were negative.

In most studies, in cultured human cells *in vitro*, positive results were obtained for inhibition of protein-DNA binding activities, inhibition of p53 binding to DNA and for induction of gene expression, induction of DNA strand breakage and sister chromatid exchange. Chromosomal aberrations were not observed in cultured human cells (IARC, 1991). [The Working Group noted the low concentrations employed.] Cobalt(II) chloride induced aneuploidy in cultured human lymphocytes.

In vivo, cobalt(II) chloride administered by intraperitoneal injection induced aneuploidy (pseudodiploidy and hyperploidy) in bone marrow and testes of Syrian hamsters, micronuclei in bone marrow in male BALB/c mice, and enhanced the micronuclei frequencies induced by the three other mutagens tested.

A gene expression mechanism is involved in several tissue and cellular responses induced by soluble cobalt (generally cobalt chloride) mimicking the pathophysiological response to hypoxia, a response which involves various genes including those coding for erythropoiesis and for growth factors for angiogenesis (Gleadle *et al.*, 1995; Steinbrech *et al.*, 2000; Beyersmann, 2002). Up-regulation of erythropoietin gene expression was observed *in vivo* after a single intraperitoneal injection of cobalt chloride (60 mg/kg bw) into rats (Göpfert *et al.*, 1995) and might be of relevance in explaining the polyglobulia noted in humans treated with high doses of cobalt (Curtis *et al.*, 1976). In Chinese hamster ovary cells, cobalt also up-regulated the expression of haeme oxygenase-1, a potent anti-oxidant and anti-inflammatory mediator which helps to maintain cellular homeostasis in response to stress and injury (Gong *et al.*, 2001).

In studies designed to explore the molecular mechanisms of gene response to hypoxia, cobalt (12 and 60 mg/kg bw as cobalt chloride) was found to up-regulate the expression of the *PDGF-B* gene in lungs and kidneys of male Sprague-Dawley rats (Bucher *et al.*, 1996). Since PDGF is an important growth factor which modulates cell proliferation and the expression of several proto-oncogenes mainly in mesenchymal cells, this effect of cobalt might explain how it may exert fibrogenic and/or carcinogenic properties, but this remains to be documented.

(d)   Other cobalt compounds

Few results are available with other cobalt(II) salts.

Molecular analysis of lung neoplasms of B6C3F$_1$ mice exposed to cobalt sulfate heptahydrate showed the presence of K-*ras* mutations with a much higher frequency (55%) of G > T transversion at codon 12 than in controls (0%). This provides suggestive evidence that cobalt sulfate heptahydrate may indirectly damage DNA by oxidative stress (National Toxicology Program, 1998).

Cobalt sulfate has been shown to induce chromosomal aberrations and aneuploidy in plant cells, chemical changes in bases in purified calf thymus DNA and in isolated human chromatin in the presence of hydrogen peroxide, and cytoskeletal perturbation of microtubules and microfilaments and p53 protein in mouse fibroblasts treated *in vitro*. Cell transformation of Syrian hamster embryo cells has been induced by cobalt sulfate *in vitro*.

A number of mammalian genes (metallothionein MT-IIA, heat-shock proteins hsp70, c-fos) are transcriptionally regulated by a *cis*-acting DNA element located in their upstream regions. This DNA element responds to various heavy metals, including cobalt, to stimulate the expression of these genes (Murata *et al.*, 1999). *MT-IIA* and *hps70* but not *c-fos* RNA transcripts were increased in HeLa S_3 cells exposed to high concentrations of cobalt sulfate (> 10 µM). Metal response element (MRE)-DNA binding activity was not inhibited by cobalt sulfate in Hela cells *in vitro* while the results for heat shock element (HSE)-DNA binding activity were inconclusive. It is unknown whether MT-IIA and hps70 induction plays a role in the pathophysiological processes involved in cobalt carcinogenesis.

Cobalt(II) acetate was found to induce cell transformation *in vitro*. Cobalt(II) acetate and cobalt(II) molybdenum(VI) oxide ($CoMoO_4$) enhanced viral transformation in Syrian hamster embryo cells. Cobalt(II) acetate was shown to induce DNA base damage in female and male Fischer 344/NCr rats. Cobalt sulfide particles were found to induce DNA strand breaks and alkali-labile sites in Chinese hamster ovary cells. Data on the induction of gene mutations in Chinese hamster cells by cobalt sulfide particles are conflicting. Cobalt sulfide was shown to induce morphological transformation in Syrian hamster embryo cells; the crystalline form of cobalt sulfide being more active than the amorphous form.

Cobalt(III) nitrate induced gene mutations in *Pisum abyssinicum* chlorophyll. Eight of 15 cobalt(III) complexes with aromatic ligands were found to be positive in a DNA repair assay and four among the eight were also mutagenic to *Salmonella typhimurium*. Cobalt(III) complexes with desferal-induced scission of double-stranded DNA, and a cobalt(III) Schiff-base complex induced inhibition of zinc-finger transcription factors.

4.5   Mechanistic considerations

It had been assumed that, as for other metals, the biological activity of cobalt-metal particles, including their genotoxic effects, were mediated by the ionic form of cobalt and could be revealed by testing soluble compounds. However, Lison *et al.* (1995) demonstrated *in vitro* that cobalt metal, and not its ionic (II) species, was thermodynamically able to reduce oxygen in ROS independently of the Fenton reaction. During this process, soluble cobalt ions are produced which have several major cellular targets for induction of genotoxic effects and may, in turn, take part in a Fenton reaction in the presence of hydrogen peroxide. Moreover, since metallic cobalt forms particles which can be inhaled, assessment of genetic effects should also take into consideration: (i) that the primary production of ROS is related to the specific surface properties of the particles or the presence of transition metals, together with other parameters such as particle size, shape and uptake; and (ii) that excessive and persistent formation of ROS by inflammatory cells can lead to secondary toxicity. Since the mechanisms leading to the genotoxic effects of metallic cobalt are complex, assessment of its mutagenic effects should not be restricted to the genetic effects of metallic cobalt alone but should be complemented by those of cobalt in association with carbides, and of cobalt salts.

The results of genotoxicity assays with cobalt salts demonstrate clearly their mutagenic potential. Recent experimental studies have contributed to better delineate the molecular mechanisms involved in the genotoxic (and carcinogenic potential) of cobalt ions. These mechanisms may conceivably apply both to soluble cobalt compounds — for example, cobalt chloride or sulfate — and also to cobalt-metal or hard-metal particles which are readily solubilized in biological media. *In vivo*, however, the bioavailability of cobalt(II) is relatively limited because these cations precipitate in the presence of physiological concentrations of phosphates $(Co_3(PO_4)_2)$; K_s: 2.5×10^{-35} at 25 °C) and bind to proteins such as albumin.

In vitro in mammalian cells, two mechanisms seem to apply :
(1)   a direct effect of cobalt(II) ions causing damage to DNA through a Fenton-like mechanism;
(2)   an indirect effect of cobalt(II) ions through inhibition of repair of DNA damage caused by endogenous events or induced by other agents.

In vitro, cobalt(II) has been shown to inhibit the excision of UV-induced pyrimidine dimers from DNA in a dose-dependent fashion. Inhibition of repair by cobalt(II) resulted in the accumulation of long-lived DNA strand breaks suggesting a block in the gap-filling stage (DNA polymerization) of repair. Ability to inhibit repair was not correlated with cytotoxicity. It has been shown that repair of X-ray-induced DNA damage is not sensitive to cobalt. All inhibitory metals inhibited closure of single-strand DNA breaks (Snyder *et al.*, 1989).

In vitro, ionic cobalt(II) was shown to inhibit nucleotide excision repair processes after ultraviolet (UV) irradiation as measured by the alkaline unwinding method. A concentration as low as 50 µM cobalt chloride inhibited the incision as well as the polymerization step of the DNA repair process in human fibroblasts treated with UV light. As the repair of DNA damage is an essential homeostatic mechanism, its inhibition may account for a mutagenic or carcinogenic effect of cobalt(II) ions. Concentrations less than 1 mM cobalt chloride did not affect the activity of bacterial fpg but significantly reduced the DNA binding activity of the mammalian damage recognition protein XPA. Competition with essential magnesium ions and binding to zinc finger domains in repair proteins have been identified as potential modes of indirect genotoxic activity of cobalt(II) ions. It has also been reported that the DNA binding activity of the p53 protein, which is a zinc-dependent mechanism, can be modulated by cobalt(II) ions (Kasten *et al.*, 1997; Palecek *et al.*, 1999; Asmuss *et al.*, 2000).

This indirect mutagenic effect of cobalt on repair enzymes is not restricted to cobalt salts but has been shown to apply also to in-vitro exposure to metallic cobalt. De Boeck *et al.* (1998) examined the effects of cobalt-metal particles using the alkaline Comet assay on methyl methanesulfonate (MMS)-treated isolated human lymphocytes. MMS induced DNA strand breaks and alkali-labile sites in the lymphocytes in a dose-dependent manner. Post-incubation of MMS-treated cells for 2 h, in the absence of cobalt, resulted in significantly less DNA damage, implying that repair took place. Post-treatment with cobalt particles at a non-genotoxic dose for 2 h, after treatment with 5.5 µg/mL MMS, resulted in higher damage values compared with post-incubation values. These results may reflect inhibition by the cobalt particles of the ongoing repair of MMS-induced DNA lesions, which had presumably reached the polymerization step. Simultaneous exposure of lymphocytes to 5.5 µg/mL MMS and 1.2 µg/mL cobalt for 2 h resulted in higher damage values, conceivably representing an interference of cobalt particles at the incision of methylated bases, allowing more alkali-labile apurinic sites to be expressed, which, in the absence of cobalt, would be repaired. The authors concluded that metallic cobalt could cause persistence of MMS-induced DNA lesions by interference during their repair.

Since the previous IARC evaluation of cobalt in 1991, additional information has been obtained on the genotoxicity of the various cobalt species.

Cobalt(II) ions have been shown to substitute for zinc in the zinc-finger domain of some important proteins, such as those controlling cell cycling and/or DNA repair processes in animal and human cells.

Cobalt-metal particles produce mutagenic effects *in vitro* by two different mechanisms:
- directly through the production of ROS resulting in DNA damage, and
- indirectly by releasing Co(II) ions which inhibit DNA repair processes.

Moreover, when cobalt-metal particles are mixed with metallic carbide particles (mainly tungsten carbide), they form a unique chemical entity which:
- produces higher amounts of ROS than cobalt alone *in vitro*,
- has a stronger mutagenic activity than cobalt alone *in vitro* in human cells, and
- is mutagenic in rat lung cells in *vivo*.

A physicochemical mechanism to explain this increased toxicity has been proposed.

In humans, a specific fibrosing alveolitis (so-called hard-metal disease) occurs in workers exposed to dusts containing metallic cobalt such as hard metal or cemented microdiamonds. Fibrosing alveolitis may be a risk factor for lung cancer in humans.

5. Summary of Data Reported and Evaluation

5.1 Exposure data

Cobalt is widely distributed in the environment, occurring in the earth's crust mainly in the form of sulfides, oxides and arsenides. Cobalt metal is used to make corrosion- and wear-resistant alloys used in aircraft engines (superalloys), in magnets (magnetic alloys) and in high-strength steels and other alloys for many applications. Cobalt metal is added to metallic carbides, especially tungsten carbide, to prepare hard metals (two-phase composites; also known as cemented carbides) for metal-working tools. Cobalt is also used to manufacture cobalt-diamond grinding tools, cobalt discs and other cutting and grinding tools made from cobalt metal. Other uses of cobalt compounds include catalysts, batteries, dyes and pigments and related applications. Occupational exposure to cobalt occurs predominantly during refining of cobalt, in the production of alloys, and in the hard-metal industry where workers may be exposed during the manufacture and maintenance of hard-metal tools and during the use of diamond-cobalt tools.

5.2   Human carcinogenicity data

Several reports addressing cancer risks among workers in hard-metal production facilities in France provide evidence of an increased lung cancer risk related to exposure to hard-metal dust containing cobalt and tungsten carbide. The risk appears to be highest among those exposed to unsintered rather than sintered hard-metal dust. There is evidence for an increasing lung cancer risk with increasing duration of exposure in analyses which took into account potential confounding by smoking and other occupational carcinogens.

An earlier and smaller study of workers exposed to cobalt and tungsten carbide in the hard-metal industry in Sweden found increased mortality from lung cancer in the full cohort, with a higher risk among those with longer duration of exposure and latency. The study provides limited confirmation due to the small number of exposed lung cancer cases, the lack of adjustment for other carcinogenic exposures and the absence of a positive relationship between intensity of exposure and lung cancer risk.

The study of workers in hard-metal factories in France also allowed estimation of lung cancer risk in relation to exposures to cobalt in the absence of tungsten carbide. A twofold increased lung cancer risk was observed. However, no exposure–response relationships were reported and the results were not adjusted for other occupational carcinogens or smoking. Another study in the cobalt production industry in France reported no increase in risk of lung cancer mortality among cobalt production workers, but the study was limited by very small numbers.

5.3   Animal carcinogenicity data

Cobalt sulfate heptahydrate as an aqueous aerosol was tested in a single study by inhalation exposure in male and female mice and rats. Increased incidences of alveolar/bronchiolar neoplasms were seen in both sexes of both species. There was also an increase in adrenal pheochromocytomas in female rats. It was uncertain whether a marginal increase in pheochromocytomas in male rats was caused by cobalt sulfate.

Cobalt metal powder was tested in two experiments in rats by intramuscular injection and in one experiment by intrathoracic injection, and in rabbits in one experiment by intra-osseous injection. All the studies revealed sarcomas at the injection site.

A finely powdered cobalt–chromium–molybdenum alloy was tested in rats by intramuscular injection and produced sarcomas at the injection site. In two other experiments in rats, coarsely- or finely-ground cobalt–chromium–molybdenum alloy implanted in muscle, or pellets of cobalt–chromium–molybdenum alloy implanted subcutaneously, did not induce sarcomas. Implantation in the rat femur of three different cobalt-containing alloys, in the form of powder, rod or compacted wire, resulted in a few local sarcomas. In another experiment, intramuscular implantation of polished rods consisting of three different cobalt-containing alloys did not produce local sarcomas. In an experiment in guinea-pigs, intramuscular implantation of a cobalt–chromium–molybdenum alloy powder did not produce local tumours.

Intraperitoneal injection of a cobalt–chromium–aluminium spinel in rats produced a few local malignant tumours, and intratracheal instillation of this spinel in rats was associated with the occurrence of a few pulmonary squamous-cell carcinomas.

Interpretation of the evidence available for the carcinogenicity of cobalt in experimental animals was difficult because many of the reports failed to include sufficient details on results of statistical analyses, on survival and on control groups. Furthermore, such statistical analyses could not be performed by the Working Group in the absence of specific information on survival including fatality due to the neoplasms. Nevertheless, in the evaluation, weight was given to the consistent occurrence of tumours at the site of administration and to the histological types of tumours observed. However, intramuscular or subcutaneous injection of relatively inert foreign materials into rats is known to result in malignant tumours at the injection site, therefore limiting the interpretation of the results.

5.4   Other relevant data

The absorption rate of inhaled cobalt-containing particles is dependent on their solubility in biological fluids and in macrophages. In humans, gastrointestinal absorption of cobalt has been reported to vary between 5 and 45% and it has been suggested that absorption is higher in women than in men. Cobalt can be absorbed through intact human skin. It does not accumulate in any specific organ, except in the lung when inhaled in the form of insoluble particles. High concentrations of cobalt in blood are found in workers exposed to cobalt, in uraemic patients and in persons taking multivitamin preparations. Most of the absorbed cobalt is excreted in the urine within days, but a certain proportion is eliminated slowly, with half-life values between 2 and 15 years. Cobalt ions bind strongly to circulating proteins, mainly albumin. Cobalt concentrations in blood and/or in urine can be used in biological monitoring to assess individual exposure. After inhalation of metallic cobalt particles with tungsten carbide, toxic effects (alveolitis, fibrosis) occur at the site of contact and deposition. These effects are caused by the particles themselves and by solubilized cobalt ions. Systemic effects outside the respiratory tract are unlikely to be due to the particles. The main non-malignant respiratory disorders caused by inhalation of metallic cobalt-containing particles are bronchial asthma (any cobalt compounds) and fibrosing alveolitis (cobalt metal mixed with tungsten carbide or with microdiamonds). Fibrosis alveolitis, also known as hard-metal lung disease, is characterized pathologically as a giant-cell interstitial pneumonia; there is no evidence that it is caused by cobalt metal alone or cobalt salts. Non-respiratory toxic effects of cobalt include stimulation of erythropoiesis, and toxicity in the thyroid and the heart. Cobalt has skin-sensitizing properties, which may lead to contact dermatitis or airborne dermatitis.

In animals, it has been demonstrated that the health status of the lung affects the rate of clearance and retention of cobalt-containing particles. Smaller particles show a higher dissolution rate than larger ones. When mixed with tungsten carbide, the absorption and subsequent excretion of intratracheally-instilled cobalt-metal particles is greatly enhanced.

In experimental animals, various cobalt compounds cause a variety of toxic effects in the respiratory tract (pulmonary oedema, acute pneumonia), thyroid, erythropoietic tissue, myocardium and reproductive organs. A mixture of cobalt-metal particles and tungsten carbide caused effects that were much more severe than those observed with cobalt metal alone. Specific surface chemistry and increased production of reactive oxygen species at the site of mutual contact between cobalt and tungsten carbide are likely to play a role in this phenomenon. Cobalt-metal particles are weak inducers of reactive oxygen species *in vitro*, but this effect is greatly enhanced by the presence of tungsten carbide particles.

Exposure by inhalation to cobalt oxide, cobalt chloride or cobalt sulfate gives rise to a spectrum of inflammatory and proliferative changes in the respiratory tract in animals. Biochemical effects include increased levels of oxidized glutathione and stimulation of the pentose phosphate pathway, both of which are indicative of oxidative stress.

Reproductive effects of cobalt chloride include teratogenic effects in mice, and growth retardation and reduced postnatal survival in rats. Decreased fertility, testicular weights and sperm concentration have also been observed in mice. Inhalation of cobalt sulfate also gave rise to decreased sperm motility and increased sperm abnormality in mice, but not in rats.

In vitro, cobalt has been shown to induce various enzymes involved in the cellular response to stress and to interfere with cell-cycle control.

The results of genotoxicity assays with a variety of cobalt salts demonstrate the mutagenic potential of these salts both *in vitro* and *in vivo*. Moreover, from experiments performed with a mixture of cobalt and tungsten carbide particles, there is strong evidence that the mixture is mutagenic *in vitro*. It was also demonstrated to be mutagenic *in vivo* in rat lung cells.

5.5 Evaluation

There is *limited evidence* in humans for the carcinogenicity of cobalt metal with tungsten carbide.

There is *inadequate evidence* in humans for the carcinogenicity of cobalt metal without tungsten carbide.

There is *sufficient evidence* in experimental animals for the carcinogenicity of cobalt sulfate.

There is *sufficient evidence* in experimental animals for the carcinogenicity of cobalt-metal powder.

There is *limited evidence* in experimental animals for the carcinogenicity of metal alloys containing cobalt.

There is *inadequate evidence* in experimental animals for the carcinogenicity of cobalt–aluminum–chromium spinel.

Overall evaluation

Cobalt metal with tungsten carbide is *probably carcinogenic to humans (Group 2A)*.

A number of working group members supported an evaluation in Group 1 because: (1) they judged the epidemiological evidence to be sufficient, leading to an overall evaluation in Group 1; and/or (2) they judged the mechanistic evidence to be strong enough to justify upgrading the default evaluation from 2A to 1. The majority of working group members, who supported the group 2A evaluation, cited the need for either sufficient evidence in humans or strong mechanistic evidence in exposed humans.

Cobalt metal without tungsten carbide is *possibly carcinogenic to humans (Group 2B)*.

Cobalt sulfate and other soluble cobalt(II) salts are *possibly carcinogenic to humans (Group 2B)*.

6. References

Abbracchio, M.P., Heck, J.D. & Costa, M. (1982) The phagocytosis and transforming activity of crystalline metal sulfide particles are related to their negative surface charge. *Carcinogenesis*, **3**, 175–180

Abraham, J.L. & Hunt, A. (1995) Environmental contamination by cobalt in the vicinity of a cemented tungsten carbide tool grinding plant. *Environ. Res.*, **69**, 67–74

ACGIH Worldwide® (2003a) *2003 Guide to Occupational Exposure Values*, Cincinnati, OH, p. 34

ACGIH Worldwide® (2003b) *2003 TLVs® and BEIs® Based on the Documentation of the Threshold Limit Values for Chemical Substances and Physical Agents & Biological Exposure Indices*, Cincinnati, OH, pp. 23, 89

ACGIH Worldwide® (2003c) *Documentation of the TLVs® and BEIs® with Other Worldwide Occupational Exposure Values CD-ROM — 2003*, Cincinnati, OH

Adachi, S., Takemoto, K., Ohshima, S., Shimizu, Y. & Takahama, M. (1991) Metal concentrations in lung tissue of subjects suffering from lung cancer. *Int. Arch. occup. environ. Health*, **63**, 193–197

Adamis, Z., Tátrai, E., Honma, K., Kárpáti, J. & Ungváry, G. (1997) A study on lung toxicity of respirable hard metal dusts in rats. *Ann. occup. Hyg.*, **41**, 515–526

Aich, P., Labiuk, S.L., Tari, L.W., Delbaere, L.J.T., Roesler, W.J., Falk, K.J., Steer, R.P. & Lee, J.S. (1999) *M*-DNA: A complex between divalent metal ions and DNA which behaves as a molecular wire. *J. mol. Biol.*, **294**, 477–485

Alexander, C.S. (1972) Cobalt-beer cardiomyopathy. A clinical and pathologic study of twenty-eight cases. *Am. J. Med.*, **53**, 395–417

Alexandersson, R. (1979) [Studies on the effects of exposure to cobalt.] *Arbete Hälsa*, **8**, 2–23 (in Swedish)

Alexandersson, R. (1988) Blood and urinary concentrations as estimators of cobalt exposure. *Arch. environ. Health*, **43**, 299–303

Alexandersson, R. & Lidums, V. (1979) [Studies on the effects of exposure to cobalt. VII. Cobalt concentrations in blood and urine as exposure indicators.] *Arbete Hälsa*, **8**, 2–23 (in Swedish)

Alimonti, A., Petrucci, F., Laurenti, F., Papoff, P. & Caroli, S. (2000) Reference values for selected trace elements in serum of term newborns from the urban area of Rome. *Clin. chim. Acta*, **292**, 163–173

A.L.M.T. Corp. (2003) *Specification Sheets: Tungsten Carbide Powder [WC Powder]: Standard, Coarse, Ultrafine*, Tokyo

Amacher, D.E. & Paillet, S.C. (1980) Induction of trifluorothymidine-resistant mutants by metal ions in L5178Y/TK$^{+/-}$ cells. *Mutat. Res.*, **78**, 279–288

An, W.G., Kanekal, M., Simon, M.C., Maltepe, E., Blagosklonny, M.V. & Neckers, L.M. (1998) Stabilization of wild-type p53 by hypoxia-inducible factor 1α. *Nature*, **392**, 405–408

Anard, D., Kirsch-Volders, M., Elhajouji, A., Belpaeme, K. & Lison, D. (1997) *In vitro* genotoxic effects of hard metal particles assessed by alkaline single cell gel and elution assays. *Carcinogenesis*, **18**, 177–184

Andersen, O. (1983) Effects of coal combustion products and metal compounds on sister chromatid exchange (SCE) in a macrophage-like cell line. *Environ. Health Perspect.*, **47**, 239–253

Angerer, J. & Heinrich, R. (1988) Cobalt. In: Seiler, H.G. & Sigel, H., *Handbook on Toxicity of Inorganic Compounds*, New York, Marcel Dekker, pp. 251–264

Angerer, J., Heinrich-Ramm, R. & Lehnert, G. (1989) Occupational exposure to cobalt and nickel: biological monitoring. *Int. J. environ. anal. Chem.*, 35, 81-88

Anon. (1989) The role of cobalt in cemented carbides. *Cobalt News*, **89**, 2–3

Anthoine, D., Petiet, G., Wurtz, M.C., Simon, B., Stefani, F. & François, M.C. (1982) [Hard metal pulmonary fibroses and their distribution in France.] *Méd. Hyg.*, **40**, 4280–4286 (in French)

Apostoli, P., Porru, S. & Alessio, L. (1994) Urinary cobalt excretion in short time occupational exposure to cobalt powders. *Sci. total Environ.*, **150**, 129–132

Arai, F., Yamamura, Y., Yoshida, M. & Kishimoto, T. (1994) Blood and urinary levels of metals (Pb, Cr, Cd, Mn, Sb, Co and Cu) in cloisonne workers. *Ind. Health*, **32**, 67–78

Araya, J., Maruyama, M., Inoue, A., Fujita, T., Kawahara, J., Sassa, K., Hayashi, R., Kawagishi, Y., Yamashita, N., Sugiyama, E. & Kobayashi, M. (2002) Inhibition of proteasome activity is involved in cobalt-induced apoptosis of human alveolar macrophages. *Am. J. Physiol. Lung Cell mol. Physiol.*, **283**, L849–L858

Arlauskas, A., Baker, R.S.U., Bonin, A.M., Tandon, R.K., Crisp, P.T. & Ellis, J. (1985) Mutagenicity of metal ions in bacteria. *Environ. Res.*, **36**, 379–388

Asmuss, M., Mullenders, L.H.F., Eker, A. & Hartwig, A. (2000) Differential effects of toxic metal compounds on the activities of Fpg and XPA, two zinc finger proteins involved in DNA repair. *Carcinogenesis*, **21**, 2097–2104

Auchincloss, J.H., Abraham, J.L., Gilbert, R., Lax, M., Henneberger, P.K., Heitzman, E.R. & Peppi, D.J. (1992) Health hazard of poorly regulated exposure during manufacture of cemented tungsten carbides and cobalt. *Br. J. ind. Med.*, **49**, 832–836

Ayala-Fierro, F., Firriolo, J.M. & Carter, D.E. (1999) Disposition, toxicity, and intestinal absorption of cobaltous chloride in male Fischer 344 rats. *J. Toxicol. environ. Health A.*, **56**, 571–591

Ball, J.C., Straccia, A.M., Young, W.C. & Aust, A.E. (2000) The formation of reactive oxygen species catalyzed by neutral, aqueous extracts of NIST ambient particulate matter and diesel engine particles. *J. Air Waste Manag. Assoc.*, **50**, 1897–1903

Balmes, J.R. (1987) Respiratory effects of hard-metal dust exposure. *Occup. Med.*, **2**, 327–344

Barany, E., Bergdahl, I.A., Schütz, A., Skerfving, S. & Oskarsson, A. (1997) Inductively coupled plasma mass spectrometry for direct multi-element analysis of diluted human blood and serum. *J. anal. Atom. Spectr.*, **12**, 1005–1009

Barceloux, D.G. (1999) Cobalt. *Clin. Toxicol.*, **37**, 201–216

Bauer, C. & Wang, A.H.-J. (1997) Bridged cobalt amine complexes induce DNA conformational changes effectively. *J. inorg. Biochem.*, **68**, 129–135

Bech, A.O. (1974) Hard metal disease and tool room grinding. *J. soc. occup. Med.*, **24**, 11–16

Bech, A.O., Kipling, M.D. & Heather, J.C. (1962) Hard metal disease. *Br. J. ind. Med.*, **19**, 239–252

Bertine, K.K. & Goldberg, E.D. (1971) Fossil fuel combustion and the major sedimentary cycle. *Science*, **173**, 233–235

Beyersmann, D. (2002) Effects of carcinogenic metals on gene expression. *Toxicol. Lett.*, **127**, 63–68

Beyersmann, D. & Hartwig, A. (1992) The genetic toxicology of cobalt. *Toxicol. appl. Pharmacol.*, **115**, 137–145

Bouros, D., Hatzakis, K., Labrakis, H. & Zeibecoglou, K. (2002) Association of malignancy with diseases causing interstitial pulmonary changes. *Chest*, **121**, 1278–1289

Bucher, J.R., Elwell, M.R., Thompson, M.B., Chou, B.J., Renne, R. & Ragan, H.A. (1990) Inhalation toxicity studies of cobalt sulfate in F344/N rats and B6C3F1 mice. *Fundam. appl. Toxicol.*, **15**, 357–372

Bucher, J.R., Hailey, J.R., Roycroft, J.R., Haseman, J.K., Sills, R.C., Grumbein, S.L., Mellick, P.W. & Chou, B.J. (1999) Inhalation toxicity and carcinogenicity studies of cobalt sulfate. *Toxicol. Sci.*, **49**, 56–67

Bucher, M., Sandner, P., Wolf, K. & Kurtz, A. (1996) Cobalt but not hypoxia stimulates PDGF gene expression in rats. *Am. J. Physiol*, **271**, E451–E457

Burgaz, S., Demircigil, G.C., Yilmazer, M., Ertas, N., Kemaloglu, Y. & Burgaz, Y. (2002) Assessment of cytogenetic damage in lymphocytes and in exfoliated nasal cells of dental laboratory technicians exposed to chromium, cobalt, and nickel. *Mutat. Res.*, **521**, 47–56

Calabrese, E.J., Canada, A.T. & Sacco, C. (1985) Trace elements and public health. *Ann. Rev. public Health*, **6**, 131–146

Camner, P., Boman, A., Johansson, A., Lundborg, M., Wahlberg, J.E. (1993) Inhalation of cobalt by sensitised guinea pigs: Effects on the lungs. *Br. J. ind. Med.*, **50**, 753–757

Cassina, G., Migliori, M., Michetti, G., Argenti, G. & Seghizzi, P. (1987) [A case of cobalt interstitial pneumonia: Pathogenetic and prognostic considerations.] *Med. Lav.*, **78**, 229–234 (in Italian)

Casto, B.C., Meyers, J. & DiPaolo, J.A. (1979) Enhancement of viral transformation for evaluation of the carcinogenic or mutagenic potential of inorganic metal salts. *Cancer Res.*, **39**, 193–198

Cereda, C., Redaelli, M.L., Canesi, M., Carniti, A. & Bianchi, S. (1994) Widia tool grinding: The importance of primary prevention measures in reducing occupational exposure to cobalt. *Sci. total Environ.*, **150**, 249–251

Chadwick, J.K., Wilson, H.K. & White, M.A. (1997) An investigation of occupational metal exposure in thermal spraying processes. *Sci. total Environ.*, **199**, 115–124

Chemical Information Services (2003) *Directory of World Chemical Producers (Online Version)*, Dallas, TX (http://www.chemicalinfo.com; accessed 18.09.2003)

Chou, I.-N. (1989) Distinct cytoskeletal injuries induced by As, Cd, Co, Cr, and Ni compounds. *Biomed. environ. Sci.*, **2**, 358–365

Christensen, J.M. (1995) Human exposure to toxic metals: factors influencing interpretation of biomonitoring results. *Sci. total Environ.*, **166**, 89–135

Christensen, J. & Mikkelsen, S. (1986) Cobalt concentration in whole blood and urine from pottery plate painters exposed to cobalt paint. In: Lakkas, T.D., ed., *Proceedings of an International Conference, Heavy Metals in the Environment, Athens, September 1985, Vol. 2*, Luxembourg, Commission of the European Communities, pp. 86–88

Christensen, J.M. & Poulsen, O.M. (1994) A 1982–1992 surveillance programme on Danish pottery painters. Biological levels and health effects following exposure to soluble or insoluble cobalt compounds in cobalt blue dyes. *Sci. total Environ.*, **150**, 95–104

Christensen, J.M., Poulsen, O.M. & Thomsen, M. (1993) A short-term cross-over study on oral administration of soluble and insoluble cobalt compounds: Sex differences in biological levels. *Int. Arch. Occup. Environ. Health*, **65**, 233–240

Coates, E.O. & Watson, J.H.L. (1971) Diffuse interstitial lung disease in tungsten carbide workers. *Ann. intern. Med.*, **75**, 709–716

Cobalt Development Institute (2003) *About Cobalt — Superalloys*, Guildford (http://www.thecdi.com/cobalt/superalloys.html; accessed 10.09.2003)

Collier, C.G., Pearce, M.J., Hodgson, A. & Ball, A. (1992) Factors affecting the *in-vitro* dissolution of cobalt oxide. *Environ. Health Perspect.*, **97**, 109–113

Coombs, M. (1996) Biological monitoring of cobalt oxide workers. *Int. Arch. occup. environ. Health*, **68**, 511–512

Cornelis, R., Heinzow, B., Herber, R.F.M., Molin Christensen, J., Paulsen, O.M., Sabbioni, E., Templeton, D.M., Thomassen Y., Vahter, M. & Vesterberg, O. (1995) Sample collection guidelines for trace elements in blood and urine. *Pure appl. Chem.*, **67**, 1575–1608

Corrin, B., Butcher, D., McAnulty, B.J., Dubois, R.M., Black, C.M., Laurent, G.J. & Harrison, N.K. (1994) Immunohistochemical localization of transforming growth factor-β_1 in the lungs of patients with systemic sclerosis, cryptogenic fibosing alveolitis and other lung disorders. *Histopathology*, **24**, 145–150

Costa, M., Heck, J.D. & Robison, S.H. (1982) Selective phagocytosis of crystalline metal sulfide particles and DNA strand breaks as a mechanism for the induction of cellular transformation. *Cancer Res.*, **42**, 2757–2763

Cugell, D.W. (1992) The hard metal diseases. *Clin. Chest Med.*, **13**, 269–279

Cugell, D.W., Morgan, W.K.C., Perkins, D.G. & Rubin, A. (1990) The respiratory effects of cobalt. *Arch. intern. Med.*, **150**, 177–183

Curtis, J.R., Goode, G.C., Herrington, J. & Urdaneta, L.E. (1976) Possible cobalt toxicity in maintenance hemodialysis patients after treatment with cobaltous chloride: A study of blood and tissue cobalt concentrations in normal subjects and patients with terminal and renal failure. *Clin. Nephrol.*, **5**, 61–65

Davison, A.G., Haslam, P.L., Corrin, B., Coutts, I.I., Dewar, A., Riding, W.D., Studdy, P.R. & Newman-Taylor, A.J. (1983) Interstitial lung disease and asthma in hard-metal workers: Bronchoalveolar lavage, ultrastructural, and analytical findings and results of bronchial provocation tests. *Thorax*, **38**, 119–128

Dawson, E.B., Evans, D.R., Harris, W.A. & Powell, L.C. (2000) Seminal plasma trace metal levels in industrial workers. *Biol. trace Elem. Res.*, **74**, 97–105

De Boeck, M., Lison, D. & Kirsch-Volders, M. (1998) Evaluation of the *in vitro* direct and indirect genotoxic effects of cobalt compounds using the alkaline comet assay. Influence of interdonor and interexperimental variability. *Carcinogenesis*, **19**, 2021–2029

De Boeck, M., Lardau, S., Buchet, J.-P., Kirsch-Volders, M. & Lison, D. (2000) Absence of significant genotoxicity in lymphocytes and urine from workers exposed to moderate levels of cobalt-containing dust: A cross-sectional study. *Environ. mol. Mutag.*, **36**, 151–160

De Boeck, M., Kirsch-Volders, M. & Lison, D. (2003a) Cobalt and antimony: Genotoxicity and carcinogenicity. *Mutat. Res.*, **533**, 135–152

De Boeck, M., Lombaert, N., De Backer, S., Finsy, R., Lison, D. & Kirsch-Volders, M. (2003b) *In vitro* genotoxic effects of different combinations of cobalt and metallic carbide particles. *Mutagenesis*, **18**, 177–186

De Boeck, M., Hoet, P., Lombaert, N., Nemery, B., Kirsch-Volders, M. & Lison, D. (2003c) *In vivo* genotoxicity of hard metal dust: Induction of micronuclei in rat type II epithelial lung cells. *Carcinogenesis*, **24**, 1793–1800

Delahant, A.B. (1955) An experimental study of the effects of rare metals on animal lungs. *Arch. ind. Health*, **12**, 116–120

Demedts, M. & Ceuppens, J.L. (1989) Respiratory diseases from hard metal or cobalt exposure. Solving the enigma. *Chest*, **95**, 2–3

Demedts, M., Gheysens, B., Nagels, J., Verbeken, E., Lauweryns, J., van den Eeckhout, A., Lahaye, D. & Gyselen, A. (1984) Cobalt lung in diamond polishers. *Am. Rev. respir. Dis.*, **130**, 130–135

Deng, J.F., Sinks, T., Elliott, D., Smith, D., Singal, M. & Fine, L. (1991) Characterisation of respiratory health and exposures at a sintered permanent magnet manufacturer. *Br. J. ind. Med.*, **48**, 609–615

Deutsche Forschungsgemeinschaft (2002) *List of MAK and BAT Values 2002* (Report No. 38 of the Commission for the Investigation of Health Hazards of Chemical Compounds in the Work Area), Weinheim, Wiley-VCH Verlag GmbH, p. 42

De Vuyst, P., Vande Weyer, R., De Coster, A., Marchandise, F.X., Dumortier, P., Ketelbant, P., Jedwab, J. & Yernault, J.C. (1986) Dental technician's pneumoconiosis. A report of two cases. *Am. Rev. respir. Dis.*, **133**, 316–320

Domingo, J.L. (1989) Cobalt in the environment and its toxicological implications. *Rev. environ, Contam. Toxicol.*, **108**, 105–132

Domingo, J.L., Llobet, J.M. & Bernat, R. (1984) A study of the effects of cobalt administered orally to rats. *Arch. Farmacol. Toxicol.*, **X**, 13–20

Domingo, J.L., Paternain, J.L., Llobet, J.M. & Corbella, J. (1985) Effects of cobalt on postnatal development and late gestation in rats upon oral administration. *Rev. Esp. Fisiol.*, **41**, 293–298

Donaldson, J.D. (2003) Cobalt and cobalt compounds. In: *Ullmann's Encyclopedia of Industrial Chemistry*, 6th Ed., Vol. 8, Weinheim, Wiley-VCH Verlag GmbH, pp. 759–793

Donaldson, J.D., Clark, S.J. & Grimes, S.M. (1986) *Cobalt in Medicine, Agriculture and the Environment*, Slough, Cobalt Development Institute

Dooms-Goossens, A.E., Debusschere, K.M., Gevers, D.M., Dupre, K.M., Degreef, H.J., Loncke, J.P. & Snauwaert, J.F. (1986) Contact dermatitis caused by airborne agents. A review and case reports. *J. Am. Acad. Dermatol.*, **15**, 1–10

Doran, A., Law, F.C., Allen, M.J. & Rushton, N. (1998) Neoplastic transformation of cells by soluble but not particulate forms of metals used in orthopaedic implants. *Biomaterials*, **19**, 751–759

Dorsit, G., Girard, R., Rousset, H., Brune, J., Wiesendanger, T., Tolot, F., Bourret, J. & Galy, P. (1970) [Pulmonary fibrosis in 3 subjects from the same factory exposed to cobalt and tungsten carbide dusts. Pulmonary disorders of the hard metal industry. A professional study.] *Sem. Hôp. Paris*, **46**, 3363–3376 (in French)

Duerksen-Hughes, P.J., Yang, J. & Ozcan, O. (1999) p53 induction as a genotoxic test for twenty-five chemicals undergoing in vivo carcinogenicity testing. *Environ. Health Perspect.*, **107**, 805–812

Edel, J., Sabbioni, E., Pietra, R., Rossi, A., Torre, M., Rizzato, G. & Fraioli, P. (1990) Trace metal lung disease: In-vitro interaction of hard metals with human lung and plasma components. *Sci. total Environ.*, **95**, 107–117

Egilsson, V., Evans, I.H. & Wilkie, D. (1979) Toxic and mutagenic effects of carcinogens on the mitochondria of *Saccharomyces cerevisiae*. *Mol. gen. Genet.*, **174**, 39–46

Einarsson, Ö., Eriksson, E., Lindstedt, G. & Wahlberg, J.E. (1979) Dissolution of cobalt from hard metal alloys by cutting fluids. *Contact Derm.*, **5**, 129–132

Elferink, J.G.R. & Deierkauf, M. (1989) Suppressive action of cobalt on exocytosis and respiratory burst in neutrophils. *Am. J. Physiol.*, **257**, C859–C864

Elinder, C.-G. & Friberg, L. (1986) Cobalt. In: Friberg, L., Nordberg, G.F. & Vouk, V.B., eds, *Handbook on the Toxicology of Metals*, 2nd Ed., Amsterdam, Elsevier, pp. 211–232

Eurotungstene Metal Powders (2003) *Product Data Sheets: Diamond Tools — Cobalt, Tungsten Based Additives, Cemented Carbides, Cemented Carbides — Tungsten Carbide, Cemented Carbides — Cobalt*, Grenoble, France

Falconbridge (2002) *Product Descriptions: Electrolytic Cobalt, S-Cobalt*, Ontario, Canada

Farah, S.B. (1983) The *in vivo* effect of cobalt chloride on chromosomes. *Rev. Brasil. Genet.*, **6**, 433–442

Fenoglio, I., Martra, G., Prandi, L., Tomatis, M., Coluccia, S. & Fubini, B. (2000) The role of mechanochemistry in the pulmonary toxicity caused by particulate minerals. *J. Mater. Synth. Proc.*, **8**, 145–153

Ferdenzi, P., Giaroli, C., Mori, P., Pedroni, C., Piccinini, R., Ricci, R., Sala, O., Veronesi, C. & Mineo, F. (1994) Cobalt powdersintering industry (stone cutting diamond wheels): A study of environmental-biological monitoring, workplace improvement and health surveillance. *Sci. total Environ.*, **150**, 245–248

Ferioli, A., Roi, R. & Alessio, L. (1987) Biological indicators for the assessment of human exposure to industrial chemicals. In: Alessio, L., Berlin, A., Boni, M. & Roi, R., eds, *CEC-Industrial Health and Safety* (EUR 11135 EN), Luxembourg, Commission of the European Communities, pp. 48–61

Ferri, F., Candela, S., Bedogni, L., Piccinini, R. & Sala, O. (1994) Exposure to cobalt in the welding process with stellite. *Sci. total Environ.*, **150**, 145–147

Figueroa, S., Gerstenhaber, B., Welch, L., Klimstra, D., Smith, G.J.W. & Beckett, W. (1992) Hard metal interstitial pulmonary disease associated with a form of welding in a metal parts coating plant. *Am. J. ind. Med.*, **21**, 363–373

Fischbein, A., Luo, J.-C.J., Solomon, S.J., Horowitz, S., Hailoo, W. & Miller, A. (1992) Clinical findings among hard metal workers. *Br. J. ind. Med.*, **49**, 17–24

Fischer, T. & Rystedt, I. (1983) Cobalt allergy in hard metal workers. *Contact Derm.*, **9**, 115–121

Fortoul, T.I., Osorio, L.S., Tovar, A.T., Salazar, D., Castilla, M.E. & Olaiz-Fernández, G. (1996) Metals in lung tissue from autopsy cases in Mexico City residents: Comparison of cases from the 1950s and the 1980s. *Environ. Health Perspect.*, **104**, 630–632

Foxmet SA (2003) *Specification Sheets: Ultra Fine Cobalt Powder, Extrafine Fine Cobalt Powder, Fine Cobalt Powder – 400 Mesh, Fine Cobalt Powder – 5M, Coarse Cobalt Powder – 'S' Grade, Coarse Cobalt Powder – 'Dgc' Grade, Fine Tungsten Carbide Powders, Fine Tungsten Carbide – Cobalt Powders, Fused Tungsten Carbide Powders*, Dondelange, Luxembourg

Frost, A.E., Keller, C.A., Brown, R.W., Noon, G.P., Short, H.D., Abraham, J.L., Pacinda, S. & Cagle, P.T. (1993) Giant cell interstitial pneumonitis. Disease recurrence in the transplanted lung. *Am. Rev. respir. Dis.*, **148**, 1401–1404

Fukunaga, M., Kurachi, Y. & Mizuguchi, Y. (1982) Action of some metal ions on yeast chromosomes. *Chem. pharm. Bull.*, **30**, 3017–3019

Garcia, F., Ortega, A., Domingo, J.L. & Corbella, J. (2001) Accumulation of metals in autopsy tissues of subjects living in Tarragona County, Spain. *J. environ. Sci. Health*, **A36**, 1767–1786

Gennart, J.P., Baleux, C., Verellen-Dumoulin, C., Buchet, J.P., De Meyer, R. & Lauwerys, R. (1993) Increased sister chromatid exchanges and tumor markers in workers exposed to elemental chromium-, cobalt- and nickel-containing dusts. *Mutat. Res.*, **299**, 55–61

Gerhardsson, L. & Nordberg, G.F. (1993) Lung cancer in smelter workers — Interactions of metals as indicated by tissue levels. *Scand. J. Work environ. Health*, **19** (Suppl. 1), 90–94

Gheysens, B., Auwerx, J., Van den Eeckhout, A. & Demedts, M. (1985) Cobalt-induced bronchial asthma in diamond polishers. *Chest*, **88**, 740–744

Gleadle, J.M., Ebert, B.L., Firth, J.D. & Ratcliffe, P.J. (1995) Regulation of angiogenic growth factor expression by hypoxia, transition metals, and chelating agents. *Am. J. Physiol.*, **268**, C1362–C1368

Gong, P., Hu, B., Stewart, D., Ellerbe, M., Figueroa, Y.G., Blank, V., Beckman, B.S. & Alam, J. (2001) Cobalt induces heme oxygenase-1 expression by a hypoxia-inducible factor-independent mechanism in Chinese hamster ovary cells: Regulation by Nrf2 and MafG transcription factors. *J. biol. Chem.*, **276**, 27018–27025

Göpfert, T., Eckardt, K.-U., Gess, B. & Kurtz, A. (1995) Cobalt exerts opposite effects on erythropoietin gene expression in rat hepatocytes in vivo and in vitro. *Am. J. Physiol.*, **269**, R995–R1001

Gori, C. & Zucconi, L. (1957) L'azione citologica indotta da un gruppo di composti inorganici su *Allium cepa*. *Caryologia*, **10**, 29–45

Hahtola, P.A., Järvenpää, R.E., Lounatmaa, K., Mattila, J.J., Rantala I., Uitti, J.A. & Sutinen, S. (2000) Hard metal alveolitis accompanied by rheumatoid arthritis. *Respiration*, **67**, 209–212

Hamilton, E.I. (1994) The geobiochemistry of cobalt. *Sci. Total Environ.*, **150**, 7–39

Hamilton-Koch, W., Snyder, R.D. & Lavelle, J.M. (1986) Metal-induced DNA damage and repair in human diploid fibroblasts and Chinese hamster ovary cells. *Chem.-biol. Interact.*, **59**, 17–28

Hanna, P.M., Kadiiska, M.B. & Mason, R.P. (1992) Oxygen-derived free radical and active oxygen complex formation from cobalt(II) chelates in vitro. *Chem. Res. Toxicol.*, **5**, 109–115

Harding, H.E. (1950) Notes on the toxicology of cobalt metal. *Br. J. ind. Med.*, **7**, 76–78

Harding, S.M. (2003) Woman with dry cough and dyspnea on exertion has clubbing, conjunctival injection, and diffuse crackles. *Chest*, **123**, 935–936

Hartung, M. (1986) [*Lungenfibrosen bei Hartmetallschleifern. Bedeutung der Cobalteinwirkung*], Sankt Augustin (Germany), Hauptverband der gewerblichen Berufsgenossenschaften

Hartung, M. & Schaller, K.-H. (1985) [Occupational medical significance of cobalt exposure in hard-metal grinding.] In: Bolt, H.M., Piekarski, C. & Rutenfranz, J., eds, *Aktuelle arbeitsmedizinische Probleme in der Schwerindustrie. Theorie und Praxis biologischer Toleranzwerte für*

Arbeitsstoffe (BAT-Werte). Bedeutung neuer Technologien für die arbeitsmedizinische Praxis. Arbeitsmedizinisches Kolloquium der gewerblichen Berufsgenossenschaften [Actual Occupational Medical Problems in Heavy Industry. Theory and Practice of Biological Tolerance Values for Industrial Substances. Significance of New Technologies for Occupational and Medical Practice. Occupational Medical Colloquium of Industrial Societies], Stuttgart, Gentner Verlag, pp. 55–63 (in German)

Hartung, M. & Spiegelhalder, B. (1982) [Zur externen und internen Belastung mit Nitrosaminen bei Hartmetallschleifern]. *Arbeitsmed. Sozialmed. Präventivmed.*, **11**, 273–275

Hartwig, A., Kasten, U., Boakye-Dankwa, K., Schlepegrell, R. & Beyersmann, D. (1990) Uptake and genotoxicity of micromolar concentrations of cobalt chloride in mammalian cells. *Toxicol. environ. Chem.*, **28**, 205–215

Hartwig, A. (1995) Current aspects in metal genotoxicity. *Biometals*, **8**, 3–11

Hartwig, A. (1998) Carcinogenicity of metal compounds: Possible role of DNA repair inhibition. *Toxicol. Lett.*, **102–103**, 235–239

Hartwig, A. (2001) Zinc finger proteins as potential targets for toxic metal ions: Differential effects on structure and function. *Antioxid. Redox Signal.*, **3**, 625–634

Hartwig, A. & Schwerdtle, T. (2002) Interactions by carcinogenic metal compounds with DNA repair processes: Toxicological implications. *Toxicol. Lett.*, **127**, 47–54

Hawkins, M. (2004) The role of the CDI in the global cobalt market. *Cobalt News*, **04/2** (April 2004), pp. 8–11

Health and Safety Executive (2002) *EH40/2002 Occupational Exposure Limits 2002*, Norwich, p. 8

Heath, J.C. (1954) Cobalt as a carcinogen. *Nature*, **173**, 822–823

Heath, J.C. (1956) The production of malignant tumours by cobalt in the rat. *Br. J. Cancer*, **10**, 668–673

Heath, J.C. (1960) The histogenesis of malignant tumours induced by cobalt in the rat. *Br. J. Cancer*, **14**, 478–482

Heath, J.C. & Daniel, M.R. (1962) The production of malignant tumours by cobalt in the rat: Intrathoracic tumours. *Br. J. Cancer*, **16**, 473–478

Heath, J.C., Freeman, M.A.R. & Swanson, S.A.V. (1971) Carcinogenic properties of wear particles from prostheses made in cobalt-chromium alloy. *Lancet*, **i**, 564–566

Hengstler, J.G., Bolm-Audorff, U., Faldum, A., Janssen, K., Reifenrath, M., Götte, W., Jung, D., Mayer-Popken, O., Fuchs, J., Gebhard, S., Bienfait, H.G., Schlink, K., Dietrich, C., Faust, D., Epe, B. & Oesch, F. (2003) Occupational exposure to heavy metals: DNA damage induction and DNA repair inhibition prove co-exposures to cadmium, cobalt and lead as more dangerous than hitherto expected. *Carcinogenesis*, **24**, 63–73

Hodge, F.G. (1993) Cobalt and cobalt compounds. In: Kroschwitz, J.I. & Howe-Grant, M., eds, *Kirk-Othmer Encyclopedia of Chemical Technology*, Vol. 6, 4th Ed., New York, John Wiley & Sons, pp. 760–777

Hoet, P.M.H., Roesems, G., Demedts, M.G. & Nemery, B. (2002) Activation of the hexose monophosphate shunt in rat type II pneumocytes as an early marker of oxidative stress caused by cobalt particles. *Arch. Toxicol.*, **76**, 1–7

Hogstedt, C. & Alexandersson, R. (1990) Mortality among hard metal workers. *Arbete Hälsa*, **21**, 1–26

Horowitz, S.F., Fischbein, A., Matza, D., Rizzo, J.N., Stern, A., Machac, J. & Solomon, S.J. (1988) Evaluation of right and left ventricular function in hard metal workers. *Br. J. ind. Med.*, **45**, 742–746

Huaux, F., Lasfargues, G., Lauwerys, R. & Lison, D. 1995, Lung toxicity of hard metal particles and production of interleukin-1, tumor necrosis factor-α, fibronectin, and cystatin-c by lung phagocytes. *Toxicol. appl. Pharmacol.*, **132**, 53–62

IARC (1991) *IARC Monographs on the Evaluation of Carcinogenic Risks to Humans*, Vol. 52, *Chlorinated Drinking-water; Chlorination By-products; Some Other Halogenated Compounds; Cobalt and Cobalt Compounds*, Lyon, pp. 363–472

Ichikawa, Y., Kusaka, Y. & Goto, S. (1985) Biological monitoring of cobalt exposure, based on cobalt concentrations in blood and urine. *Int. Arch. occup. environ. Health*, **55**, 269–276

Imbrogno, P. & Alborghetti, F. (1994) Evaluation and comparison of the levels of occupational exposure to cobalt during dry and/or wet hard metal sharpening. Environmental and biological monitoring. *Sci. total Environ.*, **150**, 259–262

Inco Ltd (2003) *Product Data Sheet: Inco Electrolytic Cobalt Rounds*, Ontario, Canada

Inoue, T., Ohta, Y., Sadaie, Y. & Kada, T. (1981) Effect of cobaltous chloride on spontaneous mutation induction in a *Bacillus subtilis* mutator strain. *Mutat. Res.*, **91**, 41–45

Iyengar, V. & Woittiez, J. (1988) Trace elements in human clinical specimens: Evaluation of literature data to identify reference values. *Clin. Chem.*, **34**, 474–481

Japan New Metals Co. Ltd (2003) *Specification Sheets: Tungsten Carbide Powders*, Osaka, Japan

Jarvis, J.Q., Hammond, E., Meier, R. & Robinson, C. (1992) Cobalt cardiomyopathy. A report of two cases from mineral assay laboratories and a review of the literature. *J. occup. Med.*, **34**, 620–626

Jasmin, G. & Riopelle, J.L. (1976) Renal carcinomas and erythrocytosis in rats following intrarenal injection of nickel subsulfide. *Lab. Invest.*, **35**, 71–78

Jayesh Group (2003) *Specification Sheets: Cobalt Metal Powder* and *Tungsten Carbide Powder*, Mumbai, India

Jensen, A.A. & Tüchsen, F. (1990) Cobalt exposure and cancer risk. *Crit. Rev. Toxicol.*, **20**, 427–437

Johansson, A., Lundborg, M., Hellström, P.-Å., Camner, P., Keyser, T.R., Kirton, S.E. & Natusch, D.F. (1980) Effect of iron, cobalt, and chromium dust on rabbit alveolar macrophages: A comparison with the effects of nickel dust. *Environ. Res.*, **21**, 165–176

Johansson, A., Camner, P., Jarstrand, C. & Wiernik, A. (1983) Rabbit alveolar macrophages after inhalation of soluble cadmium, cobalt, and copper: A comparison with the effects of soluble nickel. *Environ. Res.*, **31**, 340–354

Johansson, A., Lundborg, M., Wiernik, A., Jarstrand, C. & Camner, P. (1986) Rabbit alveolar macrophages after long-term inhalation of soluble cobalt. *Environ. Res.*, **41**, 488–496

Johansson, A., Robertson, B., & Camner, P. (1987) Nodular accumulation of type II cells and inflammatory lesions caused by inhalation of low cobalt concentrations. *Environ. Res.*, **43**, 227–243

Johansson, A., Curstedt, T., & Camner, P. (1991) Lung lesions after combined inhalation of cobalt and nickel, *Environ. Res.*, **54**, 24–38

Johnston, J.M. (1988) *Cobalt 87. A Market Research Study of Cobalt in 1987*, Slough, Cobalt Development Institute

Jordan, C., Whitman, R.D., Harbut, M. & Tanner, B. (1990) Memory deficits in workers suffering from hard metal disease. *Toxicol. Lett.*, **54**, 241–243

Jordan, C.M., Whitman, R.D. & Harbut, M. (1997) Memory deficits and industrial toxicant exposure: A comparative study of hard metal, solvent and asbestos workers. *Int. J. Neurosci.*, **90**, 113–128

Joshi, R.R. & Ganesh, K.N. (1992) Chemical cleavage of plasmid DNA by Cu(II), Ni(II) and Co(III) desferal complexes. *Biochem. biophys. Res. Commun.*, **182**, 588–592

Kada, T. & Kanematsu, N. (1978) Reduction of *N*-methyl-*N'*-nitrosoguanidine-induced mutations by cobalt chloride in *Escherichia coli. Proc. Jpn. Acad.*, **54B**, 234–237

Kadiiska, M.B., Maples, K.R. & Mason, R.P. (1989) A comparison of cobalt(II) and iron(II) hydroxyl and superoxide free radical formation. *Arch. Biochem. Biophys.*, **275**, 98–111

Kanematsu, N., Hara, M. & Kada, T. (1980) Rec assay and mutagenicity studies on metal compounds. *Mutat. Res.*, **77**, 109–116

Kaplun, Z.S. & Mezencewa, N.W. (1960) Experimentalstudie über die toxische Wirkung von Staub bei der Erzeugung von Sintermetallen. *J. Hyg. Epidemiol. Microbiol. Immunol.*, **4**, 390–399

Kasprzak, K.S., Zastawny, T.H., North, S.L., Riggs, C.W., Diwan, B.A., Rice, J.M. & Dizdaroglu, M. (1994) Oxidative DNA base damage in renal, hepatic, and pulmonary chromatin of rats after intraperitoneal injection of cobalt(II) acetate. *Chem. Res. Toxicol.*, **7**, 329–335

Kasten, U., Hartwig, A. & Beyersmann, D. (1992) Mechanisms of cobalt(II) uptake into V79 Chinese hamster cells. *Arch. Toxicol.*, **66**, 592–597

Kasten, U., Mullenders, L.H.F. & Hartwig, A. (1997) Cobalt(II) inhibits the incision and the polymerization step of nucleotide excision repair in human fibroblasts. *Mutat. Res.*, **383**, 81–89

Keane, M.J., Hornsby-Myers, J.L., Stephens, J.W., Harrison, J.C., Myers, J.R. & Wallace, W.E. (2002) Characterization of hard metal dusts from sintering and detonation coating processes and comparative hydroxyl radical production. *Chem. Res. Toxicol.*, **15**, 1010–1016

Kennametal (2003) *Product Data Sheets: Macrocrystalline WC Powder, Conventional Carburized WC, Cast Carbide Vacuum Fused KF110, Chill Cast Carbide, Kenface Sintered Tungsten Carbide/Cobalt Hardmetal*, Latrobe, PA, USA

Kennedy, S.M., Chan-Yeung, M., Marion, S., Lea, J. & Teschke, K. (1995) Maintenance of stellite and tungsten carbide saw tips: Respiratory health and exposure–response evaluations. *Occup. environ. Med.*, **52**, 185–191

Kerckaert, G.A., LeBoeuf, R.A. & Isfort, R.J. (1996) Use of the Syrian hamster embryo cell transformation assay for determining the carcinogenic potential of heavy metal compounds. *Fundam. appl. Toxicol.*, **34**, 67–72

Kerfoot, E.J., Fredrick, W.G. & Domeier, E. (1975) Cobalt metal inhalation studies in miniature swine. *Am. ind. Hyg. Assoc. J.*, **36**, 17–25

Kharab, P. & Singh, I. (1985) Genotoxic effects of potassium dichromate, sodium arsenite, cobalt chloride and lead nitrate in diploid yeast. *Mutat. Res.*, **155**, 117–120

Kharab, P. & Singh, I. (1987) Induction of respiratory deficiency in yeast by salts of chromium, arsenic, cobalt and lead. *Indian J. exp. Biol.*, **25**, 141–142

Kirk, W.S. (1985) Cobalt. In: *Mineral Facts and Problems, 1985 Edition* (Preprint from Bulletin 675), Washington DC, Bureau of Mines, US Department of the Interior, pp. 1–8

Kirsch-Volders, M. & Lison, D. (2003) Occupational exposure to heavy metals: DNA damage induction and DNA repair inhibition prove co-exposures to cadmium, cobalt and lead as more dangerous than hitherto expected (Letter to the Editor). *Carcinogenesis*, **24**, 1853–1854

Kitahara, J., Yamanaka, K., Kato, K., Lee, Y.-W., Klein, C.B. & Costa, M. (1996) Mutagenicity of cobalt and reactive oxygen producers. *Mutat. Res.*, **370**, 133–140

Kitamura, H., Yoshimura, Y., Tozawa, T., & Koshi, K. (1980) Effects of cemented tungsten carbide dust on rat lungs following intratracheal injection of saline suspension. *Acta pathol. jpn.*, **30**, 241–253

Kraus, T., Schramel, P., Schaller, K.H., Zöbelein, P., Weber, A. & Angerer, J. (2001) Exposure assessment in the hard metal manufacturing industry with special regard to tungsten and its compounds. *Occup. environ. Med.*, **58**, 631–634

Kreyling, W.G., Godleski, J.J., Kariya, S.T., Rose, R.M. & Brain, J.D. (1990) In-vitro dissolution of uniform cobalt oxide particles by human and canine alveolar macrophages. *Am. J. Respir. Cell Mol. Biol.*, **2**, 413–422

Kreyling, W.G., Cox, C., Ferron, G.A. & Oberdörster, G. (1993) Lung clearance in Long-Evans rats after inhalation of porous, monodisperse cobalt oxide particles. *Exp. Lung Res.*, **19**, 445–467

Kriss, J.P., Carnes, W.H. & Gross, R.T. (1955) Hypothyroidism and thyroid hyperplasia in patients treated with cobalt. *Hyperplasia Thyroid*, **157**, 117–121

Kristiansen, J., Christensen, J.M., Iversen, B.S. & Sabbioni, E. (1997) Toxic trace element reference levels in blood and urine: Influence of gender and lifestyle factors. *Sci. total Environ.*, **204**, 147–160

Kumagai, S., Kusaka, Y. & Goto, S. (1996) Cobalt exposure level and variability in the hard metal industry of Japan. *Am. ind. Hyg. Assoc. J.*, **57**, 365–369

Kumagai, S., Kusaka, Y. & Goto, S. (1997) Log-normality of distribution of occupational exposure concentrations to cobalt. *Ann. occup. Hyg.*, **41**, 281–286

Kusaka, Y., Yokoyama, K., Sera, Y., Yamamoto, S., Sone, S., Kyono, H., Shirakawa, T. & Goto, S. (1986) Respiratory diseases in hard metal workers: An occupational hygiene study in a factory. *Br. J. ind. Med.*, **43**, 474–485

Kusaka, Y., Nakano, Y., Shirakawa, T. & Morimoto, K. (1989) Lymphocyte transformation with cobalt in hard metal asthma. *Ind. Health*, **27**, 155–163

Kusaka, Y., Nakano, Y., Shirakawa, T., Fujimura, N., Kato, M. & Heki, S. (1991) Lymphocyte transformation test with nickel in hard metal asthma: Another sensitizing component of hard metal. *Ind. Health*, **29**, 153–160

Kusaka, Y., Kumagai, S., Kyono, H., Kohyama, N. & Shirakawa, T. (1992) Determination of exposure to cobalt and nickel in the atmosphere in the hard metal industry. *Ann. occup. Hyg.*, **36**, 497–507

Kusaka, Y., Iki, M., Kumagai, S. & Goto, S. (1996a) Epidemiological study of hard metal asthma. *Occup. environ. Med.*, **53**, 188–193

Kusaka, Y., Iki, M., Kumagai, S. & Goto, S. (1996b) Decreased ventilatory function in hard metal workers. *Occup. environ. Med.*, **53**, 194–199

Kyle, P.R. & Meeker, K.M. (1990) Emission rates of sulfur dioxide, trace gases and metals from Mount Erebus, Antarctica. *Geophys. Res. Lett.*, **17**, 2125–2128

Kyono, H., Kusaka, Y., Homma, K., Kubota, H. & Endo-Ichikawa, Y. (1992) Reversible lung lesions in rats due to short-term exposure to ultrafine cobalt particles. *Ind. Health*, **30**, 103–118

Lahaye, D., Demedts, M., Vanden Oever, R. & Roosels, D. (1984) Lung diseases among diamond polishers due to cobalt? *Lancet*, **i**, 156–157

Lasfargues, G., Lison, D., Maldague, P. & Lauwerys, R. (1992) Comparative study of the acute lung toxicity of pure cobalt powder and cobalt–tungsten carbide mixture in rat. *Toxicol. appl. Pharmacol.*, **112**, 41–50

Lasfargues, G., Wild, P., Moulin, J.J., Hammon, B., Rosmorduc, B., Rondeau du Noyer, C., Lavandier, M. & Moline, J. (1994) Lung cancer mortality in a French cohort of hard-metal workers. *Am. J. ind. Med.*, **26**, 585–595

Lasfargues, G., Lardot, C., Delos, M., Lauwerys, R. & Lison, D. (1995) The delayed lung responses to single and repeated intratracheal administration of pure cobalt and hard metal powder in the rat. *Environ. Res.*, **69**, 108–121

Lauwerys, R. & Lison, D. (1994) Health risks associated with cobalt exposure — An overview. *Sci. total Environ.*, **150**, 1–6

Leitão, A.C., Soares, R.A., Cardoso, J.S., Guillobel, H.C. & Caldas, L.R. (1993) Inhibition and induction of SOS response in *Escherichia coli* by cobaltous chloride. *Mutat. Res.*, **286**, 173–180

Léonard, A. & Lauwerys, R. (1990) Mutagenicity, carcinogenicity and teratogenicity of cobalt metal and cobalt compounds. *Mutat. Res.*, **239**, 17–27

Leonard, S., Gannett, P. M., Rojanasakul, Y., Schwegler-Berry, D., Castranova, V., Vallyathan, V. & Shi, X. (1998) Cobalt-mediated generation of reactive oxygen species and its possible mechanism. *J. inorg. Biochem.*, **70**, 239–244

Lewis, R.J., Sr, ed. (2001) *Hawley's Condensed Chemical Dictionary*, 14th Ed., New York, John Wiley & Sons, p. 998

Lewis, C.P.L., Demedts, M. & Nemery, B. (1991) Indices of oxidative stress in hamster lung following exposure to cobalt(II) ions: *In vivo* and *in vitro* studies. *Am. J. respir. Cell mol. Biol.*, **5**, 163–169

Lewis, C.P., Demedts, M. & Nemery, B. (1992) The role of thiol oxidation in cobalt(II)-induced toxicity in hamster lung. *Biochem. Pharmacol.*, **43**, 519–525

Lichtenstein, M.E., Bartl, F. & Pierce, R.T. (1975) Control of cobalt exposures during wet process tungsten carbide grinding. *Am. ind. Hyg. Assoc. J.*, 879–885

Lide, D.R., ed. (2003) *CRC Handbook of Chemistry and Physics*, 84th Ed., Boca Raton, FL, CRC Press, pp. 4-53, 4-54, 4-91

Liebow, A.A. (1975) Definition and classification of interstitial pneumonias in human pathology. *Prog. respir. Res.*, **8**, 1–33

Lindegren, C.C., Nagai, S. & Nagai, H. (1958) Induction of respiratory deficiency in yeast by manganese, copper, cobalt and nickel. *Nature*, **182**, 446–448

Linna, A., Oksa, P., Palmroos, P., Roto, P., Laippala, P. & Uitti, J. (2003) Respiratory health of cobalt production workers. *Am. J. ind. Med.*, **44**, 124–132

Linnainmaa, M. & Kiilunen, M. (1997) Urinary cobalt as a measure of exposure in the wet sharpening of hard metal and stellite blades. *Int. Arch. occup. environ. Health*, **69**, 193–200

Linnainmaa, M., Kangas, J. & Kalliokoski, P. (1996) Exposure to airborne metals in the manufacture and maintenance of hard metal and stellite blades. *Am. ind. Hyg. Assoc. J.*, **57**, 196–201

Lins, L.E. & Pehrsson, K. (1976) Cobalt intoxication in uraemic myocardiopathy? *Lancet*, **May 29**, 1191–1192

Lirsac, P.N., Nolibe, D. & Metivier, H. (1989) Immobilization of alveolar macrophages for measurement of *in-vitro* dissolution of aerosol particles. *Int. J. Radiat. Biol.*, **56**, 1011–1021

Lison, D. (1996) Human toxicity of cobalt-containing dust and experimental studies on the mechanism of interstitial lung disease (hard metal disease). *Crit. Rev. Toxicol.*, **26**, 585–616

Lison, D. (2000) Letter to the Editor. *Toxicol. appl. Pharmacol.*, **168**, 173

Lison, D. & Lauwerys, R. (1990) *In vitro* cytotoxic effects of cobalt-containing dusts on mouse peritoneal and rat alveolar macrophages. *Environ. Res.*, **52**, 187–198

Lison, D. & Lauwerys, R. (1991) Biological responses of isolated macrophages to cobalt metal and tungsten carbide–cobalt powders. *Pharmacol. Toxicol.*, **69**, 282–285

Lison, D. & Lauwerys, R. (1992) Study of the mechanism responsible for the elective toxicity of tungsten carbide-cobalt powder toward macrophages. *Toxicol. Lett.*, **60**, 203–210

Lison, D. & Lauwerys, R. (1993) Evaluation of the role of reactive oxygen species in the interactive toxicity of carbide–cobalt mixtures on macrophages in culture. *Arch. Toxicol.*, **67**, 347–351

Lison, D. & Lauwerys, R. (1994) Cobalt bioavailability from hard metal particles. Further evidence that cobalt alone is not responsible for the toxicity of hard metal particles. *Arch. Toxicol.*, **68**, 528–531

Lison, D. & Lauwerys, R. (1995) The interaction of cobalt metal with different carbides and other mineral particles on mouse peritoneal macrophages. *Toxicol. In Vitro*, **9**, 341–347

Lison, D., Buchet, J.P., Swennen, B., Molders, J. & Lauwerys, R. (1994) Biological monitoring of workers exposed to cobalt metal, salt, oxides, and hard metal dust. *Occup. environ. Med.*, **51**, 447–450

Lison, D., Carbonnelle, P., Mollo, L., Lauwerys, R. & Fubini, B. (1995) Physicochemical mechanism of the interaction between cobalt metal and carbide particles to generate toxic activated oxygen species. *Chem. Res. Toxicol.*, **8**, 600–606

Lison, D., De Boeck, M., Verougstraete, V. & Kirsch-Volders, M. (2001) Update on the genotoxicity and carcinogenicity of cobalt compounds. *Occup. environ. Med.*, **58**, 619–625

Little, J.A. & Sunico, R. (1958) Cobalt-induced goiter with cardiomegaly and congestive failure. *J. Pediatr.*, **53**, 284–288

Lob, M. & Hugonnaud, C. (1977) [Pulmonary pathology. (Risks of pneumoconiosis due to hard metals and beryliosis in dental technicians during the modelling of metal prostheses)]. *Arch. mal. prof.*, **38**, 543–549 (in French)

Louie, A.Y. & Meade, T.J. (1998) A cobalt complex that selectively disrupts the structure and function of zinc fingers. *Proc. natl Acad Sci USA*, **95**, 6663–6668

Lu, T.H., Pepe, J., Lambrecht, R.W. & Bonkovsky, H.L. (1996) Regulation of metallothionein gene expression. Studies in transfected primary cultures of chick embryo liver cells. *Biochimie*, **78**, 236–244

Lundborg, M., Falk, R., Johansson, A., Kreyling, W. & Camner, P. (1992) Phagolysosomal pH and dissolution of cobalt oxide particles by alveolar macrophages. *Environ. Health Perspect.*, **97**, 153–157

Lundborg, M., Johard, U., Johansson, A., Eklund, A., Falk, R., Kreyling, W. & Camner, P. (1995) Phagolysosomal morphology and dissolution of cobalt oxide particles by human and rabbit alveolar macrophages. *Exper. Lung Res.*, **21**, 51–66

Mao, Y., Liu, K.J., Jiang, J.J., & Shi, X. (1996) Generation of reactive oxygen species by Co(II) from H_2O_2 in the presence of chelators in relation to DNA damage and 2′-deoxyguanosine hydroxylation. *J. Toxicol. environ. Health*, **47**, 61–75

McDermott, F.T. (1971) Dust in the cemented carbide industry. *Am. ind. Hyg. Assoc. J.*, **32**, 188–193

McLean, J.R., McWilliams, R.S., Kaplan, J.G & Birnboim, H.C. (1982) Rapid detection of DNA strand breaks in human peripheral blood cells and animal organs following treatment with physical and chemical agents. In: Bora, K.C., Douglas, G.R. & Nestmann, E.R., eds, *Progress in Mutation Research*, Vol. 3, Amsterdam, Elsevier Biomedical Press, pp. 137–141

Meachim, G., Pedley, R.B. & Williams, D.F. (1982) A study of sarcogenicity associated with Co–Cr–Mo particles implanted in animal muscle. *J. biomed. Mat. Res.*, **16**, 407–416

Memoli, V.A., Urban, R.M., Alroy, J. & Galante, J.O. (1986) Malignant neoplasms associated with orthopedic implant materials in rats. *J. orthopaed. Res.*, **4**, 346–355

Merritt, K., Brown, S.A. & Sharkey, N.A. (1984) The binding of metal salts and corrosion products to cells and proteins *in vitro*. *J. biomed. Mater. Res.*, **18**, 1005–1015

Metal-Tech Ltd (2003) *Specification Sheets: Tungsten Carbide Powder (Grade 1 & 2)*, Beer Sheva, Israel

Meyer-Bisch, C., Pham, Q.T., Mur, J.-M., Massin, N., Moulin, J.-J., Teculescu, D., Carton, B., Pierre, F. & Baruthio, F. (1989) Respiratory hazards in hard metal workers: A cross sectional study. *Br. J. ind. Med.*, **46**, 302–309

Midtgård, U. & Binderup, M.L. (1994) *The Nordic Expert Group for Criteria Documentation of Health Risks from Chemicals*, Vol. 39, *114. Cobalt and Cobalt Compounds*, Arbete och Hälsa, Arbets Miljö Institutet, National Institute of Occupational Health

Migliori, M., Mosconi, G., Michetti, G., Belotti, L., D'Adda, F., Leghissa, P., Musitelli, O., Cassina, G., Motta, T., Seghizzi, P. & Sabbioni, E. (1994) Hard metal disease: Eight workers with interstitial lung fibrosis due to cobalt exposure. *Sci. total Environ.*, **150**, 187–196

Miller, C.W., Davis, M.W., Goldman, A. & Wyatt, J.P. (1953) Pneumoconiosis in the tungsten-carbide tool industry. *Arch. ind. Hyg. occup. Med.*, **8**, 453–465

Miller, A.C., Mog, S., McKinney, L., Luo, L., Allen, J., Xu, J. & Page, N. (2001) Neoplastic transformation of human osteoblast cells to the tumorigenic phenotype by heavy metal-tungsten alloy particles: Induction of genotoxic effects. *Carcinogenesis*, **22**, 115–125

Miller, A.C., Xu, J., Stewart, M., Prasanna, P.G. & Page, N. (2002) Potential late health effects of depleted uranium and tungsten used in armor-piercing munitions: Comparison of neoplastic transformation and genotoxicity with the known carcinogen nickel. *Mil. Med.*, **167**, 120–122

Minoia, C., Sabbioni, E., Apostoli, P., Pietra, R., Pozzoli, L., Gallorini, M., Nicolaou, G., Alessio, L. & Capodaglio, E. (1990) Trace element reference values in tissues from inhabitants of the European community. I. A study of 46 elements in urine, blood and serum of Italian subjects. *Sci. total Environ.*, **95**, 89–105

Minoia, C., Pietra, R., Sabbioni, E., Ronchi, A., Gatti, A., Cavalleri, A. & Manzo, L. (1992) Trace element reference values in tissues from inhabitants of the European Community. III. The control of preanalytical factors in the biomonitoring of trace elements in biological fluids. *Sci. total Environ.*, **120**, 63–79

Mitchell, D.F., Shankwalker, G.B. & Shazer, S. (1960) Determining the tumorigenicity of dental materials. *J. dent. Res.*, **39**, 1023–1028

Miyaki, M., Akamatsu, N., Ono, T. & Koyama, H. (1979) Mutagenicity of metal cations in cultured cells from Chinese hamster. *Mutat. Res.*, **68**, 259–263

Mochizuki, H. & Kada, T. (1982) Antimutagenic action of cobaltous chloride on Trp-P-1-induced mutations in *Salmonella typhimurium* TA98 and TA1538. *Mutat. Res.*, **95**, 145–157

Moens, L. & Dams, R. (1995) NAA and ICP-MS: A comparison between two methods for trace and ultra-trace element analysis. *J. Radioanal. nucl. Chem.*, **192**, 29–38

Mollenhauer, H.H., Corrier, D.E., Clark, D.E., Hare, M.F., Elissalde, M.H. (1985) Effects of dietary cobalt on testicular structure. *Virchows Arch. B. Cell Pathol.*, **49**, 241–248

Moorhouse, C. P., Halliwell, B., Grootveld, M. & Gutteridge, J.M.C. (1985) Cobalt(II) ion as a promoter of hydroxyl radical and possible 'crypto-hydroxyl' radical formation under physiological conditions. Differential effects of hydroxyl radical scavengers. *Biochim. biophys. Acta*, **843**, 261–268

Mosconi, G., Bacis, M., Leghissa, P., Maccarana, G., Arsuffi, E., Imbrogno, P., Airoldi, L., Caironi, M., Ravasio, G., Parigi, P.C., Polini, S. & Luzzana, G. (1994) Occupational exposure to metallic cobalt in the Province of Bergamo. Results of a 1991 survey. *Sci. total Environ.*, **150**, 121–128

Moulin, J.J., Wild, P., Mur, J.M., Fournier-Betz, M. & Mercier-Gallay, M. (1993) A mortality study of cobalt production workers: An extension of the follow-up. *Am. J. ind. Med.*, **23**, 281–288

Moulin, J.J., Wild, P., Romazini, S., Lasfargues, G., Peltier, A., Bozec, C., Deguerry, P., Pellet, F. & Perdrix, A. (1998) Lung cancer risk in hard metal workers. *Am. J. Epidemiol.*, **148**, 241–248

Mur, J.M., Moulin, J.J., Charruyer-Seinerra, M.P. & Lafitte, J. (1987) A cohort mortality among cobalt and sodium workers in an electrochemical plant. *Am. J. ind. Med.*, **12**, 75–81

Murata, M., Gong, P., Suzuki, K. & Koizumi, S. (1999) Differential metal response and regulation of human heavy metal-inducible genes. *J. cell Physiol.*, **180**, 105–113

Nackerdien, Z., Kasprzak, K.S., Rao, G., Halliwell, B. & Dizdaroglu, M. (1991) Nickel(II)- and cobalt(II)-dependent damage by hydrogen peroxide to the DNA bases in isolated human chromatin. *Cancer Res.*, **51**, 5837–5842

National Institute for Occupational Safety and Health (1981) *Criteria for Controlling Occupational Exposure to Cobalt* (DHHS (NIOSH) Publ. No. 82-107), Washington DC, pp. 1–95

National Institute of Environmental Health Sciences (2002) *Toxicological Summary for Cobalt Dust [7440-48-4]*, Research Triangle Park, NC, National Institute of Environmental Health Sciences

National Toxicology Program (1991) *Tox-5: Toxicity Studies of Cobalt Sulfate Heptahydrate in F344/N Rats and B6C3F1 Mice (Inhalation Studies) (CAS No. 10026-24-1* (NIH No. 91-3124), Research Triangle Park, NC

National Toxicology Program (1998) *Toxicology and Carcinogenesis Studies of Cobalt Sulfate Heptahydrate (CAS No. 10026-24-1) in F344/N Rats and B6C3F$_1$ Mice (Inhalation Studies)* (NTP Technical Report 471), Research Triangle Park, NC, pp. 1–268 and pp. 233–239

Nemery, B., Nagels, J., Verbeken, E., Dinsdale, D. & Demedts, M. (1990) Rapidly fatal progression of cobalt-lung in a diamond polisher. *Am. Rev. Respir. Dis.*, **141**, 1373–1378

Nemery, B., Casier, P., Roosels, D., Lahaye, D. & Demedts, M. (1992) Survey of cobalt exposure and respiratory health in diamond polishers. *Am. Rev. respir. Dis.*, **145**, 610–616

Nemery, B., Casier, P., Roosels, D., Lahaye, D. & Demedts, M. (1992) Survey of cobalt exposure and respiratory health in diamond polishers. *Am. Rev. respir. Dis.*, **145**, 610–616

Nemery, B., Verbeken, E.K. & Demedts, M. (2001a) Giant cell interstitial pneumonia (hard metal lung disease, cobalt lung). *Sem. respir. crit. Care Med.*, **22**, 435–447

Nemery, B., Bast, A., Behr, J., Borm, P.J.A., Bourke, S.J., Camus, P., De Vuyst, P., Jansen, H.M., Kinnula, V.L., Lison, D., Pelkonen, O. & Saltini, C. (2001b) Interstitial lung disease induced by exogenous agents: Factors governing susceptibility. *Eur. respir. J.*, **32** (Suppl.), 30S–42S

Newman, L.S., Maier, L.A. & Nemery, B. (1998) Interstitial lung disorders due to beryllium and cobalt. In: Schwartz, M.I. & King, T.E., Jr, eds, *Interstitial Lung Disease*, St Louis, MO, Mosby, pp. 367–392

Newton, D. & Rundo, J. (1970) The long-term retention of inhaled cobalt-60. *Health Phys.*, **21**, 377–384

Nilsson, K., Jensen, B.S. & Carlsen, L. (1985) The migration chemistry of cobalt. *Eur. appl. Res. Rept.-Nucl. Sci. Technol.*, **7**, 23–86

Nishioka, H. (1975) Mutagenic activities of metal compounds in bacteria. *Mutat. Res.*, **31**, 185–189

Ogawa, H.I., Sakata, K., Inouye, T., Jyosui, S., Niyitani, Y., Kakimoto, K., Morishita, M., Tsuruta, S. & Kato, Y. (1986) Combined mutagenicity of cobalt(II) salt and heteroaromatic compounds in *Salmonella typhimurium*. *Mutat. Res.*, **172**, 97–104

Ogawa, H.I., Shibahara, T., Iwata, H., Okada, T., Tsuruta, S., Kakimoto, K., Sakata, K., Kato, Y., Ryo, H., Itoh, T. & Fujikawa, K. (1994) Genotoxic activities *in vivo* of cobaltous chloride and other metal chlorides as assayed in the *Drosophila* wing spot test. *Mutat. Res.*, **320**, 133–140

Ogawa, H.I., Ohyama, Y., Ohsumi, Y., Kakimoto, K., Kato, Y., Shirai, Y., Nunoshiba, T. & Yamamoto, K. (1999) Cobaltous chloride-induced mutagenesis in the *supF* tRNA gene of *Escherichia coli*. *Mutagenesis*, **14**, 249–253

Ohori, N.P., Sciurba, F.C., Owens, G.R., Hodgson, M.J. & Yousem, S.A. (1989) Giant-cell interstitial pneumonia and hard-metal pneumoconiosis. A clinicopathologic study of four cases and review of the literature. *Am. J. surg. Pathol.*, **13**, 581–587

OM Group (2003) *Product Data: Cobalt Metals and Products, Cobalt Coarse Powders and Metal Forms, Cobalt Fine Powders, Cobalt Rechargeable Battery Chemicals*, Newark, NJ

O'Neil, M.J., ed. (2001) *The Merck Index*, 13th Ed., Whitehouse Station, NJ, Merck & Co., p. 426, 428

Oyama, S.T. & Kieffer, R. (1992) Carbides (survey). In: Kroschwitz, J.I. & Howe-Grant, M., eds, *Kirk-Othmer Encyclopedia of Chemical Technology*, Vol. 4, 4th Ed., New York, John Wiley & Sons, pp. 841–848

Pagano, D.A. & Zeiger, E. (1992) Conditions for detecting the mutagenicity of divalent metals in *Salmonella typhimurium*. *Environ. mol. Mutag.*, **19**, 139–146

Paksy, K., Forgács, Z. & Gáti, I. (1999) *In vitro* comparative effect of Cd^{2+}, Ni^{2+}, and Co^{2+} on mouse postblastocyst development. *Environ. Res.*, **80**, 340–347

Palecek, E., Brázdová, M., Cernocká, H., Vlk, D., Brázda, V. & Vojtešek, B. (1999) Effect of transition metals on binding of p53 protein to supercoiled DNA and to consensus sequence in DNA fragments. *Oncogene*, **18**, 3617–3625

Parkes, W.R. (1994) *Occupational Lung Disorders*, Oxford, Butterworth-Heinemann, pp. 1–892

Paternain, J.L., Domingo, J.L. & Corbella, J. (1988) Developmental toxicity of cobalt in the rat. *J. Toxicol. environ. Health*, **24**, 193–200

Paton, G.R. & Allison, A.C. (1972) Chromosome damage in human cell cultures induced by metal salts. *Mutat. Res.*, **16**, 332–336

Pedigo, N.G. & Vernon, M.W. (1993) Embryonic losses after 10-week administration of cobalt to male mice. *Reprod. Toxicol.*, **7**, 111–116

Pedigo, N.G., George, W.J. & Anderson, M.B. (1988) Effects of acute and chronic exposure to cobalt on male reproduction in mice. *Reprod. Toxicol.*, **2**, 45–53

Pellet, F., Perdrix, A., Vincent, M. & Mallion, J.-M. (1984) [Biological levels of urinary cobalt.] *Arch. Mal. prof.*, **45**, 81–85 (in French)

Posma, F.D. & Dijstelberger, S.K. (1985) Serum and urinary cobalt levels as indicators of cobalt exposure in hard metal workers. In: Lekkas, T.D., ed., *Proceedings of an International Conference, Heavy Metals in the Environment, Athens, September 1985*, Luxembourg, Commission of the European Communities, pp. 89-91

Potolicchio, I., Mosconi, G., Forni, A., Nemery, B., Seghizzi, P. & Sorrentino, R. (1997) Susceptibility to hard metal lung disease is strongly associated with the presence of glutamate 69 in HLA-DPβ chain. *Eur. J. Immunol.*, **27**, 2741–2743

Potolicchio, I., Festucci, A., Hausler, P. & Sorrentino, R. (1999) HLA-DP molecules bind cobalt: A possible explanation for the genetic association with hard metal disease. *Eur. J. Immunol.*, **29**, 2140–2147

Poulsen, O.M., Olsen, E., Christensen, J.M., Vinzent, P. & Petersen, O.H. (1995) Geltape method for measurement of work related surface contamination with cobalt containing dust: Correlation between surface contamination and airborne exposure. *Occup. environ. Med.*, **52**, 827–833

Prandi, L., Fenoglio, I., Corazzari, I. & Fubini, B. (2002) *Molecular Basis of Hard Metal Lung Disease, Third International Conference on Oxygen/Nitrogen Radicals: Cell Injury and Disease*, Morgantown, USA, June 2002 (Abstract)

Prazmo, W., Balbin, E., Baranowska, H., Ejchart, A. & Putrament, A. (1975) Manganese mutagenesis in yeast. II. Conditions of induction and characteristics of mitochondrial respiratory deficient *Saccharomyces cerevisiae* mutants induced with manganese and cobalt. *Genet. Res. Camb.*, **26**, 21–29

Putrament, A., Baranowska, H., Ejchart, A. & Jachymczyk, W. (1977) Manganese mutagenesis in yeast. VI. Mn^{2+} uptake, mitDNA replication and E^R induction: Comparison with other divalent cations. *Mol. gen. Genet.*, **151**, 69–76

Rae, T. (1975) A study on the effects of particulate metals of orthopaedic interest on murine macrophages *in vitro*. *J. Bone Joint Surg.*, **57**, 444–450

Reade Advanced Materials (1997) *Product Data Sheets: Cobalt Metal Powder, Tungsten Carbide (WC) Powder*, Providence, RI

Reber, E. & Burckhardt, P. (1970) Über Hartmetallstaublungen in der Schweiz. *Respiration*, **27**, 120–153

Reinl, W., Schnellbächer, F. & Rahm, G. (1979) Pulmonary fibrosis and inflammatory lung diseases following effect of cobalt contact mass. *Zentralbl. Arbeitsmed. Arbeitsschutz Prophyl.*, **29**, 318–324

Resende de Souza Nazareth, H. (1976) Efeito do cloreto de cobalto em não-disjunção. *Cie. Cult.*, **28**, 1472–1475

Reynolds, J.E.F., ed. (1989) Martindale, *The Extra Pharmacopoeia*, London, The Pharmaceutical Press, pp. 1260–1261, 1559

Richardson, C.L., Verna, J., Schulman, G.E., Shipp, K. & Grant, A.D. (1981) Metal mutagens and carcinogens effectively displace acridine orange from DNA as measured by fluorescence polarization. *Environ. Mutag.*, **3**, 545–553

Richeldi, L., Sorrentino, R. & Saltini, C. (1993) HLA-DPB1 glutamate 69: A genetic marker of beryllium disease. *Science*, **262**, 242–244

Rizzato, G., Fraioli, P., Sabbioni, E., Pietra, R. & Barberis, M. (1992) Multi-element follow up in biological specimens of hard metal pneumoconiosis. *Sarcoidosis*, **9**, 104–117

Rizzato, G., Fraioli, P., Sabbioni, E., Pietra, R. & Barberis, M. (1994) The differential diagnosis of hard metal lung disease. *Sci. total Environ.*, **150**, 77–83

Robison, S.H., Cantoni, O. & Costa, M. (1982) Strand breakage and decreased molecular weight of DNA induced by specific metal compounds. *Carcinogenesis*, **3**, 657–662

Rochat, T., Kaelin, R.M., Batawi, A. & Junod, A.F. (1987) Rapidly progressive interstitial lung disease in a hard metal coating worker undergoing hemodialysis. *Eur. J. respir. Dis.*, **71**, 46–51

Roesems, G., Hoet, P.H.M., Demedts, M. & Nemery, B. (1997) *In vitro* toxicity of cobalt and hard metal dust in rat and human type II pneumocytes. *Pharmacol. Toxicol.*, **81**, 74–80

Roesems, G., Hoet, P.H., Dinsdale, D., Demedts, M. & Nemery, B. (2000) *In vitro* cytotoxicity of various forms of cobalt for rat alveolar macrophages and type II pneumocytes. *Toxicol. appl. Pharmacol.*, **162**, 2–9

Rolfe, M.W., Paine, R., Davenport, R.B. & Strieter, R.M. (1992) Hard metal pneumoconiosis and the association of tumor necrosis factor-alpha. *Am. Rev. respir. Dis.*, **146**, 1600–1602

Rosenberg, D.W. (1993) Pharmacokinetics of cobalt chloride and cobalt-protoporphyrin. *Drug Metab. Dispos.*, **21**, 846–849

Rossman, T.G. (1981) Effect of metals on mutagenesis and DNA repair. *Environ. Health Perspect.*, **40**, 189–195

Rossman, T.G., Molina, M. & Meyer, L.W. (1984) The genetic toxicology of metal compounds: I. Induction of λ prophage in *E. coli* WP2$_s$(λ). *Environ. Mutag.*, **6**, 59–69

Sabbioni, E., Minoia, C., Pietra, R., Mosconi, G., Forni, A. & Scansetti, G. (1994a) Metal determinations in biological specimens of diseased and non-diseased hard metal workers. *Sci. total Environ.*, **150**, 41–54

Sabbioni, E., Mosconi, G., Minoia, C. & Seghizzi, P. (1994b) The European Congress on Cobalt and Hard Metal Disease. Conclusions, highlights and need of future studies. *Sci. total Environ.*, **150**, 263–270

Sadasivan, S. & Negi, B.S. (1990) Elemental characterization of atmospheric aerosols. *Sci. total Environ.*, **96**, 269–279

Sala, C., Mosconi, G., Bacis, M., Bernabeo, F., Bay, A. & Sala, O. (1994) Cobalt exposure in 'hard metal' and diamonds grinding tools manufacturing and in grinding processes. *Sci. total Environ.*, **150**, 111–116

Salnikow, K., Su, W., Blagosklonny, M.V. & Costa, M. (2000) Carcinogenic metals induce hypoxia-inducible factor-stimulated transcription by reactive oxygen species-independent mechanism. *Cancer Res.*, **60**, 3375–3378

Santhanam, A.T. (1992) Cemented carbides. In: Kroschwitz, J.I. & Howe-Grant, M., eds, *Kirk-Othmer Encyclopedia of Chemical Technology*, Vol. 4, 4th Ed., New York, John Wiley & Sons, pp. 848–860

Sariego Muñiz, C., Marchante-Gayón, J.M., García Alonso, J.I. & Sanz-Medel, A. (1999) Multi-elemental trace analysis of human serum by double-focusing ICP-MS. *J. anal. Atom. Spectrom.*, **14**, 193–198

Sariego Muñiz, C., Fernández-Martin, J.L., Marchante-Gayón, J.M., García Alonso, J.I., Cannata-Andía, J.B. & Sanz-Medel, A. (2001) Reference values for trace and ultratrace elements in human serum determined by double-focusing ICP-MS. *Biol. trace Elem. Res.*, **82**, 259–272

Sarkar, B. (1995) Metal replacement in DNA-binding zinc finger proteins and its relevance to mutagenicity and carcinogenicity through free radical generation. *Nutrition*, **11** (Suppl.), 646–649

Scansetti, G., Lamon, S., Talarico, S., Botta, G.C., Spinelli, P., Sulotto, F. & Fantoni, F. (1985) Urinary cobalt as a measure of exposure in the hard metal industry. *Int. Arch. occup. environ. Health*, **57**, 19–26

Scansetti, G., Botta, G.C., Spinelli, P., Reviglione, L. & Ponzetti, C. (1994) Absorption and excretion of cobalt in the hard metal industry. *Sci. total Environ.*, **150**, 141–144

Scansetti, G., Maina, G., Botta, G.C., Bambace, P. & Spinelli, P. (1998) Exposure to cobalt and nickel in the hard-metal production industry. *Int. Arch. occup. environ. Health*, **71**, 60–63

Schade, S.G., Felsher, B.F., Glader, B.E. & Conrad, M.E. (1970) *Effect of Cobalt upon Iron Absorption*, Vol. 134, *Proceedings of the Society for Experimental Biology and Medicine*, New York, Academic Press, pp. 741–743

Schepers, G.W.H. (1955a) The biological action of particulate cobalt metal. *Arch. ind. Health*, **12**, 127–133

Schepers, G.W.H. (1955b) The biological action of tungsten carbide and cobalt. *Arch. ind. Health*, **12**, 140–146

Schepers, G.W.H. (1955c) The biological action of tungsten carbide and carbon. *Arch. ind. Health*, **12**, 137–139

Schinz, H.R. & Uehlinger, E. (1942) Metals: A new principle of carcinogenesis. *Z. Krebsforsch.*, **52**, 425–437

Schmit, J.-P., Youla, M. & Gélinas, Y. (1991) Multi-element analysis of biological tissues by inductively coupled plasma mass spectrometry. *Anal. chim. Acta*, **249**, 495–501

Schrauzer, G.N. (1989) Cobalt. In: Merian, E., ed., *Metals and Their Compounds in the Environment. Occurrence, Analysis, and Biological Relevance*, Weinheim, VCH-Verlag, pp. 2-8-1–2-8-11

Schultz, P.N., Warren, G., Kosso, C. & Rogers, S. (1982) Mutagenicity of a series of hexacoordinate cobalt(III) compounds. *Mutat. Res.*, **102**, 393–400

Seghizzi, P., D'Adda, F., Borleri, D., Barbic, F. & Mosconi, G. (1994) Cobalt myocardiopathy. A critical review of literature. *Sci. total Environ.*, **150**, 105–109

Selden, A.I., Persson, B., Bornberger-Dankvardt, S.I., Winström, L.E. & Bodin, L.S. (1995) Exposure to cobalt chromium dust and lung disorders in dental technicians. *Thorax*, **50**, 769–772

Sesana, G., Cortona, G., Baj, A., Quaianni, T. & Colombo, E. (1994) Cobalt exposure in wet grinding of hard metal tools for wood manufacture. *Sci. total Environ.*, **150**, 117–119

Shedd, K.B. (2001) *Minerals Yearbook: Cobalt*, Reston, VA, US Geological Survey, pp. 20.1–20.18 (http://minerals.usgs.gov/minerals/pubs/commodity/cobalt/index.html; accessed 10.09.2003)

Shedd, K.B. (2003) *Mineral Commodity Summaries: Cobalt*, Reston, VA, US Geological Survey, pp. 52–53 (http://minerals.usgs.gov/minerals/pubs/commodity/cobalt/index.html; accessed 10.09.2003)

Sherson, D., Maltbaek, N. & Olsen, O. (1988) Small opacities among dental laboratory technicians in Copenhagen. *Br. J. ind. Med.*, **45**, 320–324

Shi, X., Dalal, N.S. & Kasprzak, K.S. (1993) Generation of free radicals from model lipid hydroperoxides and H_2O_2 by Co(II) in the presence of cysteinyl and histidyl chelators. *Chem. Res. Toxicol.*, **6**, 277–283

Shirakawa, T., Kusaka, Y., Fujimura, N., Goto, S. & Morimoto, K. (1988) The existence of specific antibodies to cobalt in hard metal asthma. *Clin. Allergy*, **18**, 451–460

Shirakawa, T., Kusaka, Y., Fujimura, N., Goto, S., Kato, M., Heki, S. & Morimoto, K. (1989) Occupational asthma from cobalt sensitivity in workers exposed to hard metal dust. *Chest*, **95**, 29–37

Shirakawa, T., Kusaka, Y., Fujimura, N., Kato, M., Heki, S. & Morimoto, K. (1990) Hard metal asthma: Cross immunological and respiratory reactivity between cobalt and nickel? *Thorax*, **45**, 267–271

Shirakawa, T., Kusaka, Y. & Morimoto, K. (1992) Specific IgE antibodies to nickel in workers with known reactivity to cobalt. *Clin. exp. Allergy*, **22**, 213–218

Simcox, N.J., Stebbins, A., Guffey, S., Atallah, R., Hibbard, R. & Camp, J. (2000) Hard metal exposures. Part 2: Prospective exposure assessment. *Appl. occup. environ. Hyg.*, **15**, 342–353

Singh, I. (1983) Induction of reverse mutation and mitotic gene conversion by some metal compounds in *Saccaromyces cerevisiae. Mutat. Res.*, **117**, 149–152

Sirover, M.A. & Loeb, L.A. (1976) Metal activation of DNA synthesis. *Biochem. biophys. Res. Commun.*, **70**, 812–817

Sjögren, I., Hillerdal, G., Andersson, A. & Zetterström, O. (1980) Hard metal lung disease: Importance of cobalt in coolants. *Thorax*, **35**, 653–659

Slauson, D.O., Lay, J.C. Castleman, W.L. & Neilsen, N.R. (1989) Acute inflammatory lung injury retards pulmonary particle clearance. *Inflammation*, **13**, 185–199

Smith, T., Edmonds, C.J. & Barnaby, C.F. (1972) Absorption and retention of cobalt in men by whole-body counting. *Health Phys.*, **22**, 359–367

Snyder, R.D., Davis, G.F. & Lachmann, P.J. (1989) Inhibition by metals of X-ray and ultraviolet-induced DNA repair in human cells. *Biol. trace Elem. Res.*, **21**, 389–398

Sorbie, J., Olatunbosun, D., Corbett, W.E.N. & Valberg, L.S. (1971) Cobalt excretion test for the assessment of body iron stores. *Can. med. Assoc. J.*, **104**, 777–782

Sprince, N.L., Chamberlin, R.I., Hales, C.A., Weber, A.L. & Kazemi, H. (1984) Respiratory disease in tungsten carbide production workers. *Chest*, **86**, 549–557

Sprince, N.L., Oliver, L.C., Eisen, E.A., Greene, R.E. & Chamberlin, R.I. (1988) Cobalt exposure and lung disease in tungsten carbide production: A cross-sectional study of current workers. *Am. Rev. respir. Dis.*, **138**, 1220–1226

Starck, H.C. (2003) *Product Data Sheets: Tungsten Carbides (WC, W2C) — Hardmetal/Cemented Carbides*, Goslar, Germany

Stea, S., Visentin, M., Granchi, D., Savarino, L., Dallari, D., Gualtieri, G., Rollo, G., Toni, A., Pizzoferrato, A. & Montanaro, L. (2000) Sister chromatid exchange in patients with joint prostheses. *J. Arthroplast.*, **15**, 772–777

Stebbins, A.I., Horstman, S.W., Daniell, W.E. & Atallah, R. (1992) Cobalt exposure in a carbide tip grinding process. *Am. ind. Hyg. Assoc. J.*, **53**, 186–192

Steinbrech, D.S., Mehrara, B.J., Saadeh, P.B., Greenwald, J.A., Spector, J.A., Gittes, G.K. & Longaker, M.T. (2000) VEGF expression in an osteoblast-like cell line is regulated by a hypoxia response mechanism. *Am. J. Physiol. Cell Physiol.*, **278**, C853–C860

Steinhoff, D. & Mohr, U. (1991) On the question of a carcinogenic action of cobalt-containing compounds. *Exp. Pathol.*, **41**, 169–174

Stoll, W.M. & Santhanam, A.T. (1992) Industrial hard carbides. In: Kroschwitz, J.I. & Howe-Grant, M., eds, *Kirk-Othmer Encyclopedia of Chemical Technology*, Vol. 4, 4th Ed., New York, John Wiley & Sons, pp. 861–878

Suardi, R., Belotti, L., Ferrari, M.T., Leghissa, P., Caironi, M., Maggi, L., Alborghetti, F., Storto, T., Silva, T. & Piazzolla, S. (1994) Health survey of workers occupationally exposed to cobalt. *Sci. total Environ.*, **150**, 197–200

Sunderman, F.W., Jr, Hopfer, S.M., Swift, T., Rezuke, W.N., Ziebka, L., Highman, P., Edwards, B., Folcik, M. & Gossling, H.R. (1989) Cobalt, chromium, and nickel concentrations in body fluids of patients with porous-coated knee or hip prostheses. *J. orthopaed. Res.*, **7**, 307–315

Suva (2003) *Grenzwerte am Arbeitsplatz 2003*, Luzern, Switzerland [Swiss OELs]

Suzuki, Y., Shimizu, H., Nagae, Y., Fukumoto, M., Okonogi, H. & Kadokura, M. (1993) Micronucleus test and erythropoiesis: Effect of cobalt on the induction of micronuclei by mutagens. *Environ. mol. Mutag.*, **22**, 101–106

Swanson, S.A.V., Freeman, M.A.R. & Heath, J.C. (1973) Laboratory tests on total joint replacement prostheses. *J. Bone Joint Surg.*, **55B**, 759–773

Swennen, B., Buchet, J.-P., Stánescu, D., Lison, D. & Lauwerys, R. (1993) Epidemiological survey of workers exposed to cobalt oxides, cobalt salts, and cobalt metal. *Br. J. ind. Med.*, **50**, 835–842

Swierenga, S.H.H., Gilman, J.P.W. & McLean, J.R. (1987) Cancer risk from inorganics. *Cancer Metastasis Rev.*, **6**, 113–154

Szakmáry, E., Ungváry, G., Hudák, A., Tátrai, E., Náray, M. & Morvai, V. (2001) Effects of cobalt sulfate on prenatal development of mice, rats, and rabbits, and on early postnatal development of rats. *J. Toxicol. environ. Health*, **A62**, 367–386

Takemoto, K., Kawai, H., Kuwahara, T., Nishina, M. & Adachi, S. (1991) Metal concentrations in human lung tissue, with special reference to age, sex, cause of death, emphysema and contamination of lung tissue. *Int. Arch. Occup. Environ. Health*, **62**, 579–586

Talakin, Y.N., Ivanova, L.A., Kostetskaya, N.I., Komissarov, V.N. & Belyaeva, I.V. (1991) [Hygienic characteristics of working conditions and health state of workers engaged in the production of cobalt salts.] *Gig. Tr. prof. Zbl.*, **1**, 10–11 (in Russian)

Tan, K.L., Lee, H.S., Poh, W.T., Ren, M.Q., Watt, F., Tang, S.M. & Eng, P. (2000) Hard metal lung disease — The first case in Singapore. *Ann. Acad. Med. Singapore*, **29**, 521–527

Thomas, I.T. & Evans, E.J. (1986) The effect of cobalt-chromium-molybdenum powder on collagen formation by fibroblasts *in vitro*. *Biomaterials*, **7**, 301–304

Thomassen, Y., Nieboer, E., Ellingsen, D., Hetland, S., Norseth, T., Odland, J.Ø., Romanova, N., Chernova, S. & Tchachtchine, V.P. (1999) Characterisation of workers' exposure in a Russian nickel refinery. *J. environ. Monit.*, **1**, 15–22

Tolot, F., Girard, R., Dortit, G., Tabourin, G., Galy, P. & Bourret, J. (1970) [Pulmonary manifestations of 'hard metals': Irritative disorders and fibrosis (survey and clinical observations).] *Arch. Mal. prof.*, **31**, 453–470 (in French)

Tozawa, T., Kitamura, H., Koshi, K., Ikemi, Y., Ambe, K. & Kitamura, H. (1981) [Experimental pneumoconiosis induced by cemented tungsten and sequential concentrations of cobalt and tungsten in the lungs of the rat.] *Jpn J. Ind. Health [Sangyo Igaku]*, **23**, 216–226 (in Japanese)

Tso, W.-W. & Fung, W.-P. (1981) Mutagenicity of metallic cations. *Toxicol. Lett.*, **8**, 195–200

Tüchsen, F., Jensen, M.V., Villadsen, E. & Lynge, E. (1996) Incidence of lung cancer among cobalt-exposed women. *Scand. J. Work Environ. Health*, **22**, 444–450

Umicore Specialty Metals (2002) *Technical Data Sheets: High Purity Cobalt, Cobalt Powders 400 Mesh/100 Mesh, 5M Cobalt Powder, Extra Fine Cobalt Powder, Half Micron Cobalt Powder, Submicron Size Cobalt Powder, Ultrafine Cobalt Powder*, Olen, Belgium

Valberg, L.S., Ludwig, J. & Olatunbosun, D. (1969) Alteration in cobalt absorption in patients with disorders of iron metabolism. *Gastroenterology*, **56**, 241–251

Van Cutsem, E.J., Ceuppens, J.L., Lacquet, L.M. & Demedts, M. (1987) Combined asthma and alveolitis induced by cobalt in a diamond polisher. *Eur. J. respir. Dis.*, **70**, 54–61

Van Den Eeckhout, A., Verbeken, E. & Demedts, M. (1988) [Pulmonary pathology due to cobalt and heavy metals.] *Rev. Mal. respir.*, **5**, 201–207 (in French)

Van den Oever, R., Roosels, D., Douwen, M., Vanderkeel, J. & Lahaye, D. (1990) Exposure of diamond polishers to cobalt. *Ann. occup. Hyg.*, **34**, 609–614

Van Goethem, F., Lison, D. & Kirsch-Volders, M. (1997) Comparative evaluation of the in vitro micronucleus test and the alkaline single cell gel electrophoresis assay for the detection of DNA damaging agents: Genotoxic effects of cobalt powder, tungsten carbide and cobalt-tungsten carbide. *Mutat. Res.*, **392**, 31–43

Veien, N.K. & Svejgaard, E. (1978) Lymphocyte transformation in patients with cobalt dermatitis. *Br. J. Dermatol.*, **99**, 191–196

Versieck, J. & Cornelis, R. (1980) Normal levels of trace elements in human blood plasma or serum. *Anal. chim. Acta*, **116**, 217–254

Vollmann, J. (1938) Animal experiments with intraosseous arsenic, chromium and cobalt implants. *Schweiz. Z. Allg. Pathol. Bakteriol.*, **1**, 440–443 (in German)

Von Rosen, G. (1964) Mutations induced by the action of metal ions in *Pisum*. II: Further investigations on the mutagenic action of metal ions and comparison with the activity of ionizing radiation. *Hereditas*, **51**, 89–134

Voroshilin, S.I., Plotko, E.G., Fink, T.V. & Nikiforova, V.J. (1978) [Cytogenetic effect of inorganic compounds of tungsten, zinc, cadmium and cobalt on animal and human somatic cells]. *Tsitol. Genet.*, **12**, 241–243 (in Russian)

Wahlberg, J.E. & Boman, A. (1978) Sensitization and testing of guinea pigs with cobalt chloride. *Contact Dermatitis*, **4**, 128–132

Wang, X., Yokoi, I., Liu, J., & Mori, A. (1993) Cobalt(II) and nickel(II) ions as promoters of free radicals *in vivo*: Detected directly using electron spin resonance spectrometry in circulating blood in rats. *Arch. Biochem. Biophys.*, **306**, 402–406

Wedrychowski, A., Schmidt, W.N. & Hnilica, L.S. (1986) DNA-protein crosslinking by heavy metals in Novikoff hepatoma. *Arch. Biochem. Biophys.*, **251**, 397–402

Wehner, A.P., Busch, R.H., Olson, R.J. & Craig, D.K. (1977) Chronic inhalation of cobalt oxide and cigarette smoke by hamsters. *Am. ind. Hyg. Assoc. J.*, **38**, 338–346

Wetterhahn, K.J. (1981) The role of metals in carcinogenesis: Biochemistry and metabolism. *Environ. Health Perspect.*, **40**, 233–252

White, M.A. (1999) A comparison of inductively coupled plasma mass spectrometry with electrothermal atomic absorption spectrophotometry for the determination of trace elements in blood and urine from nonoccupationally exposed populations. *J. trace Elem. Med. Biol.*, **13**, 93–101

White, M.A. & Dyne, D. (1994) Biological monitoring of occupational cobalt exposure in the United Kingdom. *Sci. total Environ.*, **150**, 209–213

White, M.A. & Sabbioni, E. (1998) Trace element values in tissues from inhabitants of the European Union. X. A study of 13 elements in blood and urine of a United Kingdom population. *Sci. total Environ.*, **216**, 253–270

Wide, M. (1984) Effect of short-term exposure to five industrial metals on the embryonic and fetal development of the mouse. *Environ. Res.*, **33**, 47–53

Wiethege, T., Wesch, H., Wegener, K., Müller, K.-M., Mehlhorn, J., Spiethoof, A., Schömig, D., Hollstein, M., Bartsch, H. & the German Uranium Miner Study, Research Group Pathology (1999) Germanium uranium miner study — Pathological and molecular genetic findings. *Radiat. Res.*, **152**, S52–S55

Wild, P., Perdrix, A., Romazini, S., Moulin, J.J. & Pellet, F. (2000) Lung cancer mortality in a site producing hard metals. *Occup. environ. Med.*, **57**, 568–573

Wilk-Rivard, E. & Szeinuk, J. (2001) Occupational asthma with paroxysmal atrial fibrillation in a diamond polisher. *Environ. Health Perspect.*, **109**, 1303–1306

Wong, P.K. (1988) Mutagenicity of heavy metals. *Bull. environ. Contam. Toxicol.*, **40**, 597–603

Yesilada, E. (2001) Genotoxicity testing of some metals in the *Drosophila* wing somatic mutation and recombination test. *Bull. environ. Contam. Toxicol.*, **66**, 464–469

Yokoiyama, A., Kada, T. & Kuroda, Y. (1990) Antimutagenic action of cobaltous chloride on radiation-induced mutations in cultured Chinese hamster cells. *Mutat. Res.*, **245**, 99–105

Zanetti, G. & Fubini, B. (1997) Surface interaction between metallic cobalt and tungsten carbide particles as a primary cause of hard metal lung disease. *J. Mater. Chem.*, **7**, 1647–1654

Zeiger, E., Anderson, B., Haworth, S., Lawlor, T. & Mortelmans, K. (1992) Salmonella mutagenicity tests: V. Results from the testing of 311 chemicals. *Environ mol. Mutagen.*, **19** (Suppl. 21), 2–141

Zhang, Q., Kusaka, Y., Sato, K., Nakakuki, K., Kohyama, N. & Donaldson, K. (1998) Differences in the extent of inflammation caused by intratracheal exposure to three ultrafine metals: Role of free radicals. *J. Toxicol. environ. Health*, **A53**, 423–438

Zou, W., Yan, M., Xu, W., Huo, H., Sun, L., Zheng, Z., & Liu, X. (2001) Cobalt chloride induces PC12 cells apoptosis through reactive oxygen species and accompanied by AP-1 activation. *J. neurosci. Res.*, **64**, 646–653

Zou, W., Zeng, J., Zhuo, M., Xu, W., Sun, L., Wang, J. & Liu, X. (2002) Involvement of caspase-3 and p38 mitogen-activated protein kinase in cobalt chloride-induced apoptosis in PC12 cells. *J. neurosci. Res.*, **67**, 837–843

MONOGRAPHS ON GALLIUM ARSENIDE
AND INDIUM PHOSPHIDE

INTRODUCTION TO THE MONOGRAPHS ON GALLIUM ARSENIDE AND INDIUM PHOSPHIDE

Studies of Cancer in Humans

Two compounds evaluated in this volume of the *IARC Monographs*, gallium arsenide and indium phosphide, are predominantly used in the semiconductor industry. However, only parts of the workforce in this industry are exposed to these compounds, and there is also potential for exposure to several other carcinogens in this industry. As none of the studies of cancer in the semiconductor industry were informative with regard to gallium arsenide or indium phosphide, these studies are reported in this introduction.

Two analyses of a small cohort of workers at a semiconductor manufacturing facility in the West Midlands in England have been published (Sorahan *et al.*, 1985, 1992). These were prompted by the observation (reported by Sorahan *et al.*, 1985) of several instances of skin cancer among the workers. A cohort of 1807 (1526 women) workers first employed in or before 1970, initially followed up until 1982 (Sorahan *et al.*, 1985) and further until 1989 for mortality and 1988 for cancer incidence, was compiled (Sorahan *et al.*, 1992). The overall SMR for all causes was 0.72 (107 observed; 95% CI, 0.59–0.87); the standardized registration ratio (SRR) for all cancer registration was 0.96 [93 observed; 95% CI, 0.77–1.18]. The SRR for cancer of the respiratory system was 0.97 [11 observed; 95% CI, 0.48–1.74], that for melanoma was 2.00, [three observed; 95% CI, 0.41–5.84] and that for non-melanoma skin cancer was 1.52 [13 observed; 95% CI, 0.81–2.59] (Table 1).

McElvenny *et al.* (2003) analysed a cohort of 4388 current and former employees (2126 men, 2262 women) at a semiconductor manufacturing facility in Scotland who were first employed at the facility between 1970 (date of start of the operations) and 30 April 1999. Employees were followed up until the end of 2000 for mortality and 1998 for cancer registration analyses. Death and cancer registration rates in the employees were compared with those for Scotland as a whole, with and without adjustment for socio-economic status (using the distribution of the Carstairs deprivation index among current employees). The mean length of follow-up was only 12.5 years. The results, adjusted for socioeconomic status, are presented in Table 1. The study found an excess of lung cancer among women but not men. Small excesses were observed for stomach cancer in women

Table 1. Cohort studies of cancer in the semiconductor industry

Reference, facilities	Cohort characteristics	Cancer site	No. of cases	Relative risk (95% CI)	Comments
Sorahan et al. (1992) 1 factory	1526 female and 181 male workers first employed in or before 1970, follow-up for mortality to 1989 and for cancer incidence to 1988	All cause mortality	107	0.72 [0.59–0.87][a]	
		Registration, all cancer	93	0.96 [0.77–1.18][a]	
		Respiratory cancer	11	0.97 [0.48–1.74][a]	
		Melanoma	3	2.00 [0.42–5.92][a]	
		Non-melanoma skin cancer	13	1.52 [0.81–2.59][a]	
McElvenny et al. (2003) 1 factory	4388 employees (2126 men, 2262 women) first employed between 1970 and 1999, follow-up for mortality to 2000 and cancer incidence to 1998	All cause mortality			Adjusted for socioeconomic status
		Men	27	0.40 (0.27–0.59)	
		Women	44	0.75 (0.54–1.01)	
		All cancer mortality			
		Men	6	0.47 (0.17–1.02)	
		Women	23	1.10 (0.69–1.64)	
		All cancer registration			
		Men	25	0.99 (0.64–1.47)	
		Women	54	1.11 (0.83–1.45)	
		Lung cancer			
		Men	2	0.56 (0.07–2.02)	
		Women	11	2.73 (1.36–4.88)	
		Melanoma			
		Men	2	1.86 (0.23–6.71)	
		Women	2	0.88 (0.11–3.19)	
		Non-melanoma skin cancer			
		Men	4	1.04 (0.28–2.65)	
		Women	6	1.25 (0.46–2.72)	
		Stomach			
		Men	0		
		Women	3	4.38 (0.90–12.81)	

[a] 95% CI calculated by Working Group

and brain cancer in men (three deaths; unadjusted SMR, 4.01; 95% CI, 0.83–11.72). The SRR for lung cancer in women for up to 10 years since first employment was 3.90 (five registrations; 95% CI, 1.27–9.11), which was higher than that for 10 or more years since first employment (six registrations; SRR, 2.18; 95% CI, 0.80–4.75). [No data on smoking habits were available; however, these results are adjusted for socioeconomic status, which is highly correlated with smoking habits.]

References

McElvenny, D.M., Darnton, A.J., Hodgson, J.T., Clarfke, S.D., Elliott, R.C. & Osman, J. (2003) Investigation of cancer incidence and mortality at a Scottish semiconductor manufacturing facility. *Occup. Med.*, **53**, 419–430

Sorahan, T., Waterhouse, J.A.H., McKiernan, M.J. & Aston, R.H.R. (1985) Cancer incidence and cancer mortality in a cohort of semiconductor workers. *Br. J. ind. Med.*, **42**, 546–550

Sorahan, T., Pope, D.J. & McKiernan, M.J. (1992) Cancer incidence and cancer mortality in a cohort of semiconductor workers: an update. *Br. J. ind. Med.*, **49**, 215–216

GALLIUM ARSENIDE

1. Exposure Data

1.1 Chemical and physical data

1.1.1 *Nomenclature*

Chem. Abstr. Serv. Reg. No.: 1303-00-0
Deleted CAS Reg. No.: 12254-95-4, 106495-92-5, 116443-03-9, 385800-12-4
Chem. Abstr. Serv. Name: Gallium arsenide (GaAs)
IUPAC Systematic Name: Gallium arsenide
Synonyms: Gallium monoarsenide

1.1.2 *Molecular formula and relative molecular mass*

GaAs   Relative molecular mass: 144.6

1.1.3 *Chemical and physical properties of the pure substance*

(a) *Description*: Grey, cubic crystals (Lide, 2003)
(b) *Melting-point*: 1238 °C (Lide, 2003)
(c) *Density*: 5.3176 g/cm³ (Lide, 2003)
(d) *Solubility*: Insoluble in water (Wafer Technology Ltd, 1997); slightly soluble in 0.1 M phosphate buffer at pH 7.4 (Webb *et al.*, 1984)
(e) *Stability*: Decomposes with evolution of arsenic vapour at temperatures above 480 °C (Wafer Technology Ltd, 1997)
(f) *Reactivity*: Reacts with strong acid reducing agents to produce arsine gas (Wafer Technology Ltd, 1997)

1.1.4 *Technical products and impurities*

Purity requirements for the raw materials used to produce gallium arsenide are stringent. For optoelectronic devices (light-emitting diodes (LEDs), laser diodes, photo-detectors, solar cells), the gallium and arsenic must be at least 99.9999% pure; for integrated circuits, a purity of 99.99999% is required. These purity levels are referred to by several names: 99.9999%-pure gallium is often called 6-nines, 6N or optoelectronic grade, while 99.99999%-pure gallium is called 7-nines, 7N, semi-insulating (SI) or integrated circuit (IC) grade.

For 7N gallium, the total of the impurities must be < 100 µg/kg. In addition to the challenge of consistently producing material with such high purity, there are difficulties in detecting the small quantity of impurities. Certain impurities cause more problems than others during gallium arsenide production. Those of most concern are calcium, carbon, copper, iron, magnesium, manganese, nickel, selenium, silicon, sulfur, tellurium and tin. Generally, these elements should be present in concentrations < 1 µg/kg in both the gallium and the arsenic. Lead, mercury and zinc should be present in concentrations < 5 µg/kg. Although aluminum, chlorine and sodium are often present, the concentrations of each should be < 10 µg/kg (Kramer, 1988). Some companies have even more stringent requirements (Recapture Metals, 2003).

1.1.5   *Analysis*

The monitoring of occupational exposure to gallium arsenide can only be based on measurements of arsenic or gallium concentrations in workplace air or in human tissues or body fluids (biological monitoring), because there is no analytical method capable of measuring gallium arsenide per se in the above media.

(*a*)   *Determination of gallium*

Monitoring of exposure to gallium arsenide by determination of gallium has so far not been used due to the limited availability of analytical methods with sufficiently low detection limits.

Instrumental neutron activation analysis (INAA) has been employed successfully for the determination of gallium in ambient air (Kucera *et al.*, 1999) and can therefore be used for the determination of gallium in workplace air. A newly developed Mist-UV sampling system coupled with ICP–MS analysis allows the determination of volatile gallium compounds in the atmosphere (Ito & Shooter, 2002).

Several methods have been designed and tested for the determination of gallium in biological materials, mainly in the analysis of spiked samples. These include spectrophotometry (Beltrán Lucena *et al.*, 1994), fluorescence spectrometry (Requena *et al.*, 1983; Afonso *et al*,. 1985; Ureña *et al.*, 1985; Ureña Pozo *et al.*, 1987; Cano Pavón *et al.*, 1988; Salgado *et al.*, 1988; Cano Pavón *et al.*, 1990; Sanchez Rojas & Cano Pavón, 1995), electrothermal AAS (Nakamura *et al.*, 1982; Ma *et al.*, 1999) and F-AAS with preconcentration (Anthemidis *et al.*, 2003).

Data on gallium concentrations in human tissues and body fluids are scarce. Scansetti (1992) reported a mean blood concentration in healthy donors of 3 µg/L.

(b) Determination of arsenic

Until now, only measurements of arsenic have been used for monitoring exposure to gallium arsenide because occupational exposure limits for arsenic have been established in many countries and the analytical methods available for its determination are more sensitive than those for gallium.

(i) Workplace air monitoring

Yamauchi *et al.* (1989) measured arsenic concentrations in air in gallium-arsenide plants using a method similar to those used to determine inorganic arsenic in workplace air (Hakala & Pyy, 1995; Jakubowski *et al.*, 1998; Apostoli *et al.*, 1999). Airborne respirable particulate matter is collected by drawing air into a stationary or personal sampler through a membrane filter made of polycarbonate, cellulose ester and/or teflon. The filter containing the collected air particulates is digested in various concentrated mineral acids (chlorhydric acid, nitric acid, sulfuric acid, perchloric acid) or a mixture thereof and the arsenic concentration in the digest is determined by hydride generation AAS (Hakala & Pyy, 1995; Jakubowski *et al.*, 1998) or ICP-MS (Apostoli *et al.*, 1999). Non-destructive determination of the arsenic content on a filter can also be achieved using INAA, with a detection limit of about $0.5 \, \text{ng/m}^3$ (Kucera *et al.*, 1999).

(ii) Biological monitoring

Preliminary considerations

Biomonitoring of exposure to gallium arsenide by measuring arsenic in human tissues or body fluids has several limitations. Firstly, gallium arsenide has a low solubility and is poorly absorbed in the gastrointestinal tract, resulting in rapid elimination of the compound in the faeces (Webb *et al.*, 1984; Yamauchi *et al.*, 1986) (see also Section 4.1.2). Secondly, determining the total concentration of arsenic in body fluids, e.g. in urine, is not an optimum measure of occupational exposure to arsenic. Gallium arsenide is partly dissociated *in vivo* into inorganic arsenic and gallium (Webb *et al.*, 1984; Yamauchi *et al.*, 1986). Inorganic arsenic is methylated in the human body to monomethylarsonic acid (MMA^V) and dimethylarsinic acid (DMA^V), which are readily excreted in urine (Vahter, 2002). However, arsenic in seafood is present predominantly in the forms of arsenobetaine and arsenocholine, which do not undergo biotransformation and are also excreted in urine (Apostoli *et al.*, 1999; Vahter, 2002). Hence, the concentration of total arsenic in body fluids is dependent on the concentration and the species of dietary arsenic. Therefore, it is preferable to use speciation analysis in biomonitoring of arsenic exposure; alternatively, subjects enrolled in occupational health studies should refrain from consuming seafood for a couple of days before the analysis of their body fluids.

Similarly, determination of arsenic concentration in hair is associated with the problem of distinguishing external contamination from the endogenous content of arsenic in this tissue (Yamauchi *et al.*, 1989).

Analytical methods

Numerous methods (reviewed recently in IARC, 2004) are available for the determination of the total concentration of arsenic and its species in blood, serum, urine and other biological materials. The methods most commonly used include electrothermal AAS, AAS with hydride generation, ICP-AES, ICP-MS, atomic fluorescence spectrometry (AFS) and INAA. The advantages and shortcomings of these and other less frequently used techniques, such as spectrophotometric and electroanalytical methods, have been reviewed comprehensively (Burguera & Burguera, 1997; IARC, 2004). The analytical methods used for arsenic speciation are based on the combination of a powerful separation process and an adequate element-specific detection, using so-called hyphenated analytical techniques. The methods most frequently employed for separation and preconcentration involve solvent extraction, including solid-phase extraction (Yalcin & Le, 2001; Yu *et al.*, 2003), precipitation and coprecipitation, ion-exchange chromatography (IEC), capillary electrophoresis (Greschonig *et al.*, 1998), gas chromatography and high-performance liquid chromatography (HPLC). The element-specific detection is performed using the same analytical techniques as for determination of total arsenic. On-line coupling of some separation techniques, usually HPLC or IEC, with the most sensitive detection methods (AAS, AFS, ICP-AES, ICP-MS) is frequently used. A detection limit at the nanogram level has been achieved by AAS coupled with hydride generation before detection (Burguera & Burguera, 1997).

(iii)	*Reference values for occupationally non-exposed populations*

Concentrations of total arsenic found in human blood, serum, plasma and urine have been reviewed (Iyengar *et al.*, 1978; Versieck & Cornelis, 1980; Heydorn, 1984; Iyengar & Woittiez, 1988). The data suggest that there are regional differences and short-term effects of dietary intake, especially seafood. After exclusion of values suspected of such variations, the following median reference values were given: whole blood, 5 μg/L (Iyengar & Woittiez, 1988); serum or plasma, 1–3.5 μg/L (Heydorn, 1984; Iyengar & Woittiez, 1988); and urine, 20 μg/L (Iyengar & Woittiez, 1988). The data available for the individual arsenic species are still insufficient to make reliable estimates of reference values, presumably because of methodological problems.

## 1.2	Production and use

### 1.2.1	*Production*

Gallium occurs in very small concentrations in many rocks and ores of other metals. Most gallium is produced as a by-product of processing bauxite, and the remainder is produced from zinc-processing residues. Only part of the gallium present in bauxite and zinc ores is recoverable, and the factors controlling the recovery are proprietary. Therefore, an estimate of current reserves cannot be made. The world bauxite reserve base is so large that much of it will not be mined for many decades; hence, most of the gallium in the bauxite reserve base cannot be considered to be available in the short term (Kramer, 2003).

Estimates of primary production of gallium in the world between 1995 and 2002 have varied between 35 and 100 tonnes. In 2003, about 64 tonnes were produced, with China, Germany, Japan and the Russian Federation being the major producers; countries with smaller output included Hungary, Kazakhstan, Slovakia and the Ukraine. Refined gallium production in 2003 was estimated to be about 83 tonnes, including some scrap refining. France was the largest producer of refined gallium, using crude gallium produced in Germany as feed material. Japan and the USA are two other large gallium-refining countries (Kramer, 1996–2004).

Demand for gallium in the USA in 2003 was satisfied by imports, mainly low-purity material from China, Kazakhstan and the Russian Federation and smaller amounts of high-purity material from France. In addition, in 2002, the USA imported an estimated 120 tonnes of doped and undoped gallium arsenide wafers, mainly from Finland, Germany, Italy and Japan (Kramer, 2002, 2003).

Consumption of high-purity gallium in Japan in 2002 was estimated to be 108 tonnes, including domestic production of 8 tonnes, imports of 55 tonnes and scrap recycling of 45 tonnes (Kramer, 2003).

The technology of gallium arsenide processing has been reviewed in detail (Harrison, 1986; Kitsunai & Yuki, 1994). Gallium arsenide can be obtained by direct combination of the elements at high temperature and pressure; it can also be prepared, mainly as a thin film, by numerous exchange reactions in the vapour phase (Sabot & Lauvray, 1994).

Gallium arsenide single crystals are more difficult to fabricate than those of silicon. With silicon, only one component needs to be controlled, whereas with gallium arsenide, a 1:1 ratio of gallium atoms to arsenic atoms must be maintained. At the same time, arsenic volatilizes at the temperatures needed to grow crystals. To prevent loss of arsenic, which would result in the formation of an undesirable gallium-rich crystal, gallium arsenide single-crystal ingots are grown in an enclosed environment. Two basic methods are used to fabricate gallium arsenide ingots: the boat-growth, horizontal Bridgeman or gradient-freeze technique and the liquid-encapsulated Czochralski technique. Ingots produced by the horizontal Bridgeman method are D-shaped and have a typical cross-sectional area of about 2 in^2 [13 cm^2]. In contrast, single-crystal ingots grown by the liquid-encapsulated Czochralski method are round and are generally 3 in [7.5 cm] in diameter, with a cross-sectional area of about 7 in^2 [45 cm^2] (Kramer, 1988; Kitsunai & Yuki, 1994). Ingots grown by the horizontal Bridgeman method are cleaned in chemical baths of aqua regia and isopropyl alcohol, and sandblasted using an abrasive material such as silicon carbide or calcined alumina.

The crystalline orientation of the gallium arsenide ingot is checked by X-ray diffraction and the ends are cut off with a diamond blade saw. The ingots are shaped by grinding the edges and then sliced into wafers along the proper crystalline axis. Wafers pass through several stages of surface preparation, polishing and testing before they are ready for device manufacture or epitaxial growth. [Epitaxy is a method for growing single crystals in which chemical reactions produce thin layers of materials whose lattice structures are identical to that of the substrate on which they are deposited.] Pure gallium arsenide is semi-insulating and, in order to conduct electricity, a small number of atoms of another element must be incorporated into the crystal structure; this is called doping. Doping is accomplished by either ion implantation or epitaxial growth (Harrison, 1986; Kramer, 1988).

Because of the low yield in processing gallium arsenide for optoelectronic devices or integrated circuits, substantial quantities of scrap are generated during the various processing stages. This scrap has varying gallium (from < 1 to 99.99%) and impurity contents, depending on the processing step from which it results. In processing gallium arsenide scrap, the material is crushed, if necessary, and then dissolved in a hot acidic solution. This acidic solution is neutralized with a caustic solution to precipitate the gallium as gallium hydroxide, which is filtered from the solution and washed. The gallium hydroxide filter cake is redissolved in a caustic solution and electrolysed to recover 3N to 4N gallium metal. This metal may be refined to 6N or 7N gallium by conventional purification techniques (Kramer, 1988).

Available information indicates that gallium arsenide is produced by three companies in Taiwan, China, two companies in Japan, and one company each in China, the Ukraine and the USA (Chemical Information Services, 2003).

1.2.2   Use

Gallium arsenide has light-emitting properties, high electron mobility, electromagnetic properties and photovoltaic properties. As a semiconductor, it has several unique material properties which can be utilized in high speed semi-conductor devices, high power microwave and millimetre-wave devices, and optoelectronic devices including fibreoptic sources and detectors. Its advantages as a material for high speed devices are high electron mobility and saturation velocity, and relatively easy growth of semi-insulating substrates which render low parasitics and good device isolation. Other useful properties are controllable band gap by alloying, desirable ionization and optical absorption properties. Gallium arsenide has certain advantages over other semiconductor materials: (1) faster operation with lower power consumption, (2) better resistance to radiation and, most importantly, (3) it may be used to convert electrical into optical signals (Chakrabarti, 1992; Greber, 2003).

In 2002, more than 95% of gallium consumed in the USA was in the form of gallium arsenide for optoelectronic devices and integrated circuits. Analogue ICs were the largest single application for gallium, representing 65% of gallium demand (Kramer, 2003). ICs are used in defence applications, high-performance computers and telecommunications. The developments in gallium arsenide IC technology have been reviewed (Welch *et al.*, 1985; Chakrabarti, 1992). About 34% of the gallium consumed was used in optoelectronic devices, which include LEDs, laser diodes, photodetectors and solar cells. Optoelectronic devices are used in applications such as aerospace, consumer goods, industrial components, medical equipment and telecommunications. The remaining 1% was used in research and development, specialty alloys and other applications (Kramer, 2003).

Many manufacturers have introduced new LEDs based on gallium arsenide technology which offer improvements over current LEDs. In many cases, the new LEDs are brighter, last longer and/or can be used in new applications (Kramer, 2002).

Gallium arsenide wafer manufacturers and some electrical companies produce gallium arsenide epitaxial-growth wafers and LED drips. Vapour-phase epitaxy or liquid-phase epitaxy is used to grow gallium arsenide layers for most LEDs. The super-bright red LEDs are manufactured using liquid-phase epitaxy to grow aluminum–gallium–arsenide on gallium arsenide substrates. Epitaxial growth based on metal–organic chemical vapour deposition (MOCVD) technology is used in manufacturing some types of infrared LEDs used in optocouplers. MOCVD is also used to grow a gallium arsenide layer (buffer layer) on gallium arsenide substrates for low-cost optic fibres dedicated to local area computer networks. Gallium arsenide-based laser diodes are manufactured using liquid-phase epitaxy, MOCVD and molecular beam epitaxy technologies (Kitsunai & Yuki, 1994; Sabot & Lauvray, 1994).

For analogue ICs, the requirements for epitaxy grow at the same rate as frequencies increase. An epitaxial gallium arsenide layer is also required for most microwave devices with frequencies over 20 GHz. Photovoltaic applications require gallium arsenide wafers and epitaxial layers. Night-vision system devices use an epitaxial layer of gallium arsenide applied to one end of a photomultiplier to enhance infrared images. Gallium arsenide epitaxial-growth wafers are is also used in optical ICs and magnetoelectric transducers (Sabot & Lauvray, 1994).

1.3   Occurrence and exposure

1.3.1   *Natural occurrence*

Gallium arsenide does not occur naturally. Gallium is present in the earth's crust at 5–15 mg/kg and is recovered as a by-product of the extraction of aluminum and zinc from their ores (Beliles, 1994; Sabot & Lauvray, 1994).

Arsenic concentration in the earth's crust is generally < 2 mg/kg, but may be elevated in zones of active or extinct volcanic activity (IARC, 2004).

1.3.2   *Occupational exposure*

Exposure to gallium arsenide occurs predominantly in the microelectronics industry where workers are involved in the production of gallium arsenide crystals, ingots and wafers, in grinding and sawing operations, in device fabrication, and in sandblasting and clean-up activities (Webb *et al.*, 1984; Harrison, 1986). The National Institute for Occupational Safety and Health (NIOSH) estimated that in 1981 the microelectronics industry employed approximately 180 000 workers in the USA, with over 500 plants manufacturing semiconductors (National Institute for Occupational Safety and Health, 1985).

Exposure to gallium arsenide can only be monitored by determining arsenic concentrations. Several reports describe the assessment of exposure to arsenic during gallium arsenide production and use (Harrison, 1986; Yamauchi *et al.*, 1989; Sheehy & Jones, 1993). Harrison (1986) reported short-term exposure concentrations of arsenic measured at two facilities during epitaxial vacuum servicing and beadblasting of 0.29 and 2.5 mg/m^3, respectively.

Sheehy and Jones (1993) conducted more thorough workplace assessments of total arsenic exposure during 1986–87 by collecting personal breathing zone and workplace air samples at various stages of gallium arsenide production in three different plants. In areas where arsine gas was used, arsine concentrations were also measured. In general, arsenic concentrations in air in personal breathing zones were found to be < 5 µg/m^3 in each of the three plants. However, concentrations in air samples collected from personal breathing zones of individuals responsible for cleaning activities in the crystal-growth area were as high as 2.7 mg/m^3. Wipe samples collected from various work sites showed mean concentrations up to 970 µg/100 cm^2. The authors noted that in two of the three plants monitored, 30–70% of the arsenic collected in air in personal breathing zones passed through the filters and was collected on charcoal tubes, implying that a large portion of the exposure to arsenic was due to arsine gas. The authors concluded that in order to determine exposure to arsenic during gallium arsenide production, both particulate and gaseous arsenic should be monitored.

Yamauchi *et al.* (1989) measured inorganic arsenic, MMA^V, DMA^V and trimethyl-arsenic compounds in the urine and hair of workers involved in various stages of gallium arsenide crystal and wafer production. Total arsenic concentration in workplace air ranged from 2 to 24 µg/m^3. For workers in these areas, the mean concentration of total arsenic in hair was significantly greater than that in the controls and ranged from 1.11 to 6.28 µg arsenic per g of hair, with inorganic arsenic contributing 85–99.6% of total arsenic. [The Working Group noted discrepancies between the means and ranges of concentrations of arsenic species in hair, and between text and table in the percentage of inorganic arsenic over total arsenic.] There was no difference in DMA^V concentrations in hair between workers and controls (approximately 0.03 µg/g arsenic), and MMA^V and trimethylarsenic compounds were not detected in either group. Of the arsenic species detected in urine, trimethylarsenic compounds were the most abundant, followed by DMA^V, inorganic arsenic and MMA^V. There was no difference between pre- and postwork concentrations for any of the arsenic species analysed. The authors suggested that the high concentrations observed were possibly due to the high consumption of seafood containing arsenic (arsenobetaine and arseno-choline) by workers in Japan. They concluded that urinary arsenic could be used as a bio-marker of exposure only if speciation analyses are performed (see also Section 1.1.5(*b*)(ii)); determination of arsenic in hair, on the other hand, was suggested for environmental monitoring of arsenic.

A study was conducted to examine the relationship between total arsenic concentrations in hair of employees in a semiconductor fabrication facility and their job responsibility (de Peyster & Silvers, 1995). Airborne arsenic was found in areas where equipment was cleaned but not in administrative areas. The highest arsenic concentration found in the study (15 μg/m³) was in an air sample collected over a period of 2 h in the breathing zone of an employee cleaning a source housing in an area with local exhaust ventilation. A concentration of 2 μg/m³ was found during the remainder of the cleaning period (~53 min). Maintenance workers who were regularly assigned to cleaning equipment, and therefore presumed to have the highest potential exposure, had a mean concentration of arsenic in hair of 0.042 μg/g. This was slightly higher than the mean of 0.033 μg/g observed in controls working in administrative areas, but the difference was not statistically significant. Maintenance workers who only occasionally cleaned and maintained arsenic-contaminated equipment had a mean arsenic concentration in hair of 0.034 μg/g. The highest mean concentration of arsenic in hair, 0.044 μg/g, was found in the group of supervisors and engineers. However, the highest concentrations in this group (0.076 and 0.106 μg/g) were observed in two heavy smokers. When smokers were eliminated from the analysis, means increased according to levels of presumed occupational exposure. Sex, tap-water consumption and dietary habits may also have affected arsenic concentrations in hair.

1.4   Regulations and guidelines

The only occupational exposure limit for gallium arsenide in the available literature was reported by NIOSH. NIOSH recommended a ceiling value of 0.002 mg/m³ for gallium arsenide (ACGIH Worldwide®, 2003). No occupational exposure limits have been set for gallium.

Occupational exposure limits and guidelines for arsenic in some countries are presented in Table 1. Regulations and guidelines for arsenic in drinking-water were summarized recently by IARC (IARC, 2004).

2.   Studies of Cancer in Humans

See Introduction to the Monographs on Gallium Arsenide and Indium Phosphide.

Table 1. Occupational exposure limits and guidelines for arsenic (elemental and inorganic)

Country or region	Concentration (mg/m^3)	Interpretation[a]	Carcinogen classification
Australia	0.05	TWA	1[b]
Belgium	0.1	TWA	Ca[c]
Canada			
Alberta	0.2	TWA	
Quebec	0.6	STEL	
	0.1	TWA	
China	0.01	TWA	
	0.02	STEL	
Finland	0.01	TWA	
Germany		MAK	1[d]
Hong Kong SAR	0.01	TWA	A1[e]
Ireland	0.1	TWA	Ca1[f]
Japan	0.003	TWA	1[g]
Malaysia	0.01	TWA	
Netherlands	0.05	TWA	
	0.1	STEL	
New Zealand	0.05	TWA	A1[e]
Norway	0.01	TWA	Ca[h]
Poland	0.01	TWA	Rc[i]
South Africa	0.1	TWA	
Sweden	0.01 (new facilities or alteration of old ones)	TWA	Ca[j]
	0.03	TWA	Ca
UK	0.1	TWA (MEL)	
USA[l]			
ACGIH	0.01	TWA (TLV)	A1[e]
NIOSH	0.002	Ceiling (REL)	Ca[k]
OSHA	0.01	TWA (PEL)	Ca[k]

From ACGIH Worldwide® (2003)

[a] TWA, time-weighted average; STEL, short-term exposure limit; MAK, maximum allowed concentration; MEL, maximum exposure limit; TLV, threshold limit value; REL, recommended exposure limit; PEL, permissible exposure limit

[b] Established human carcinogen

[c] Carcinogen

[d] Substance which causes cancer in man

[e] Confirmed human carcinogen

[f] Substance known to be carcinogenic to humans

[g] Carcinogenic to humans

[h] Potential cancer-causing agent

[I] Agent carcinogen to humans

[j] Substance is carcinogenic.

[k] Carcinogen

[l] ACGIH, American Conference of Governmental Industrial Hygienists; NIOSH, National Institute for Occupational Safety and Health; OSHA, Occupational Health and Safety Administration

3. Studies of Cancer in Experimental Animals

3.1 Inhalation exposure

3.1.1 *Mouse*

In a study undertaken by the National Toxicology Program (2000), groups of 50 male and 50 female B6C3F$_1$ mice, 6 weeks of age, were exposed by inhalation to gallium arsenide particulate (purity, > 98%; MMAD, 0.9–1.0 μm; GSD, 1.8–1.9 μm) at concentrations of 0, 0.1, 0.5 or 1 mg/m^3 for 6 h per day, 5 days per week, for 105 weeks (males) or 106 weeks (females). No adverse effects on survival were observed in exposed males or females compared with chamber controls (survival rates: 35/50 (control), 38/50 (low dose), 34/50 (mid dose) and 34/50 (high dose) in males and 36/50, 34/50, 31/50 or 29/50 in females, respectively; mean survival times: 687, 707, 684 or 701 days in males and 699, 699, 665 or 682 days in females, respectively). There was no evidence of carcinogenic activity in male or female mice exposed to gallium arsenide; however, exposure did result in the development of a spectrum of inflammatory and proliferative lesions of the respiratory tract of mice (National Toxicology Program, 2000) (see Section 4.3).

3.1.2 *Rat*

In a study undertaken by the National Toxicology Program (2000), groups of 50 male and 50 female Fischer 344/N rats, 6 weeks of age, were exposed by inhalation to gallium arsenide particulate (purity, > 98%; MMAD, 0.9–1.0 μm; GSD, 1.8–1.9 μm) at concentrations of 0, 0.01, 0.1 or 1 mg/m^3 for 6 h per day, 5 days per week, for 105 weeks. No adverse effects on survival were observed in treated males or females compared with chamber controls (survival rates: 13/50 (control), 13/50 (low dose), 15/50 (mid dose) and 13/50 (high dose) in males and 19/50, 17/50, 21/50 or 11/50 in females, respectively; mean survival times: 651, 627, 656 or 636 days in males and 666, 659, 644 or 626 days in females, respectively). Mean body weights were generally decreased in males exposed to the high dose throughout the study and slightly decreased in females exposed to the same dose during the second year compared with chamber controls. Although there was no evidence of carcinogenic activity in male rats exposed to gallium arsenide, exposure did result in the development of a spectrum of inflammatory and proliferative lesions of the respiratory tract (see Section 4.3). A clear neoplastic response was observed in the lung and the adrenal medulla of female rats. Increased incidence of mononuclear cell leukaemia was also observed. However, exposure to gallium arsenide did not cause an increased incidence of neoplasms in other tissues. The incidence of neoplasms and non-neoplastic lesions in female rats is reported in Table 2.

Table 2. Incidence of neoplasms and non-neoplastic lesions in female rats in a 2-year inhalation study of gallium arsenide

	No. of rats exposed to gallium arsenide at concentrations (mg/m^3) of			
	0 (chamber control)	0.01	0.1	1.0
Lung				
Total no. examined	50	50	50	50
No. with:				
Cyst, squamous	0	0	1 (4.0)	0
Hyperplasia, atypical	0	0	9^b (2.2)	16^b (2.2)
Inflammation, chronic active	11 (1.1)^a	46^b (1.5)	49^b (2.8)	50^b (3.7)
Metaplasia, squamous	0	0	2 (2.5)	1 (2.0)
Proteinosis	1 (1.0)	24^b (1.0)	47^b (2.2)	49^b (3.8)
Alveolar epithelium, hyperplasia	14 (1.5)	9 (1.6)	17 (2.1)	14 (2.3)
Alveolar epithelium, metaplasia	0	1 (1.0)	36^b (2.4)	41^b (2.6)
Alveolar/bronchiolar adenoma				
Overall rate	0	0	2	7^b
Alveolar/bronchiolar carcinoma				
Overall rate	0	0	2	3
Alveolar/bronchiolar adenoma or carcinoma				
Overall rate	0	0	4	9^b
Squamous-cell carcinoma	0	0	0	1
Adrenal medulla				
Total no. examined	50	49	50	49
No. with:				
Hyperplasia	16 (2.0)	11 (1.8)	16 (1.8)	12 (2.5)
Benign pheochromocytoma	4	5	6	13^b
Malignant pheochromocytoma	0	1	0	0
Mononuclear cell leukaemia				
Overall rate	22	21	18	33^c

From National Toxicology Program (2000)

^{a} Average severity grade of lesions in affected animals: 1, minimal; 2, mild; 3, moderate; 4, marked

^{b} Significantly different ($p \leq 0.01$) from the chamber control group by the Poly-3 test

^{c} Significantly different ($p \leq 0.05$) from the chamber control group by the Poly-3 test

In female rats, exposure to gallium arsenide caused a broad spectrum of proliferative, non-proliferative, and inflammatory lesions in the lungs, including a concentration-related increase in the incidence of alveolar/bronchiolar adenoma, and alveolar/bronchiolar adenoma and carcinoma (combined). Benign and malignant neoplasms of the lung

occurred in an exposure concentration-related manner in female rats. An increased incidence of atypical hyperplasia of the alveolar epithelium was observed in both male and
female rats. Most lesions identified as atypical epithelial hyperplasia were irregular, often
multiple, lesions that occurred at the edges of foci of chronic active inflammation. The
incidence of alveolar epithelial metaplasia was significantly increased in females exposed
to 0.1 or 1.0 mg/m^3 gallium arsenide. Alveolar epithelial metaplasia generally occurred
within or adjacent to foci of chronic active inflammation and was characterized by
replacement of normal alveolar epithelial cells (type I cells) with ciliated cuboidal to
columnar epithelial cells. The incidences of chronic active inflammation and alveolar
proteinosis were significantly increased in all exposed females, and severity of these
lesions increased with increasing exposure concentration. Gallium arsenide particles were
observed in the alveolar spaces and in macrophages, primarily in animals exposed to the
higher concentrations.

Squamous metaplasia was present in a few gallium arsenide-exposed males and
females and was usually associated with foci of chronic active inflammation. In one male
in the high-dose group and one female in the mid-dose group, the squamous epithelium
formed large cystic lesions diagnosed as squamous cysts. Although squamous epithelium
is not a component of the normal lung, it often develops as a response to pulmonary injury
associated with inhalation of irritants, especially particulates. One female in the high-dose
group had an invasive squamous-cell carcinoma. The incidence of benign pheochromocytoma occurred in a dose-related manner in females and the incidence in females
exposed to 1.0 mg/m^3 gallium arsenide was significantly increased compared to the
chamber controls. Relative to chamber controls, the incidence of mononuclear cell leukaemia was significantly increased in females exposed to 1.0 mg/m^3. Mononuclear cell
leukaemia is a common spontaneous neoplasm in Fischer 344/N rats and presents charac-
teristically as a large granular lymphocytic leukaemia (National Toxicology Program,
2000).

3.2   Intratracheal instillation

Hamster

In a study by Ohyama and colleagues (1988), groups of 33 male 6-week old Syrian
golden hamsters received weekly intratracheal instillations of 0 or 0.25 mg/animal gallium
arsenide in 200 µL phosphate buffer [particle size and purity of vehicle not provided] for
15 weeks and were observed for 111–730 days. Gallium arsenide instillations significantly
reduced survival (by 50%) at 1 year (mean survival time, 399 days versus 517 days in
controls) and caused an increased incidence of alveolar cell hyperplasia (14/30) compared
with controls (5/30). [The Working Group noted the low dose used, the short exposure
duration, the small number of animals and the high mortality in the first year.] However,
histopathological examination (larynx, trachea, lungs, liver, spleen, gastric tract, kidneys,
bladder, and other tissues not further specified) of 30 hamsters that had died or been killed

gave no indication of an increased incidence of neoplasms (Ohyama *et al.*, 1988). [The Working Group noted the inadequate reporting of the study and also judged the study design inadequate for carcinogenic effect determination.]

4. Other Data Relevant to an Evaluation of Carcinogenicity and its Mechanisms

4.1   Deposition, retention, clearance and metabolism

4.1.1   *Humans*

Data on excretion of gallium have been collected from cancer patients who had received radioactive gallium for radiotherapy or scintigraphy. There was a wide variation in urinary excretion but most subjects excreted about half of the given dose during the 4 days following administration and the major part during the first 8 h. Brucer *et al.* (1953) reported autopsy data from patients who had received intravenous radioactive gallium (^{72}Ga) which showed accumulation in the lung.

Studies by Krakoff *et al.* (1979) showed that an intravenous injection of gallium nitrate at a concentration of 10 mg/mL to patients with advanced cancer was followed by biphasic clearance with half-lives of 87 min and 24.5 h, respectively. Imaging studies using ^{67}Ga have shown that this element localizes in several major tumour categories (Edwards & Hayes, 1969, 1970; Edwards *et al.*, 1970; Winchell *et al.*, 1970; Ha *et al.*, 2000; Nishiyama *et al.*, 2002).

Following exposure, gallium is known to be transported in blood bound to transferrin and to be capable of up-regulating the transferrin receptor (Chitambar & Zivkovic, 1987; Drobyski *et al.*, 1996; Jiang *et al.*, 2002). The gallium–transferrin complex hence appears to be the primary mechanism by which the gallium ion is presented to the target cellular system.

4.1.2   *Experimental systems*

(*a*)   *In-vitro solubility and dissolution in body fluids*

Although the solubility of gallium arsenide in pure water is very low (see Section 1), its dissolution in body fluids is greatly enhanced by endogenous chelating molecules. When incubated in artificial body fluid (Gamble's solution), gallium arsenide progressively releases both gallium and arsenic. A selective leaching appears to take place, probably by chelating components of the solution, whereby more arsenic than gallium is found in solution. The gallium arsenide particle surface is enriched in arsenic, which migrates from the bulk, and which is ultimately oxidized to arsenic oxide (Pierson *et al.*, 1989). When dissolution of gallium arsenide was tested *in vitro* in phosphate buffer and various acids and bases, the amount of dissolved arsenic was highest in phosphate buffer (Yamauchi *et al.*,

1986). These observations help to explain how arsenic may be released from inhaled gallium arsenide particles.

(b)   *Respiratory system deposition, retention and clearance of gallium compounds*

(i)   *Inhalation studies*

Gallium arsenide

Greenspan *et al.* (1991) studied the clearance of inhaled gallium arsenide in male Fischer 344 rats exposed to 0.1, 1.0, 10, 37 and 75 mg/m³ gallium arsenide (MMAD, 1.2 µm) for 6 h per day on 5 days per week for 13 weeks. The half-life of clearance from the lung was found to be 17 days for both arsenic and gallium. The findings differ from those obtained using intratracheal instillation which often results in preferential clearance of arsenic over gallium (see below).

The National Toxicology Program (2000) reported results from studies in groups of 10 male and 10 female Fischer 344/N rats exposed to particulate aerosols of gallium arsenide (MMAD, 0.81–1.60 µm) by inhalation of concentrations of 0, 0.1, 1, 10, 37 or 75 mg/m³ for 6 h per day on 5 days per week for 14 weeks. Tissue burden was evaluated at days 23, 45 and 93. Lung weights increased with increasing exposure concentration in males exposed to 1 mg/m³ or more when examined on days 23 and 45 and in all exposed groups at week 14. In addition, lung weights of exposed rats continued to increase to a greater extent throughout the study compared with those of chamber controls. The percentages of gallium and arsenic in the lung relative to the total lung burden of gallium arsenide were similar at all exposure concentrations throughout the study. The deposition and clearance rates in the lung for gallium and arsenic were similar within each exposure group. Lung clearance half-lives decreased for gallium, from 56 days in rats exposed to 1 mg/m³ to 20 days in the highest exposure group (75 mg/m³). Corresponding values for arsenic were 31 and 19 days.

A 2-year study was subsequently performed in rats exposed to 0.01, 0.1 or 1.0 mg/m³ gallium arsenide using the same experimental conditions as above. Lung weights measured at months 1, 2, 4, 6, 12 and 18 were increased to a greater extent in all male rats exposed to 0.1 or 1.0 mg/m³ throughout the study than did lung weights of chamber controls and of the group exposed to 0.01 mg/m³. The percentages of gallium and arsenic in the lung relative to the total lung burden were similar at all exposure concentrations throughout the study because the deposition and clearance rates in the lung for gallium and arsenic were similar within each exposed group. Deposition rates for gallium and arsenic increased with increasing exposure concentration. Lung clearance half-lives of gallium in the group exposed to 1.0 mg/m³ were considerably less (37 days) than those for the groups exposed to 0.1 (96 days) or 0.01 mg/m³ (133 days). Lung clearance half-lives of arsenic were similar to those of gallium. The gallium lung tissue burdens at 18 months were 1.60, 13.86 and 22.87 µg/g for groups exposed to 0.01, 0.1 and 1.0 mg/m³, respectively. Gallium concentrations in whole blood, serum and testes and arsenic concentrations in serum and testes were above the limits of detection only at the higher exposure concentrations and at the later time points in the study. The mean gallium concentration in whole blood was 0.05 μg/g at 18 months in the highest exposure group; corresponding values were 0.08 μg/g in serum and 1.5 μg/g in testes (National Toxicology Program, 2000).

Gallium oxide

Wolff *et al.* (1984) studied the deposition and retention of single doses of inhaled aggregate radiolabelled gallium oxide ($^{67}Ga_2O_3$) test particles (MMAD, 0.1 μm) in beagle dogs, Fischer 344 rats and CD-1 mice using a 30-min nose-only exposure. In dogs, total gallium deposition was 39 ± 19% (mean ± SD) of the administered dose, pulmonary deposition was 25%, bronchial deposition was 7% and nasopharyngeal deposition was 7%. Corresponding values in rats were 11, 5 and 9% for pulmonary, bronchial and nasopharyngeal deposition, respectively. Pulmonary deposition in mice was estimated to be 15–20% of the administered dose. Whole-body retention was measured and in dogs the long-term plateau represented more than 70% of the particles, compared with 38% and 28% for rats and mice, respectively. The half-life of the long-term component of clearance was 75 ± 19 days for mice, 65 ± 17 days for rats and 52 ± 25 days for dogs.

Wolff *et al.* (1989) presented results of modelling accumulation of particles in rat lung during chronic nose-only inhalation exposure of Fischer 344 rats to 23 mg/m^3 gallium oxide for 2 h per day on 5 days a week for 4 weeks. Impaired clearance occurred early after accumulation of a low burden of the particles. A half-life in the order of 170 days was observed rather than the 65-day half-life reported earlier (Wolff *et al.*, 1984; see above). This impairment of clearance might influence toxicity and the local dose of particles of low solubility in experimental studies.

Battelle Pacific Northwest Laboratories (1990a) carried out 13-week inhalation studies of gallium(III) oxide in male Fischer 344 rats exposed to 0, 0.12, 0.48, 4.8, 24 or 48 mg/m^3 gallium oxide particles (MMAD, ~0.9 μm). Gallium exposure concentrations were approximately equimolar to those used in the studies of gallium arsenide cited above (Greenspan *et al.*, 1991; National Toxicology Program, 2000). As observed with gallium arsenide, following inhalation of gallium oxide, blood and urinary concentrations of gallium were found to be extremely low and only detectable in animals exposed to 24 and 48 mg/m^3 throughout the study. The results indicated that gallium oxide, like gallium arsenide, is not readily absorbed and that, when absorbed, it is rapidly cleared from the blood and either excreted or sequestered in the tissues. Considerable concentrations of gallium were detected in the faeces. Lung burdens increased with increasing exposure concentration. However, when normalized to exposure concentration, accumulation in the lung during the study increased as exposure concentrations increased. Overload may have occurred at gallium oxide concentrations of 24 mg/m^3 and above; this would be in line with the results of Wolff *et al.* (1989).

(ii) *Instillation studies with gallium arsenide*

Webb *et al.* (1984) investigated absorption, excretion and pulmonary retention of gallium arsenide after intratracheal instillation doses of 10, 30 and 100 mg/kg bw (mean volume particle diameter, 12.7 μm) in male Fischer 344 rats. At day 14, gallium was not detected in the blood and urine at any dosage but was retained in the lungs; arsenic retention (measured by F-AAS) ranged from 17 to 32% of the doses given while gallium retention (measured also by F-AAS) ranged from 23 to 42%. In a later study, Webb *et al.* (1986) exposed male Fischer 344 rats to gallium arsenide (100 mg/kg bw) and gallium trioxide (65 mg/kg bw) (equimolar for gallium) by intratracheal instillation (mean volume particle diameters, 12.7 μm and 16.4 μm, respectively). The mean retention of gallium in the lung at day 14 was fairly similar for the two compounds (44% and 36% for gallium arsenide and gallium trioxide, respectively). Webb *et al.* (1987) showed that smaller gallium arsenide particles (mean volume particle diameter, 5.82 μm) had an increased in-vivo dissolution rate and there was increased severity of pulmonary lesions in male Fischer 344 rats after intratracheal instillation of a suspension containing 100 mg/kg bw. Clearance from lung was faster for arsenic (half-life, 4.8 days) than for gallium (half-life, 13.2 days).

Rosner and Carter (1987) studied metabolism and excretion after intratracheal instillation of 5 mg/kg bw gallium arsenide (mean volume particle diameter, 5.8 μm) in Syrian golden hamsters. Blood arsenic concentrations increased from 0.185 ± 0.041 ppm (2.4 μM) after day 1 to 0.279 ± 0.021 ppm (3.7 μM) on day 2. Blood concentrations of arsenic peaked at day 2 after dosing, indicating continued absorption. Of the arsenic, 5% was excreted in the urine during the first 4 days after gallium arsenide instillation compared with 48% after exposure to soluble arsenic compounds. Arsenic derived from gallium arsenide was converted into arsenate (As^{III}), arsenite (As^V) and a major metabolite dimethyl arsinic acid, and rapidly excreted. Twenty-seven per cent of the arsenic derived from gallium arsenide were excreted in the faeces the first day after the instillation; this was probably due to lung clearance into gastrointestinal tract after expectoration.

Omura *et al.* (1996a) exposed hamsters to 7.7 mg/kg bw gallium arsenide, 7.7 mg/kg bw indium arsenide or 1.3 mg/kg bw arsenic trioxide by intratracheal instillation twice a week, 14–16 times. Arsenic concentrations in serum on the day after the last instillation were 0.64 μM after gallium arsenide, 0.34 μM after indium arsenide and 1.31 μM after arsenic trioxide. Serum concentrations of gallium and indium were about 20 μM. The results indicated a high retention of both gallium and indium compared with that of arsenic which might be of importance in toxicity from long-term exposure.

Gallium arsenide might in itself impair lung clearance. Aizawa *et al.* (1993) used magnetometric evaluation to study the effects of gallium arsenide on clearance of iron oxide test particles in rabbits. Instillation of 30 mg or 300 mg gallium arsenide per animal in 2 mL saline significantly impaired clearance at 14, 21 and 28 days after exposure. However, although the effect was clear, the dose was high. Impaired clearance might be caused by gallium arsenide itself or by dissolved arsenic-induced inflammation.

(c)   *Gastrointestinal exposure to gallium*

(i)   *Oral and intraperitoneal studies*

Yamauchi *et al.* (1986) studied metabolism and excretion of gallium arsenide (mean volume particle diameter, 14 μm) in Syrian golden hamsters exposed to single doses of 10, 100 or 1000 mg/kg bw in phosphate buffer administered orally through a stomach tube and 100 mg/kg bw intraperitoneally. Urinary excretion of arsenic during the following 120 h was 0.15, 0.11 and 0.05% of the high, medium and low oral doses, respectively, and 0.29% of the intraperitoneal dose. During the same time period, faecal excretion of arsenic was around 80% of the oral doses and 0.38% of the intraperitoneal dose.

Flora *et al.* (1997) exposed groups of male albino rats to single oral doses of 500, 1000 or 2000 mg/kg bw gallium arsenide. Blood was collected at 24 h, and on days 7 and 15 following exposure. Urinary samples were taken at 24 h. Animals were killed on days 1, 7 and 15 and heart tissue was collected. Blood and heart tissue concentrations of gallium and arsenic were determined using GF-AAS and were found to peak at day 7. In a later study, Flora *et al.* (1998) exposed male Wistar albino rats to single doses of 100, 200 or 500 mg/kg bw gallium arsenide or vehicle (control) by gastric intubation. Concentrations of gallium and arsenic were measured at 24 h, and on days 7 and 21 following administration and peaked at day 7 in the blood, liver and kidney but continued to increase up to day 21 in the spleen.

(ii)   *Intravenous injection of gallium-67: tracer studies*

Sasaki *et al.* (1982) studied differences in the liver retention of ^{67}Ga (as gallium citrate) administered intravenously in controls and rats fed with the liver carcinogen 3'-methyl-4-di-methylaminobenzene for 20 weeks. They observed that the accumulation of ^{67}Ga in the carcinogen-fed animals at 20 weeks was about 2.3 times greater (per gram of liver) than in the controls. This increase correlated with increases in γ-glutamyl transpeptidase and glucose-6-phosphatase activities at late stages during hepatocarcinogenesis. The most marked change in ^{67}Ga accumulation occurred in the nuclear/whole cell (800 × g) liver fraction suggesting that ^{67}Ga may bind to components in this fraction, induced by 3'-methyl-4-dimethylaminobenzene.

4.1.3   *Data relevant to an evaluation of gallium arsenide as an arsenic compound*

(a)   *Metabolism of the arsenic oxides*

Radabaugh and coworkers (2002) recently characterized arsenate reductase enzyme and identified it as a purine nucleoside phosphorylase, an ubiquitous enzyme that required dihydrolipoic acid for maximum reduction of arsenate As^{V} to arsenite As^{III} in mammals. [The valences of different forms of arsenic and their metabolites are indicated by super-script roman numerals such as it is reported in scientific publications.] The As^{III} formed may then be methylated to MMA^{V} and to DMA^{V} by methyl transferases which have been partially characterized (Zakharyan *et al.*, 1995; Wildfang *et al.*, 1998; Styblo *et al.*, 1999).

In mice, the highest methylating activity occurred in testes followed by kidney, liver and lung (Healy *et al.*, 1998). The analogous enzymatic reduction of MMA^V to monomethyl-arsonous acid (MMA^{III}) was also demonstrated in hamster; MMA^V reductase-specific activities have been shown in all organs (Sampayo-Reyes *et al.*, 2000).

(b)   *Variation in arsenic methylation between species*

Most human organs can metabolize arsenic by oxidation/reduction reactions, methylation and protein binding. However, there is a pronounced species difference in this metabolism. Arsenic is strongly retained in rat erythrocytes but not in those of other species. The unique disposition of arsenic in rats may be due to the pronounced biliary excretion of MMA^{III} and erythrocyte of DMA^{III} (Gregus *et al.*, 2000; Shiobara *et al.*, 2001) which may explain the lower toxicity of arsenic in rats. Thus, previous scientific committees have stated that they did not recommend rats for arsenic oxide disposition studies (National Academy of Sciences, 1977; Aposhian, 1997). Most experimental animals excrete very little MMA [valence not specified] in urine compared to humans (Vahter, 1999) and some animal species, in particular guinea-pigs and several non-human primates, are unable to methylate arsenic at all (Healy *et al.*, 1997; Vahter, 1999; Wildfang *et al.*, 2001). The effect of the inability to methylate As^{III} compounds on toxicity following repeated dosing is unknown but methylation has long been considered the primary mechanism of detoxification of arsenic in mammals (Buchet *et al.*, 1981). However, non-methylator animals were not found to be more sensitive to the acute effect of arsenic than methylators in the few tests that have been performed. The toxic response of non-methylators needs to be examined in more detail. At present, the most toxic arsenic species is thought to be the MMA^{III} (Petrick *et al.*, 2000; Styblo *et al.*, 2000; Petrick *et al.*, 2001), leading to the view that this methylation should be considered as bioactivation of the metalloid rather than detoxification.

Arsenic detoxification mechanisms other than methylation have been poorly investigated. The fact that man is more than 10 times more sensitive to the effect of arsenic oxides when compared to all other animal species is remarkable. The explanation of this difference in sensitivity is important in order to understand the mechanism of action of arsenic (see IARC, 2004).

4.2   Toxic effects

4.2.1   *Humans*

There are no published reports specific to the toxicity of gallium arsenide in humans.

4.2.2   *Experimental systems*

   (*a*)   *Gallium arsenide and gallium oxide*

      (i)   *Non-neoplastic and pre-neoplastic effects in the respiratory*
            tract

   Results of studies undertaken by the National Toxicology Program (2000) (see also
Section 3.1) confirmed that the respiratory tract was the primary site of toxicity, indicated
by a spectrum of inflammatory and proliferative lesions of the lung. As described in Sections
3.1.1 and 3.1.2, and in Table 2, groups of 50 male and 50 female B6C3F$_1$ mice and groups
of 50 male and 50 female Fischer 344/N rats, 6 weeks of age, were exposed by inhalation
to gallium arsenide particulate (purity, > 98%; MMAD, 0.8–1.0 µm; GSD, 1.8–1.9 µm) at
concentrations of 0, 0.1, 0.5 or 1 mg/m³ for mice and 0, 0.01, 0.1 and 1 mg/m³ for rats, for
6 h per day on 5 days per week for 105 or 106 weeks. In mice, non-neoplastic effects were
observed in the lung (which included focal suppurative inflammation, focal chronic
inflammation, histiocyte infiltration, hyperplasia of the alveolar epithelium, proteinosis of
the alveoli and tracheobronchial lymph nodes). The non-neoplastic effects observed in the
lung of exposed rats included atypical hyperplasia, active chronic inflammation, proteinosis
and metaplasia of the alveolar epithelium in both sexes. In male rats, hyperplasia of the
alveolar epithelium of the lung and chronic active inflammation, squamous metaplasia and
hyperplasia of the epiglottis and the larynx were observed (National Toxicology Program,
2000).

   The most prominent toxic effect of gallium arsenide after a single intratracheal instillation to rats is pulmonary inflammation (Webb *et al.*, 1987; Goering *et al.*, 1988). Histopathological changes and changes in tissue concentrations of protein, lipid, and DNA have been
observed (Webb *et al.*, 1986). The effects caused by gallium arsenide (100 mg/kg bw) were
compared with those elicited by equimolar gallium oxide (65 mg/kg bw) and maximally-
tolerated amounts of (17 mg/kg bw, 0.25 equimolar) arsenious (III) acid (Webb *et al.*, 1986).
Two weeks after exposure to gallium arsenide, increases in lipid concentrations, comparable
to those observed following exposure to equimolar gallium, and increases in protein concentrations similar to those found after exposure to arsenious acid were observed. DNA concentrations were significantly increased after exposure to gallium arsenide but not to the same
magnitude as those seen after arsenious acid exposure (arsenious acid was given at 0.25
times the molar dose of gallium arsenide). Only exposure to arsenious acid resulted in
increases in 4-hydroxyproline, an indicator of a fibrotic process. Lung wet weights, lung wet
weight/body weight and lung dry weights were all increased after instillation of gallium
arsenide but not after instillation of gallium oxide or arsenious acid. Goering *et al.* (1988)
reported similar histopathological changes in the lungs of rats treated with gallium arsenide
in the same conditions.

   In a 16-day inhalation study (National Toxicology Program, 2000) of rats exposed to
gallium arsenide at concentrations of 0, 1, 10, 37, 75 or 150 mg/m³, statistically-significant
increases in the weights of lungs and liver relative to body weight were noted in animals
exposed to concentrations of 1 mg/m³ and greater. These effects were noted only for lungs

following exposure to 0.1 mg/m³ and above in a 14-week study. When the studies were repeated in mice, only the lungs were found to show increases relative to body weights.

(ii)   *Haematological effects*

A study (National Toxicology Program, 2000; see Section 4.1.2) of mice and rats exposed to gallium arsenide at chamber concentrations of 0, 0.1, 1, 10, 37 or 75 mg/m³ for 14 weeks, showed statistically-significant decreases in haematocrit and haemoglobin concentrations, and increased numbers of erythrocytes and reticulocytes at 14 weeks in both species exposed to 37 and 75 mg/m³. Statistically-significant decreases in leucocyte numbers were noted in rats exposed to the two highest doses, whereas increases in leucocyte numbers were observed in mice exposed to the three highest doses. Zinc protoporphyrin/ haeme ratios increased in male and female mice exposed to the two highest doses while methaemoglobin increased only in female rats.

Effects on the haem biosynthetic pathway

In the 14-week exposure study cited above (National Toxicology Program, 2000), concentrations of δ-aminolevulinic acid (ALA) and porphobilinogen were not increased in urine of rats exposed by inhalation to gallium arsenide, suggesting that the effect of the porphyria, as it relates to haeme synthesis, was marginal.

Goering and colleagues (1988) observed systemic effects after intratracheal administration of 50, 100 and 200 mg/kg bw gallium arsenide to rats. Activity of δ-aminolevulinic acid dehydratase (ALAD) in blood and urinary excretion of δ-aminolevulinic acid (ALA) were examined. A dose-dependent inhibition of ALAD activity in blood and an increase in excretion of ALA in urine were observed with a maximum response 3–6 days after exposure. A urinary porphyrin excretion pattern characteristic of arsenic exposure (Woods & Fowler, 1978) was also observed in these animals (Bakewell *et al.*, 1988).

In-vitro studies with gallium nitrate, sodium arsenite and sodium arsenate showed that 75 μM gallium nitrate inhibited the activity of blood ALAD and 2 μM gallium nitrate inhibited liver and kidney ALAD. The inorganic arsenic compounds inhibited ALAD in blood at much higher concentrations (15 mM, 200-fold) (Goering *et al.*, 1988). Subsequent in-vivo and in-vitro studies on ALAD in blood, liver and kidney showed that the mechanism of gallium inhibition involves zinc displacement from the sulfhydryl group of the enzyme active site (Goering & Rehm, 1990).

(iii)   *Immunological effects*

A variety of changes have been reported in animals exposed to gallium arsenide including inhibition of T-cell proliferation and suppression of immunological functions at locations distal to a single exposure site (Sikorski *et al.*, 1989; Burns *et al.*, 1991; Burns & Munson, 1993; Hartmann & McCoy, 1996). The effects included decreases in both humoral and cellular antibody response. The dissolution of gallium arsenide to form gallium and arsenic oxides may be the origin of the effects; arsenic has been shown to be the primary immunosuppressive component of gallium arsenide (Burns *et al.*, 1991), but it was unclear whether all the immunological effects reported were caused by dissolved arsenic.

(b)   Other gallium compounds

(i)   In vitro

Studies by Chitambar and Seligman (1986), Chitambar and co-workers (1988, 1990, 1991) and Narasimhan *et al.* (1992) have shown that transferrin-gallium exerts its toxic effects at the molecular level by inhibiting ribonucleotide reductase, specifically by displacing iron from the M2 subunit of this enzyme.

(ii)   In vivo

Early studies by Dudley and Levine (1949) demonstrated the acute renal toxicity of gallium lactate 3 or 4 days after its intravenous injection in rats. Studies by Hart *et al.* (1971) and Adamson *et al.* (1975) further extended the database on the renal toxicity of gallium nitrate; a limiting factor in its use in the treatment of tumours.

4.3   Reproductive and developmental effects

4.3.1   *Humans*

There have been several studies that have reported that workers in the semiconductor industry experience increased rates of spontaneous abortion, but the evidence is inconclusive (Elliot *et al.*, 1999). No single metal has been denoted as a more possible causative agent than any other because of the complex chemical exposures, and other factors, encountered in these environments (Fowler & Sexton, 2002).

4.3.2   *Animals*

(a)   Testicular function changes

(i)   Gallium arsenide

Testicular toxicity has been reported in rats and hamsters after intratracheal administration of 7.7 mg/kg bw gallium arsenide twice a week for a total of 8 weeks (Omura *et al.*, 1996a,b). A significant decrease in sperm count and in the proportion of morphologically-abnormal sperm were found in the epididymis in the gallium arsenide-treated rats. In hamsters, gallium arsenide caused testicular spermatid retention and epididymal sperm reduction. Animals treated with arsenic trioxide (1.3 mg/kg) or indium arsenide (7.7 mg/kg bw) did not show any testicular toxicities. The arsenic concentrations in serum of gallium arsenide-treated rats were almost twice those found in arsenic trioxide-treated rats. In addition, the molar concentration of gallium was found to be 10–20-fold higher than that of arsenic in gallium arsenide-treated rats (Omura *et al.*, 1996a). In contrast, the arsenic concentrations in serum of gallium arsenide-treated hamsters were less than half of those found in arsenic trioxide-treated hamsters. Moreover, the molar concentration of gallium was 32 times higher than that of arsenic in gallium arsenide-treated hamsters. Therefore gallium may play a main role in the testicular toxicity in hamsters (Omura *et al.*, 1996b).

Similar testicular toxicities were observed in 14-week and 2-year gallium arsenide inhalation studies (National Toxicology Program, 2000). The effects included decreases in epididymal weights and sperm motility in both rats and mice exposed to 37 and 75 mg/m^3 in the 14-week study. Decreases in epididymal weights and an epididymal hypospermia were also observed in mice exposed to 10 mg/m^3. Decreased testicular weights, genital atrophy and interstitial hyperplasia were observed in rats exposed to 1 mg/m^3 of gallium arsenide in the 2-year study.

(ii)   *Gallium oxide*

In a 13-week study of gallium oxide in male rats and mice, exposure to concentrations of 0, 0.16, 0.64, 6.4, 32 or 64 mg/m^3 were found to have no effect on male rat reproductive parameters. However, exposure to gallium oxide at 32 mg/m^3 or greater caused decreases in cauda epididymis and testis weights. Decreases in epididymal sperm motility and concentration were observed in animals exposed to 64 mg/m^3. Testicular degeneration and increased cellular debris in the epididymis were observed in mice exposed to gallium oxide at 64 mg/m^3 (Battelle Pacific Northwest Laboratories, 1990a,b).

(b)   *Effects on estrous cycles, gestation and foetal development*

In a 13-week study of gallium oxide in female rats and mice, there was no effect of exposure to concentrations of 0.16–64 mg/m^3 on the estrous cycles of either animal species (Battelle Pacific Northwest Laboratories, 1990a,b).

Studies to assess the developmental toxicity of gallium arsenide were performed with Sprague-Dawley rats and Swiss mice exposed to 0, 10, 37 or 75 mg/m^3 gallium arsenide by inhalation 6 h per day, 7 days per week. Rats were exposed on gestation days 4 through 19. There were no signs of maternal toxicity. Minimal effects on the fetuses were noted, including a marginal reduction in body weight in the group exposed to 75 mg/m^3 and concentration-dependent reduced ossification of the sternebrae. There was a non-significant increase in the incidence of incompletely ossified vertebral centra. Mice were exposed on gestation days 4 through 17. Considerable fetal and maternal toxicity was seen in groups exposed to 37 and 75 mg/m^3 gallium arsenide, with 50% of the female animals found dead or moribund. Most exposed females were hypoactive, had laboured breathing and failed to gain weight. The number of resorptions per litter was significantly increased and occurred earlier, while the number of corpora lutea per dam and the number of live fetuses per litter were significantly decreased. Fetal weights were reduced in all exposed groups. Although not statistically significant, various skeletal malformations were observed including cleft palate, encephalocele, and vertebral defects (Battelle Pacific Northwest Laboratories, 1990c; Mast *et al.*, 1991).

4.4   Genetic and related effects (see Table 3)

Gallium arsenide (10 000 μg/plate) was not mutagenic in *Salmonella typhimurium* strains TA97, TA98, TA100, TA102 or TA1535, with or without induced rat or hamster liver S9 enzymes (Zeiger *et al.*, 1992). No increase in the frequency of micronucleated normochromatic erythrocytes was seen in peripheral blood samples from male or female B6C3F$_1$ mice exposed to gallium arsenide by inhalation in concentrations up to 75 mg/m^3, during a 14-week study (National Toxicology Program, 2000). The majority of these experiments were carried out assuming arsenite (As^{III}) was the toxic species; however, there is evidence that it is not. It appears that dimethyl arsinous acid may be a carcinogen but that the most toxic arsenic species may be MMA^{III} (see Section 4.1.3). It is believed that many studies have assigned a toxic dose to arsenate but the effect was actually the result of the reduction of arsenate (As^V) to arsenite (As^{III}) (Carter *et al.*, 1999, 2003). It is also of concern that experiments with arsenate using cells have been done without consideration of the concentration of phosphate, an arsenate uptake inhibitor (Huang & Lee, 1996).

4.5   Mechanistic considerations

The hypothesis used to interpret the carcinogenesis results appears to accept the finding that gallium arsenide causes cancer in female rats and that the non-neoplastic hyperplasia is a precursor to neoplasms. The lung effects appear to be 'point of contact' effects. The mechanism of lung cancer fits with a highly toxic compound which kills many different cells without killing the host organism. This leads to regenerative cell proliferation that magnifies any errors in DNA replication and results in enough errors to make organ neoplastic changes in the lung. Some systemic effects were found to be sex-specific and, therefore, a selectivity of response between males and females is not surprising.

It is clear that there is partial dissolution of gallium arsenide particles *in vivo* and that while the majority of a dose of gallium arsenide remains in the lung, there is redistribution of solubilized gallium and arsenic to other organ systems. This results in a variety of toxic effects including inhibition of haeme biosynthesis in a number of organ systems, testicular damage and impaired immune function. Some of the biochemical effects, such as inhibition of haeme pathway enzymes such as ALAD, appear to be relatively specific. However, more pronounced cellular changes in target organ systems such as the kidney, testes, or immune system may be the result of gallium or arsenic or combined exposure to these elements. Further mechanistic research is needed to elucidate the primary underlying roles played by these elements in organ systems outside the lungs.

There is evidence from in-vitro test systems that ionic gallium, such as the gallium transferrin complex, may influence the carcinogenic process by inducing apoptosis at low doses and producing necrosis at high doses in cancer cell lines (Jiang *et al.*, 2002).

Table 3. Genetic and related effects of gallium arsenide

Test system	Result[a]		Dose[b] (LED or HID)	Reference
	Without exogenous metabolic system	With exogenous metabolic system		
Salmonella typhimurium TA97, TA98, TA100, TA108, TA1535, reverse mutation	–	–	10 000 µg/plate	*Zeiger et al.* (1992)
Formation of micronuclei in binucleates, cytochalasin-B assay, Syrian hamster embryo cells *in vitro*	–		10 µg/mL	*Gibson et al.* (1997)
Formation of micronuclei, B6C3F$_1$ mice erythrocytes in peripheral blood *in vivo*	–		75 µg/m^3 inhalation (14 wk)	National Toxicology Program (2000)
Cell transformation, Syrian hamster embryo cells *in vitro*	+		0.5 µg/mL	*Kerckaert et al.* (1996)

[a] +, positive; –, negative
[b] LED, lowest effective dose; HID, highest ineffective dose

5. Summary of Data Reported and Evaluation

5.1   Exposure data

Gallium arsenide is extensively used in the microelectronics industry because of its photovoltaic properties. Gallium arsenide is produced as high purity single crystals and cut into wafers and other shapes which are used primarily for integrated circuits and opto-electronic devices. Exposure to gallium arsenide occurs predominantly in the micro-electronics industry where workers are involved in the production of gallium arsenide crystals, ingots and wafers, grinding and sawing operations, device fabrication and sand-blasting and clean-up activities.

5.2   Human carcinogenicity data

See Introduction to the Monographs on Gallium Arsenide and Indium Phosphide.

5.3   Animal carcinogenicity data

Gallium arsenide was tested for carcinogenicity in a single study by chronic inhalation exposure in mice and rats. In female rats exposed to the highest concentration, significantly increased incidences of alveolar/bronchiolar neoplasms, benign pheochromocytoma of the adrenal medulla and mononuclear-cell leukaemia were observed. There was no evidence of carcinogenic activity in male rats, or in male or female mice.

Gallium arsenide was tested by intratracheal instillation in male hamsters and showed no carcinogenic response. However, due to inadequacies in design and reporting, the study did not contribute to this evaluation.

5.4   Other relevant data

Gallium arsenide has low solubility. There is in-vitro and in-vivo evidence that gallium arsenide releases gallium and arsenic moieties.

Uptake from the gastrointestinal tract is low. In inhalation studies, lung retention of inhaled gallium arsenide has been shown to be influenced by toxic effects from gallium arsenide itself. Tissue burdens are highest in the lung. Concentrations of gallium and arsenic in blood and serum remain low in long-term inhalation studies. Concentrations of gallium in testes show evidence of accumulation, but at a much lower level than in the lung. After intratracheal instillation of gallium arsenide, data indicate slower elimination and higher serum concentrations of gallium compared with arsenic.

The most prominent toxic effect of gallium arsenide is pulmonary inflammation, which may occur after a single intratracheal dose. Gallium arsenide and gallium nitrate inhibit the activity of δ-aminolevulinic acid dehydratase.

Immunological effects of exposure to gallium arsenide include inhibition of T-cell proliferation and decrease of both humoral and cellular immune response. These effects are partly due to the arsenic moiety.

Testicular toxicity was observed in rats and hamsters exposed to gallium arsenide by intratracheal administration, while animals treated with arsenic trioxide and indium arsenide did not show these effects. In inhalation studies with gallium arsenide, decreased epididymal weights and reduced sperm mobility were observed. A number of reproductive toxic effects were reported following exposure of pregnant rodents to gallium arsenide. These effects were more severe in mice than in rats.

Based on limited data, gallium arsenide does not show genotoxic activity.

5.5   Evaluation

There is *inadequate evidence* in humans for the carcinogenicity of gallium arsenide.

There is *limited evidence* in experimental animals for the carcinogenicity of gallium arsenide.

Overall evaluation

Gallium arsenide is *carcinogenic to humans (Group 1)*.

The Working Group noted that there were no data on cancer in humans and that gallium arsenide is, at best, a weak carcinogen in experimental animals. In reaching an overall evaluation of *Group 1*, the Working Group noted the potential for gallium arsenide to cause cancer through two separate mechanisms of action. Once in the body, gallium arsenide releases a small amount of its arsenic, which behaves as inorganic arsenic at the sites where it is distributed. (Arsenic and arsenic compounds have been evaluated as IARC Group 1, carcinogenic to humans.) At the same time, the gallium moiety may be responsible for the lung cancers observed in the study in female rats, due to the apparent resistance of rats to the carcinogenic potential of arsenic that is manifest in humans. The similarity of toxicochemical responses observed in subchronic studies with gallium arsenide and gallium oxide adds weight to the finding that the gallium moiety is active and suggests that a carcinogenic response might be observed with other gallium compounds. The observed findings may also be a result of the combination of the two moieties.

6. References

ACGIH Worldwide® (2003) *Documentation of the TLVs® and BEIs® with Other Worldwide Occupational Exposure Values — CD-ROM — 2003*, Cincinnati, OH

Adamson, R.H., Canellos, G.P. & Sieber, S.M. (1975) Studies on the antitumor activity of gallium nitrate (NSC-15200) and other group IIIa metal salts. *Cancer Chemother. Rep.*, **59**, 599–610

Afonso, A.M., Santana, J.J. & García Montelongo, F.J. (1985) Pyrocatechol-1-aldehyde 2-benzothiazolylhydrazone as reagent for the spectrofluorimetric determination of nanogram amounts of gallium in urine and blood serum. *Anal. Lett.*, **A18**, 1003–1012

Aizawa, Y., Takata, T., Karube, H., Tatsumi, H., Inokuchi, N., Kotani, M. & Chiyotani, K. (1993) Magnetometric evaluation of the effects of gallium arsenide on the clearance and relaxation of iron particles. *Ind. Health*, **31**, 143–153

Anthemidis, A.N., Zachariadis, G.A. & Stratis, J.A. (2003) Gallium trace on-line preconcentration/separation and determination using a polyurethane foam mini-column and flame atomic absorption spectrometry. Application in aluminum alloys, natural waters and urine. *Talanta*, **60**, 929–936

Aposhian, H.V. (1997) Enzymatic methylation of arsenic species and other new approaches to arsenic toxicity. *Annu. Rev. Pharmacol. Toxicol.*, **37**, 397–419

Apostoli, P., Bartoli, D., Alessio, L. & Buchet, J.P. (1999) Biological monitoring of occupational exposure to inorganic arsenic. *Occup. environ. Med.*, **56**, 825–832

Bakewell, W.E., Goering, P.L., Moorman, M.P. & Fowler, B.A. (1988) Arsine (AsH₃) and gallium arsenide (GAAS)-induced alterations in heme metabolism (Abstract). *Toxicologist*, **8**, 20

Battelle Pacific Northwest Laboratories (1990a) *Thirteen-week Subchronic Inhalation Toxicity Study Report of Gallium Oxide in Rats*, Final report (NIH No. N01-ES-85211)

Battelle Pacific Northwest Laboratories (1990b) *Thirteen-week Subchronic Inhalation Toxicity Study Report of Gallium Oxide in Mice*, Final report (NIH No. N01-ES-85211)

Battelle Pacific Northwest Laboratories (1990c) *Inhalation Developmental Toxicology Studies: Gallium Arsenide in Mice and Rats*, Final report (NIH No. N01-ES-70153)

Beliles, R.P. (1994) The metals. In: Clayton, G.D. & Clayton, F.E., eds, *Patty's Industrial Hygiene and Toxicology*, 4th Ed., Vol. 2C, John Wiley & Sons, New York, pp. 1879–2013

Beltrán Lucena, R., Morales, E. & Gomez-Ariza, J.L. (1994) Spectrophotometric determination of gallium in biological materials at nanogram levels with thiocarbohydrazone derivatives. *Farmaco*, **49**, 291–295

Brucer, M., Andrews, G.A. & Bruner, H.D. (1953) A study of gallium. *Radiology*, **61**, 534–613

Buchet, J.P., Lauwerys, R. & Roels, H. (1981) Comparison of the urinary excretion of arsenic metabolites after a single oral dose of sodium arsenite, monomethylarsonate, or dimethylarsinate in man. *Int. arch. occup. environ. Health*, **48**, 71–79

Burguera, M. & Burguera, J.L. (1997) Analytical methodology for speciation of arsenic in environmental and biological samples. *Talanta*, **44**, 1581–1604

Burns, L.A. & Munson, A.E. (1993) Gallium arsenide selectively inhibits T cell proliferation and alters expression of CD25 (IL-2R/p55). *J. Pharmacol. exp. Ther.*, **265**, 178–186

Burns, L.A., Sikorski, E.E., Saady, J.J. & Munson, A.E. (1991) Evidence for arsenic as the immuno-suppressive component of gallium arsenide. *Toxicol. appl. Pharmacol.*, **110**, 157–169

Cano Pavón, J.M., Ureña Pozo, E. & Garcia de Torres, A. (1988) Spectrofluorimetric determination of gallium with salicylaldehyde carbohydrazone and its application to the analysis of biological samples and alloys. *Analyst*, **113**, 443–445

Cano Pavón, J.M., Garcia de Torres, A. & Ureña Pozo, M.E. (1990) Simultaneous determination of gallium and aluminium in biological samples by conventional luminescence and derivative synchronous fluorescence spectrometry. *Talanta*, **37**, 385–391

Carter, D.E., Peraza, M.A., Ayala-Fierro, F., Casarez, E., Barber, D.S. & Winski, S.L. (1999) Arsenic metabolism after pulmonary exposure. In: Chappell, W.R., Abernathy, C.O. & Calderon, R.L., eds, *Arsenic Exposure and Health Effects*, Amsterdam, Elsevier, pp. 299–309

Carter, D.E., Aposhian, H.V. & Gandolfi, A.J. (2003) The metabolism of inorganic arsenic oxides, gallium arsenide and arsine: a toxico-chemical review. *Toxicol. appl. Pharmacol.*, **193**, 309–334

Chakrabarti, N.B. (1992) GaAs integrated circuits. *J. Inst. Electron. Telecommun. Eng.*, **38**, 163–178

Chemical Information Services (2003) *Directory of World Chemical Producers (Version 2003)*, Dallas, TX (http://www.chemicalinfo.com, accessed 18.09.2003)

Chitambar, C.R. & Seligman, P.A. (1986) Effects of different transferrin forms on transferrin receptor expression, iron uptake, and cellular proliferation of human leukemic H160 cells. Mechanisms responsible for the specific cytotoxicity of transferrin-gallium. *J. clin. Invest.*, **78**, 1538–1546

Chitambar, C.R. & Zivkovic, Z. (1987) Uptake of gallium-67 by human leukemic cells: Demonstration of transferrin receptor-dependent and transferrin-independent mechanisms. *Cancer Res.*, **47**, 3939–3934

Chitambar, C.R., Matthaeus, W.G., Antholine, W.E., Graff, K. & O'Brien, W.J. (1988) Inhibition of leukemic HL60 cell growth by transferrin-gallium: Effects on ribonucleotide reductase and demonstration of drug synergy with hydroxyurea. *Blood*, **72**, 1930–1936

Chitambar, C.R., Zivkovic-Gilgenbach, Z., Narasimhan, J. & Antholine, W.E. (1990) Development of drug resistance to gallium nitrate through modulation of cellular iron uptake. *Cancer Res.*, **50**, 4468–4472

Chitambar, C.R., Narasimhan, J., Guy, J., Sem, D.S. & O'Brien, W.J. (1991) Inhibition of ribo-nucleotide reductase by gallium in murine leukemic L1210 cells. *Cancer Res.*, **51**, 6199–6201

Drobyski, W.R., Ul-Haq, R., Majewski, D. & Chitambar, C.R. (1996) Modulation of in-vitro and in-vivo T-cell responses by transferrin-gallium and gallium nitrate. *Blood*, **88**, 3056–3064

Dudley, H.C. & Levine, M.D. (1949) Studies of the toxic action of gallium. *J. Pharmacol. exp. Ther.*, **95**, 487–493

Edwards, C.L. & Hayes, R.L. (1969) Tumor scanning with ^{67}Ga-citrate. *J. nucl. Med.*, **10**, 103–105

Edwards, C.L. & Hayes, R.L. (1970) Scanning malignant neoplasms with gallium 67. *J. am. med. Assoc.*, **212**, 1182–1190

Edwards, C.L., Nelson, B. & Hayes, R.L. (1970) Localization of gallium in human tumors. *Clin. Res.*, **18**, 89

Flora, S.J.S., Dube, S.N., Vijayaraghavan, R. & Pant, S.C. (1997) Changes in certain hematological and physiological variables following single gallium arsenide exposure in rats. *Biol. Trace Elem. Res.*, **58**, 197–208

Flora, S.J.S., Kumar, P., Kannan, G.M. & Rai, G.P. (1998) Acute oral gallium arsenide exposure and changes in certain hematological, hepatic, renal and immunological indices at different time intervals in male Wistar rats. *Toxicol. Lett.*, **94**, 103–113

Fowler, B.A. & Sexton, M.J. (2002) *Chapter 18. Semiconductors.* In: Bibudhendra Sarkar, ed., *Heavy Metals in the Environment*, New York, Marcel Dekker, Inc., pp. 631–645

Gibson, D.P., Brauninger, R., Shaffi, H.S., Kerckaert, G.A., Leboeuf, R.A., Isfort, R.J. & Aardema, M.J. (1997) Induction of micronuclei in Syrian hamster embryo cells: Comparison to results in the SHE cell transformation assay for National Toxicology Program test chemicals. *Mutat. Res.*, **392**, 61–70

Goering, P.L. & Rehm, S. (1990) Inhibition of liver, kidney, and erythrocyte δ-aminolevulinic acid dehydratase (porphobilinogen synthase) by gallium in the rat. *Environ. Res.*, **53**, 135–151

Goering, P.L., Maronpot, R.R. & Fowler, B.A. (1988) Effect of intratracheal gallium arsenide administration on δ-aminolevulinic acid dehydratase in rats: relationship to urinary excretion of aminolevulinic acid. *Toxicol. appl. Pharmacol.*, **92**, 179–193

Greber, J.F. (2003) Gallium and gallium compounds. In: *Ullmann's Encyclopedia of Industrial Chemistry*, 6th rev. Ed., Vol. 15, Weinheim, Wiley-VCH Verlag GmbH & Co., pp. 235–240

Greenspan, B.J., Dill, J.A., Mast, T.J., Chou, B.J., Stoney, K.H., Morrissey, R. & Roycroft, J. (1991) Lung clearance of inhaled gallium arsenide. *Toxicologist*, **11**, 234

Gregus, Z., Gyurasics, A. & Csanaky, I. (2000) Biliary and urinary excretion of inorganic arsenic: Monomethylarsonous acid as a major biliary metabolite in rats. *Toxicol. Sci.*, **56**, 18–25

Greschonig, H., Schmid, M.G. & Gübitz, G. (1998) Capillary electrophoretic separation of inorganic and organic arsenic compounds. *Fresenius J. anal. Chem.*, **362**, 218–223

Ha, C.S., Choe, J.-G., Kong, J.S., Allen, P.K., Oh, Y.K., Cox, J.D. & Edmund, E.K. (2000) Agreement rates among single photon emission computed tomography using gallium-67, computed axial tomography and lymphangiography for Hodgkin disease and correlation of image findings with clinical outcome. *Cancer*, **89**, 1371–1379

Hakala, E. & Pyy, L. (1995) Assessment of exposure to inorganic arsenic by determining the arsenic species excreted in urine. *Toxicol. Lett.*, **77**, 249–258

Harrison, R.J. (1986) Gallium arsenide. In: LaDou, J., ed., *Occupational Medicine: The Microelectronics Industry*, Vol. 1, Philadelphia, PA, Hanley & Belfus, pp. 49–58

Hart, M.M., Smith, C.F., Yancey, S.T. & Adamson, R.H. (1971) Toxicity and antitumor activity of gallium nitrate and periodically related metal salts. *J. natl Cancer Inst.*, **47**, 1121–1127

Hartmann, C.B. & McCoy, K.L. (1996) Gallium arsenide augments antigen processing by peritoneal macrophages for CD4[+] helper T cell stimulation. *Toxicol. appl. Pharmacol.*, **141**, 365–372

Healy, S.M., Zakharyan, R.A. & Aposhian, H.V. (1997) Enzymatic methylation of arsenic compounds: IV. In vitro and in vivo deficiency of the methylation of arsenite and monomethylarsonic acid in the guinea pig. *Mutat. Res.*, **386**, 229–239

Healy, S.M., Casarez, E.A., Ayala-Fierro, F. & Aposhian, H.V. (1998) Enzymatic methylation of arsenic compounds. *Toxicol. appl. Pharmacol.*, **148**, 65–70

Heydorn, K. (1984) *Neutron Activation Analysis for Clinical Trace Element Research*, Vol. II, Boca Raton, FL, CRC Press, p. 72

Huang, R.-N. & Lee, T.-C. (1996) Cellular uptake of trivalent arsenite and pentavalent arsenate in KB cells cultured in phosphate-free medium. *Toxicol. appl. Pharmacol.*, **136**, 243–249

IARC (2004) *IARC Monographs on the Evaluation of Carcinogenic Risks to Humans*, Vol. 84, *Some Drinking-Water Disinfectants and Contaminants, including Arsenic*, Lyon, IARCPress

Ito, M. & Shooter, D. (2002) Detection and determination of volatile metal compounds in the atmosphere by a Mist-UV sampling system. *Atmos. Environ.*, **36**, 1499–1508

Iyengar, V. & Woittiez, J. (1988) Trace elements in human clinical specimens: Evaluation of literature data to identify reference values. *Clin. Chem.*, **34**, 474–481

Iyengar, G.V., Kollmer, W.E. & Bowen, H.J.M. (1978) *The Elemental Composition of Human Tissues and Body fluids, A Compilation of Values for Adults*, Weinheim, Verlag Chemie

Jakubowski, M., Trzcinka-Ochocka, M., Razniewska, G. & Matczak, W. (1998) Biological monitoring of occupational exposure to arsenic by determining urinary content of inorganic arsenic and its methylated metabolites. *Int. Arch. occup. environ. Health*, **71**, S29–S32

Jiang, X.P., Wang, F., Yang, D.C., Elliott, R.L. & Head, J.F. (2002) Induction of apoptosis by iron depletion in the human breast cancer MCF-7 cell line and the 13762NF rat mammary adenocarcinoma *in vivo*. *Anticancer Res.*, **22**, 2685–2692

Kerckaert, G.A., Brauninger, R., LeBoeuf, R.A. & Isfort, R.J. (1996) Use of the Syrian hamster embryo cell transformation assay for carcinogenicity prediction of chemicals currently being tested by the National Toxicology Program in rodent bioassays. *Environ. Health Perspect.*, **104** (Suppl. 5), 1075–1084

Kitsunai, M. & Yuki, T. (1994) Review: How gallium arsenide wafers are made. *Appl. organometal. Chem.*, **8**, 167–174

Krakoff, I.H., Newman, R.A. & Goldberg, R.S. (1979) Clinical toxicologic and pharmacologic studies of gallium nitrate. Cancer, 44, 1722–1727

Kramer, D.A. (1988) *Gallium and Gallium Arsenide: Supply, Technology, and Uses.* Information Circular 9208, Washington DC, US Department of the Interior, Bureau of Mines

Kramer, D.A. (1996–2004) *Mineral Commodity Summary: Gallium*, Reston, VA, US Geological Survey (http://minerals.usgs.gov/minerals/pubs/commodity/gallium/index.html; accessed 20.09.2003)

Kramer, D.A. (2002) *Minerals Yearbook: Gallium*, Reston, VA, US Geological Survey (http://minerals.usgs.gov/minerals/pubs/commodity/gallium/index.html; accessed 20.09.2003)

Kramer, D.A. (2003) *Mineral Commodity Summary: Gallium*, Reston, VA, US Geological Survey (http://minerals.usgs.gov/minerals/pubs/commodity/gallium/index.html; accessed 20.09.2003)

Kucera, J., Havránek, V., Smolík, J., Schwarz, J., Vesely, V., Kugler, J., Sykorová, I. & Šantroch, J. (1999) INAA and PIXE of atmospheric and combustion aerosols. *Biol. trace Elem. Res.*, **71–72**, 233–245

Lide, D.R., ed. (2003) *CRC Handbook of Chemistry and Physics*, 84th Ed., Boca Raton, FL, CRC Press LLC, p. 4-58

Ma, D., Okamoto, Y., Kumamaru, T. & Iwamoto, E. (1999) Determination of gallium by graphite furnace atomic absorption spectrometry with combined use of a tungsten-coated L'vov platform tube and a chemical modification technique. *Anal. Chim. Acta*, **390**, 201–206

Mast, T.J., Dill, J.A., Greenspan, B.J., Evanoff, J.J., Morrissey, R.E. & Schwetz, B.A. (1991) The development toxicity of inhaled gallium arsenide in rodents (Abstract). *Teratology*, **43**, 455–456

Nakamura, K., Fujimori, M., Tsuchiya, H. & Orii, H. (1982) Determination of gallium in biological materials by electrothermal atomic absorption spectrometry. *Anal. Chim. Acta*, **138**, 129–136

Narasimhan, J., Antholine, W.E. & Chitambar, C.R. (1992) Effect of gallium on the tyrosyl radical of the iron-dependent M2 subunit of ribonucleotide reductase. *Biochem. Pharmacol.*, **44**, 2403–2408

National Academy of Sciences (1977) *Medical and Biological Effects of Environmental Pollutants, Arsenic*, Washington, DC, National Research Council

National Institute for Occupational Safety and Health (1985) *Technical Report: Hazard Assessment of the Electronic Component Manufacturing Industry*, US Department of Health and Human Services, Cincinnati, OH

National Toxicology Program (2000) *Toxicology and Carcinogenesis Studies of Gallium Arsenide (CAS No. 1303-00-0) in F344/N Rats and B6C3F$_1$ Mice (Inhalation Studies)* (NTP Technical Report 492), Research Triangle Park, NC

Nishiyama, Y., Yamamoto, Y., Fukunaga, K., Satoh, K. & Ohkawa, M. (2002) Ga-67 scintigraphy in patients with breast lymphoma. *Clin. nucl. Med.*, **27**, 101–104

Ohyama, S. Ishinishi, S., Hisanaga, A. & Yamamoto, A. (1988) Comparative chronic toxicity, including tumorigenicity, of gallium arsenide and arsenic trioxide intratracheally instilled into hamsters. *Appl. organometall. Chem.*, **2**, 333–337

Omura, M., Hirata, M., Tanaka, A., Zhao, M., Makita, Y., Inoue, N., Gotoh, K. & Ishinishi, N. (1996a) Testicular toxicity evaluation of arsenic-containing binary compound semiconductors, gallium arsenide and indium arsenide, in hamsters. *Toxicol. Lett.*, **89**, 123–129

Omura, M., Tanaka, A., Hirata, M., Zhao, M., Makita, Y., Inoue, N., Gotoh, K. & Ishinishi, N. (1996b) Testicular toxicity of gallium arsenide, indium arsenide and arsenic oxide in rats by repetitive intratracheal instillation. *Fundam. appl. Toxicol.*, **32**, 72–78

Petrick, J.S., Ayala-Fierro, F., Cullen, W.R., Carter, D.E. & Aposhian, H.V. (2000) Monomethylarsonous acid (MMA[III]) is more toxic than arsenite in Chang human hepatocytes. *Toxicol. appl. Pharmacol.*, **163**, 203–207

Petrick, J.S., Jagadish, B., Mash, E.A. & Aposhian, H.V. (2001) Monomethylarsonous acid (MMA[III]) and arsenite: LD$_{50}$ in hamsters and in vitro inhibition of pyruvate dehydrogenase. *Chem. Res. Toxicol.*, **14**, 651–656

de Peyster, A. & Silvers, J.A. (1995) Arsenic levels in hair of workers in a semiconductor fabrication facility. *Am. ind. Hyg. Assoc. J.*, **56**, 377–383

Pierson, B., Van Wagenen, S., Nebesny, K.W., Fernando, Q., Scott, N. & Carter, D.E. (1989) Dissolution of crystalline gallium arsenide in aqueous solutions containing complexing agents. *Am. ind. Hyg. Assoc. J.*, **50**, 455–459

Radabaugh, T.R., Sampayo-Reyes, A., Zakharyan, R.A. & Aposhian, H.V. (2002) Arsenate reductase II. Purine nucleoside phosphorylase in the presence of dihydrolipoic acid is a route of reduction of arsenate to arsenite in mammalian systems. *Chem. Res. Toxicol.*, **15**, 692–698

Recapture Metals (2003) *High Purity Gallium: Product Specifications*, Blanding, UT (http://www.recapturemetals.com; accessed 19.09.2003)

Requena, E., Laserna, J.J., Navas, A. & Sánchez, F.G (1983) Pyridine-2-aldehyde 2-furoylhydrazone as a fluorogenic reagent for the determination of nanogram amounts of gallium. *Analyst*, **108**, 933–938

Rosner, M.H. & Carter, D.E. (1987) Metabolism and excretion of gallium arsenide and arsenic oxides by hamsters following intratracheal instillation. *Fundam. appl. Toxicol.*, **9**, 730–737

Sabot, J.L. & Lauvray, H. (1994) Gallium and gallium compounds. In: Kroschwitz, J.I. & Howe-Grant, M., eds, *Kirk-Othmer Encyclopedia of Chemical Technology*, 4th Ed., Vol. 12, New York, John Wiley & Sons, pp. 299–317

Salgado, M., Bosch Ojeda, C., García de Torres, A. & Cano Pavón, J.M. (1988) Di-2-pyridyl ketone 2-furoylhydrazone as a reagent for the fluorimetric determination of low concentrations of gallium and its application to biological samples. *Analyst*, **113**, 1283–1285

Sampayo-Reyes, A., Zakharyan, R.A., Healy, S.M. & Aposhian, H.V. (2000) Monomethylarsonic acid reductase and monomethylarsonous acid in hamster tissue. *Chem. Res. Toxicol.*, **13**, 1181–1186

Sanchez Rojas, F. & Cano Pavón, J.M. (1995) Spectrofluorimetric determination of gallium with N-(3-hydroxy-2-pyridyl)salicylaldimine and its application to the analysis of biological samples. *Anales de Quimica*, **91**, 537–539

Sasaki, T., Kojima, S. & Kubodera, A. (1982) Changes of ^{67}Ga-citrate accumulation in the rat liver during feeding with chemical carcinogen 3'-methyl-4-dimethylaminoazobenzene. *Kaku Igaku*, **19**, 201–208 (in Japanese)

Scansetti, G. (1992) Exposure to metals that have recently come into use. *Sci. total Environ.*, **120**, 85–91

Sheehy, J.W. & Jones, J.H. (1993) Assessment of arsenic exposures and controls in gallium arsenide production. *Am. ind. Hyg. Assoc. J.*, **54**, 61–69

Shiobara, Y., Ogra, Y. & Suzuki, K.T. (2001) Animal species difference in the uptake of dimethyl-arsinous acid (DMA^{III}) by red blood cells. *Chem. Res. Toxicol.*, **14**, 1446–1452

Sikorski, E.E., McCay, J.A., White, K.L., Jr, Bradley, S.G. & Munson, A.E. (1989) Immunotoxicity of the semiconductor gallium arsenide in female B6C3F1 mice. *Fundam. appl. Toxicol.*, **13**, 843–858

Styblo, M., Del Razo, L.M., LeCluyse, E.L., Hamilton, G.A., Wang, C., Cullen, W.R. & Thomas, D.J. (1999) Metabolism of arsenic in primary cultures of human and rat hepatocytes. *Chem. Res. Toxicol.*, **12**, 560–565

Styblo, M., Del Razo, L.M., Vega, L., Germolec, D.R., LeCluyse, E.L., Hamilton, G.A., Reed, W., Wang, C., Cullen, W.R. & Thomas, D.J. (2000) Comparative toxicity of trivalent and penta-valent inorganic and methylated arsenicals in rat and human cells. *Arch. Toxicol.*, **74**, 289–299

Ureña, E., Garcia de Torres, A., Cano Pavón, J.M. & Gómez Ariza, J.L. (1985) Determination of traces of gallium in biological materials by fluorometry. *Anal. Chem.*, **57**, 2309–2311

Ureña Pozo, M.E., Garcia de Torres, A. & Cano Pavón, J.M. (1987) Simultaneous determination of gallium and zinc in biological samples, wine, drinking water, and wastewater by derivative synchronous fluorescence spectrometry. *Anal. Chem.*, **59**, 1129–1133

Vahter, M. (1999) Methylation of inorganic arsenic in different mammalian species and population groups. *Science Progress*, **82**, 69–88

Vahter, M. (2002) Mechanisms of arsenic biotransformation. *Toxicology*, **181–182**, 211–217

Versieck, J. & Cornelis, R. (1980) Normal levels of trace elements in human blood plasma or serum. *Anal. chim. Acta*, **116**, 217–254

Wafer Technology Ltd (1997) *MSDS for Gallium Arsenide*, Milton Keynes [http://www.wafertech.co.uk/msds/msds_gaas.html]

Webb, D.R., Sipes, I.G. & Carter, D.E. (1984) *In vitro* solubility and *in vivo* toxicity of gallium arsenide. *Toxicol. appl. Pharmacol.*, **76**, 96–104

Webb, D.R., Wilson, S.E. & Carter, D.E. (1986) Comparative pulmonary toxicity of gallium arsenide, gallium(III) oxide, or arsenic(III) oxide intratracheally instilled into rats. *Toxicol. appl. Pharmacol.*, **82**, 405–416

Webb, D.R., Wilson, S.E. & Carter, D.E. (1987) Pulmonary clearance and toxicity of respirable gallium arsenide particulates intratracheally instilled into rats. *Am. ind. Hyg. Assoc. J.*, **48**, 660–667

Welch, B.M., Eden, R.C. & Lee, F.S. (1985) GaAs digital integrated circuit technology. In: Howes, M.J. & Morgan, D.V., eds, *Gallium Arsenide*, New York, John Wiley & Sons, pp. 517–573

Wildfang, E., Zakharyan, R.A. & Aposhian, H.V. (1998) Enzymatic methylation of arsenic compounds. V. Characterization of hamster liver arsenite and methylarsonic acid methyltransferase activities *in vitro. Toxicol. appl. Pharmacol.*, **152**, 366–375

Wildfang, E., Radabaugh, T.R. & Aposhian, H.V. (2001) Enzymatic methylation of arsenic compounds. IX. Liver arsenite methyltransferase and arsenate reductase activities in primates. *Toxicology*, **168**, 213–221

Winchell, H.S., Sanchez, P.D., Watanabe, C.K., Hollander, L., Anger, H.O. & McRae, J. (1970) Visualization of tumors in humans using ^{67}Ga-citrate and the Anger whole-body scanner, scintillation camera and tomographic scanner. *J. nucl. Med.*, **11**, 459–466

Wolff, R.K., Kanapilly, G.M., Gray, R.H. & McClellan, R.O. (1984) Deposition and retention of inhaled aggregate ^{67}Ga$_2$O$_3$ particles in beagle dogs, Fischer-344 rats, and CD-1 mice. *Am. ind. Hyg. Assoc. J.*, **45**, 377–381

Wolff, R.K., Griffith, W.C., Jr, Cuddihy, R.G., Snipes, M.B., Henderson, R.F., Mauderly, J.L. & McClellan, R.O. (1989) Modeling accumulations of particles in lung during chronic inhalation exposures that lead to impaired clearance. *Health Phys.*, **57**, 61–68

Woods, J.S. & Fowler, B.A. (1978) Altered regulation of mammalian hepatic heme biosynthesis and urinary porphyrin excretion during prolonged exposure to sodium arsenate. *Toxicol. appl. Pharmacol.*, **43**, 361–371

Yalcin, S. & Le, X.C. (2001) Speciation of arsenic using solid phase extraction cartridges. *J. environ. Monitoring*, **3**, 81–85

Yamauchi, H., Takahashi, K. & Yamamura, Y. (1986) Metabolism and excretion of orally and intraperitoneally administered gallium arsenide in the hamster. *Toxicology*, **40**, 237–246

Yamauchi, H., Takahashi, K., Mashiko, M. & Yamamura, Y. (1989) Biological monitoring of arsenic exposure of gallium arsenide- and inorganic arsenic-exposed workers by determination of inorganic arsenic and its metabolites in urine and hair. *Am. ind. Hyg. Assoc. J.*, **50**, 606–612

Yu, C., Cai, Q., Guo, Z.-X., Yang, Z. & Khoo, S.B. (2003) Inductively coupled plasma mass spectrometry study of the retention behavior of arsenic species on various solid phase extraction cartridges and its application in arsenic speciation. *Spectrochimica Acta, Part B*, **58**, 1335–1349

Zakharyan, R., Wu, Y., Bogdan, G.M. & Aposhian, H.V. (1995) Enzymatic methylation of arsenic compounds: Assay, partial purification, and properties of arsenite methyltransferase and monomethylarsonic acid methyltransferase of rabbit liver. *Chem. Res. Toxicol.*, **8**, 1029–1038

Zeiger, E., Anderson, B., Haworth, S., Lawlor, T. & Mortelmans, K. (1992) Salmonella mutagenicity tests: V. Results from the testing of 311 chemicals. *Environ. mol. Mutagen.*, **19** (Suppl. 21), 2–141

INDIUM PHOSPHIDE

1. Exposure Data

1.1   Chemical and physical data

1.1.1   *Nomenclature*

Chem. Abstr. Serv. Reg. No.: 22398-80-7
Deleted CAS Reg. No.: 1312-40-9, 99658-38-5, 312691-22-8
Chem. Abstr. Serv. Name: Indium phosphide (InP)
IUPAC Systematic Name: Indium phosphide
Synonyms: Indium monophosphide

1.1.2   *Molecular formula and relative molecular mass*

InP   Relative molecular mass: 145.79

1.1.3   *Chemical and physical properties of the pure substance*

(*a*)   *Description*: Black cubic crystals (Lide, 2003)
(*b*)   *Melting-point*: 1062 °C (Lide, 2003)
(*c*)   *Density*: 4.81 g/cm^3 (Lide, 2003)
(*d*)   *Solubility*: Slightly soluble in acids (Lide, 2003)
(*e*)   *Reactivity*: Can react with moisture or acids to liberate phosphine (PH_3); when heated to decomposition, it may emit toxic fumes of PO_x (ESPI, 1994)

1.1.4   *Technical products and impurities*

No data were available to the Working Group.

1.1.5   *Analysis*

Occupational exposure to indium phosphide can be determined by measurement of the indium concentration in workplace air or by biological monitoring of indium. No analytical methods are available for determination of indium phosphide *per se*. Determination of phosphorus cannot provide the required information on occupational exposure.

(*a*)   *Workplace air monitoring*

The respirable fraction of airborne indium, collected by drawing air through a membrane filter in a stationary or personal sampler, can be determined by nondestructive, INAA. This technique has been applied to the determination of indium concentration in ambient air particulates (Kucera *et al.*, 1999). Using irradiation with epithermal neutrons, indium concentrations have also been determined in arctic aerosols (Landsberger *et al.*, 1992).

(*b*)   *Biological monitoring*

Analytical methods capable of determining low concentrations of indium in biological matrices are largely lacking. The sensitivity of those methods commonly used for indium determination in geological and environmental samples, such as hydride generation atomic absorption spectrometry (Busheina & Headridge, 1982; Liao & Li, 1993), GF-AAS with preconcentration (Minamisawa *et al.*, 2003), electrothermal atomization laser-excited AFS (Aucélio *et al.*, 1998) and fluorimetric determination with HPLC (Uehara *et al.*, 1997), is usually not sufficient for measuring indium in biological materials. Recently, indium concentrations in the body fluids of workers exposed to partially respirable particles containing unspecified indium compounds were evaluated by graphite-furnace atomic absorption spectrophotometry (Miyaki *et al.*, 2003). The detection limits of indium in blood, serum and urine were found to be 0.7 µg/L, 0.4 µg/L and 0.4 µg/L, respectively. It was possible to determine indium concentrations in blood, serum and urine of the exposed workers, but those in control subjects were below the limits of detection.

Data on indium concentrations in the body fluids of occupationally non-exposed persons are insufficient to allow a reliable estimate of reference values.

1.2   Production and use

1.2.1   *Production*

Indium is recovered from fumes, dusts, slags, residues and alloys from zinc and lead–zinc smelting. The source material itself, a reduction bullion, flue dust or electrolytic slime intermediate, is leached with sulfuric or hydrochloric acid, the solutions are concentrated if necessary, and crude indium is recovered as ≥ 99% metal. This impure indium is then refined to 99.99%, 99.999%, 99.9999% or to higher grades by a variety of classical chemical and electrochemical processes (Slattery, 1995; Felix, 2003).

Indium combines with several non-metallic elements, including phosphorus, to form semiconducting compounds. Indium phosphide is prepared by direct combination of the highly-purified elements at elevated temperature and pressure under controlled conditions.

Indium phosphide is also obtained by thermal decomposition of a mixture of a trialkyl indium compound and phosphine (PH_3) (Slattery, 1995; Felix, 2003).

Single crystals of indium phosphide for the manufacture of semiconductor wafers are prepared by the liquid encapsulated Czochralski method. In this method, a single crystal is pulled through a boric oxide liquid encapsulant starting from a single crystal seed and a melt of polycrystalline indium phosphide. For specifications that require doping, the dopant (Fe, S, Sn or Zn), is added to the melt before extrusion of the single crystal. High pressure is applied inside the chamber to prevent decomposition of the indium phosphide. The single crystal is shaped into a cylinder of the appropriate diameter by grinding. The crystal is then sliced into wafers (InPACT, 2003).

World production of indium was constant at approximately 200 tonnes/year between 1995 and 1999, and rapidly increased to over 300 tonnes in 2000. Major producers of indium in 2002 and production levels (tonnes) were: China (85), France (65), Japan (60), Canada (45), Belgium (40), the Russian Federation (15), Peru (5) and other countries (20) (Jorgenson, 1997–2003, 2002; McCutcheon, 2001).

Available information indicates that indium phosphide is produced by two companies in Taiwan (China) and one company each in Japan and the USA (Chemical Information Services, 2003).

1.2.2   *Use*

Indium phosphide is a semiconductor and is probably the best understood semiconductor after silicon and gallium arsenide. Indium phosphide is used primarily for the fabrication of optoelectronic devices, because it is operating at high efficiency and high power. It is also used in the fabrication of laser diodes, LEDs, heterojunction bipolar transistors for optoelectronic integration, and in solar cells. Indium phosphide is also used in certain niche areas such as high-performance ICs. The use of indium phosphide in field effect transistor ICs is being driven by two application areas: microelectronics, where indium-aluminium arsenide/indium-gallium arsenide/indium phosphide-based high-electron mobility transistors are used in millimetre-wave frequencies; and optoelectronics, where indium phosphide-based field effect transistors are incorporated into long-wavelength optoelectronic components such as lasers and photodetectors in optoelectronic ICs (Materials Database, 2003; Szweda, 2003).

One of the key advantages of indium phosphide is its potential for the fabrication of very small devices. Because indium phosphide and its ternary (InGaAs) and quaternary (InGaAsP) derivatives have relatively higher refractive indices than those of other optical materials, these compounds allow for devices with much sharper and smaller bends. As their energy band gap is also closer to light energy, electro-optical effects are stronger than those in other materials (which again translates into shorter distances and lower drive voltages). As a result, extremely small devices can be produced: dice are typically < 5 mm and for many functions (e.g. lasers, modulators) they are 1 mm or less (Reade Advanced Materials, 1997; CyOptics, 2002).

Japan accounts for about 56% of the consumption of available indium phosphide wafers, and Europe and the USA for 22% each (Bliss, 2001). In 1998, it was estimated that the use of indium for semiconductor applications worldwide was 19 tonnes (McCutcheon, 2001).

1.3   Occurrence and exposure

1.3.1   *Natural occurrence*

Indium phosphide does not occur naturally. Indium is present in the earth's crust at concentrations of 50–200 µg/kg and is recovered primarily as a by-product of zinc smelting. Indium is also found in trace amounts in association with sulfide ores of iron, tin, lead, cobalt, bismuth and copper (Beliles, 1994; Slattery, 1995; Blazka, 1998; Slattery, 1999).

1.3.2   *Occupational exposure*

Exposure to indium phosphide occurs predominantly in the microelectronics industry where workers are involved in the production of indium phosphide crystals, ingots and wafers, in grinding and sawing operations, in device fabrication and in clean-up activities. NIOSH estimated that in 1981 approximately 180 000 workers were employed in the micro-electronics industry in the USA, with over 500 plants manufacturing semiconductors (NIOSH, 1985). No assessment of occupational exposure is available specifically for indium phosphide.

In a study of workplace exposure to unspecified indium compounds at a factory in Japan (Miyaki *et al.*, 2003), concentrations of indium in blood and urine were determined for workers exposed ($n = 107$) and those not exposed ($n = 24$) to water-insoluble partially-respirable indium-containing particles in workplace air. Concentrations reported (geometric mean ± GSD) in blood were 4.09 ± 7.15 µg/L and 0.45 ± 1.73 µg/L in exposed and non-exposed workers, respectively, and in urine were 0.93 ± 4.26 µg/L and < 0.4 µg/L for exposed and non-exposed workers, respectively.

1.3.3   *Environmental exposure*

There are no data available on environmental exposure to indium phosphide.

Indium has been detected in air (43 ng/m^3), seawater (20 µg/L) and rainwater (0.59 µg/L). Indium concentrations of up to 10 µg/kg have been detected in beef and pork, and up to 15 mg/kg in algae, fish and shellfish from contaminated water near smelters. The average daily human intake of indium is estimated to be 8–10 µg/day and is regarded as minimal (Fowler, 1986; Scansetti, 1992; Blazka, 1998).

1.4   Regulations and guidelines

Occupational exposure limits have not been established specifically for indium phosphide. Table 1 presents occupational exposure limits and guidelines from several countries for indium and indium compounds in workplace air.

Table 1. Occupational exposure limits and guidelines for indium and indium compounds

Country or region	Concentration (mg/m^3) (as indium)	Interpretation[a]
Australia	0.1	TWA
Belgium	0.1	TWA
Canada		
Alberta	0.1	TWA
	0.3	STEL
Quebec	0.1	TWA
China	0.1	TWA
	0.3	STEL
Finland	0.1	TWA
Ireland	0.1	TWA
	0.3	STEL
Malaysia	0.1	TWA
Mexico	0.1	TWA
	0.3	STEL
Netherlands	0.1	TWA
New Zealand	0.1	TWA
Norway	0.1	TWA
South Africa	0.1	TWA
	0.3	STEL
Spain	0.1	TWA
Sweden	0.1	TWA
Switzerland	0.1	TWA
UK	0.1	TWA (MEL)
	0.3	STEL
USA[b]		
ACGIH	0.1	TWA (TLV)
NIOSH	0.1	TWA (REL)

From Työsuojelusäädöksiä (2002); ACGIH Worldwide® (2003); Suva (2003)
[a] TWA, time-weighted average; STEL, short-term exposure limit; MEL, maximum exposure limit; TLV, threshold limit value; REL, recommended exposure limit
[b] ACGIH, American Conference of Governmental Industrial Hygienists; NIOSH, National Institute for Occupational Safety and Health

2.   Studies of Cancer in Humans

See Introduction to the Monographs on Gallium Arsenide and Indium Phosphide.

3.   Studies of Cancer in Experimental Animals

3.1   Inhalation exposure

3.1.1   *Mouse*

In a study undertaken by the National Toxicology Program (2001), groups of 60 male and 60 female B6C3F$_1$ mice, 6 weeks of age, were exposed to particulate aerosols of indium phosphide (purity, > 99%; MMAD, 1.2 μm; GSD, 1.7–1.8 μm) at concentrations of 0, 0.03, 0.1 or 0.3 mg/m^3 for 6 h per day on 5 days per week for 22 weeks (0.1 and 0.3 mg/m^3) or 105 weeks (0 and 0.03 mg/m^3). An interim sacrifice of 10 males and 10 females per group after 3 months showed increased lung weights and lung lesions in animals exposed to 0.1 or 0.3 mg/m^3. The changes were considered sufficiently severe that exposure was discontinued in these groups and the animals were maintained on filtered air from the termination of exposure at week 22 until the end of the study. Survival rates were decreased in exposed males and females compared with chamber controls (survival rates: 37/50 (control), 24/50 (low dose), 29/50 (mid dose) or 27/50 (high dose) in males and 42/50, 13/50, 33/50 or 21/50 in females, respectively; mean survival times: 711, 660, 685 or 679 days in males and 713, 655, 712 or 654 days in females, respectively). Mean body weights were decreased in males exposed to 0.03 and 0.3 mg/m^3 and in all exposed females compared with chamber controls. Incidences of neoplasms and non-neoplastic lesions are reported in Tables 2 and 3.

There was an increased incidence of lung neoplasia in male and female mice exposed to indium phosphide. Alveolar/bronchiolar adenomas and many of the alveolar/bronchiolar carcinomas resembled those which arise spontaneously. However, exposure to indium phosphide did not cause increased incidences of neoplasms in other tissues. The lung carcinomas were distinguished from adenomas by local invasion, metastasis and/or greater anaplasia and/or pleomorphism of component cells. Some of the carcinomas differed somewhat from spontaneous carcinomas. Carcinomas in mice exposed to indium phosphide were very anaplastic with papillary and sclerosing patterns; several appeared to have spread outside the lungs into the mediastinum and some to distant metastases. A few appeared to have extensive intrapulmonary spread which in several instances was diagnosed as multiple carcinoma. Alveolar epithelial hyperplasia in the lung is generally considered to be a precursor to neoplasia in the mouse but was not significantly increased in male or female mice exposed to indium phosphide. There were increased incidences of chronic active inflammation, alveolar proteinosis and foreign bodies (indium phosphide

Table 2. Incidence of neoplasms and non-neoplastic lesions of the lung and associated lymph nodes in mice in a 2-year inhalation study of indium phosphide

Lesions observed	No. of mice exposed to indium phosphide at concentrations (mg/m^3) of			
	0 (chamber control)	0.03	0.1[a]	0.3[a]
Males				
Lung				
Total no. examined	50	50	50	50
No. with:				
Alveolar epithelium, hyperplasia	2 (1.5)[b]	5 (2.4)	3 (2.7)	7 (2.1)
Chronic active inflammation	2 (1.0)	50[c] (2.9)	45[c] (1.6)	46[c] (2.1)
Alveolus, proteinosis	0	14[c] (1.0)	0	10[c] (1.0)
Foreign body (indium phosphide particles)	0	49[c] (1.0)	42[c]	49[c]
Serosa, fibrosis	0	50[c] (3.5)	49[c] (2.0)	50[c] (2.4)
Alveolar/bronchiolar adenoma, multiple	1	2	0	3
Alveolar/bronchiolar adenoma (includes multiple)	13	9	7	13
Alveolar/bronchiolar carcinoma, multiple	1	8[d]	3	14
Alveolar/bronchiolar carcinoma (includes multiple)	6	15[c]	22[c]	13[d]
Alveolar/bronchiolar adenoma or carcinoma (includes multiple)	18	23	24	21
Pleural mesothelium, hyperplasia	0	19[c] (2.1)	4 (2.0)	6[d] (1.5)
Lymph node, bronchial				
Total no. examined	35	48	45	48
No. with:				
Hyperplasia	2 (2.5)	36[c] (2.3)	22[c] (2.0)	22[c] (2.0)
Foreign body (indium phosphide particles)	0	43[c] (1.0)	40[c] (1.0)	40[c] (1.0)
Lymph node, mediastinal				
Total no. examined	40	49	45	48
No. with:				
Hyperplasia	0	34[c] (2.5)	17[c] (2.1)	27[c] (2.2)
Foreign body (indium phosphide particles)	0	24[c] (1.0)	14[c] (1.0)	25[c] (1.0)

Table 2 (contd)

Lesions observed	No. of mice exposed to indium phosphide at concentrations (mg/m^3) of			
	0 (chamber control)	0.03	0.1^a	0.3^a
Females				
Lung				
Total no. examined	50	50	50	50
No. with:				
Alveolar epithelium, hyperplasia	0	1 (2.0)	1 (3.0)	2 (2.0)
Chronic active inflammation	2 (2.5)	49^c (2.9)	45^c (1.7)	50^c (2.1)
Alveolus, proteinosis	0	31^c (1.1)	0	8^c (1.4)
Foreign body	0	49^c (1.0)	36^c	49^c
Serosa, fibrosis	0	50^c (3.8)	47^c (1.8)	49^c (2.5)
Alveolar/bronchiolar adenoma, multiple	0	0	1	2
Alveolar/bronchiolar adenoma (includes multiple)	3	6	10^d	7
Alveolar/bronchiolar carcinoma, multiple	0	1	0	0
Alveolar/bronchiolar carcinoma (includes multiple)	1	6	5	7
Alveolar/bronchiolar adenoma or carcinoma (includes multiple)	4	11^d	15^d	14^c
Pleural mesothelium, hyperplasia	0	16^c (1.8)	3 (1.7)	13^c (1.9)
Lymph node, bronchial				
Total no. examined	36	50	48	50
No. with:				
Hyperplasia	5 (1.8)	42^c (2.8)	31^c (2.2)	28^c (2.2)
Foreign body (indium phosphide particles)	0	44^c (1.0)	33^c (1.0)	40^c (1.0)
Lymph node, mediastinal				
Total no. examined	42	48	46	49
No. with:				
Hyperplasia	2 (2.0)	40^c (3.0)	11^c (2.2)	29^c (2.6)
Foreign body	0	20^c (1.0)	7^c (1.0)	16^c (1.0)

From National Toxicology Program (2001)
[a] Exposure stopped after 22 weeks.
[b] Average severity grade of lesions in affected animals: 1, minimal; 2, mild; 3, moderate; 4, marked
[c] Significantly different ($p \leq 0.01$) from the chamber control group by the Poly-3 test
[d] Significantly different ($p \leq 0.05$) from the chamber control group by the Poly-3 test

Table 3. Incidence of neoplasms and non-neoplastic lesions of the liver in mice in a 2-year inhalation study of indium phosphide

Lesions observed	No. of mice exposed to indium phosphide at concentrations (mg/m^3) of			
	0 (chamber control)	0.03	0.1[a]	0.3[a]
Males				
Liver				
No. examined microscopically	50	50	50	50
Eosinophilic focus	10	16[b]	19[b]	18[b]
Hepatocellular adenoma, multiple	8	13	10	14
Hepatocellular adenoma (includes multiple)	17	24	23	32
Hepatocellular carcinoma, multiple	1	7[b]	10[c]	5
Hepatocellular carcinoma (includes multiple)	11	22[b]	23[b]	16
Hepatoblastoma				
Hepatocellular adenoma, hepatocellular	0	1	0	0
carcinomas, or hepatoblastoma (includes multiple)	26	40	37	39
Females				
Liver				
No. examined microscopically	50	50	50	50
Eosinophilic focus	6	9	4	12[b]
Hepatocellular adenoma, multiple	12	14	18	14
Hepatocellular adenoma (includes multiple)	2	4	1	2
Hepatocellular carcinoma, multiple	6	17[c]	8	10
Hepatoblastoma	0	0	0	1
Hepatocellular adenoma, hepatocellular carcinomas, or hepatoblastoma (includes multiple)	18	28[c]	24	23

From National Toxicology Program (2001)

[a] Exposure stopped after 22 weeks.

[b] Significantly different ($p \leq 0.05$) from the chamber control group by the Poly-3 test

[c] Significantly different ($p \leq 0.01$) from the chamber control group by the Poly-3 test

particles) in the lungs of exposed mice. A prominent feature of the inflammatory process was the presence of pleural fibrosis (serosal fibrosis). Usually, these fibrotic areas were associated with areas of inflammation. Pulmonary interstitial fibrosis was an uncommon finding in control animals. The incidence of visceral pleural mesothelial hyperplasia was increased in males and females exposed to 0.03 and 0.3 mg/m³ indium phosphide. Usually in association with chronic inflammation and fibrosis, the pleural mesothelium from many animals was hypertrophic and/or hyperplastic. Normal visceral mesothelium is a single layer of flattened epithelium, whereas affected mesothelium ranged from a single layer of plump (hypertrophic) cells to several layers of rounded cells (hyperplasia). In the more severe cases, the proliferations formed papillary fronds that projected into the pleural cavity.

There were increased incidences of hepatocellular adenoma and carcinoma in males and females. The incidence of multiple hepatocellular tumours per animal was increased in exposed groups. The incidence of eosinophilic foci was increased in all groups of exposed males and in females exposed to 0.3 mg/m³. Foci of hepatocellular alteration, hepatocellular adenoma, and hepatocellular carcinoma are thought to represent a spectrum that constitutes the progression of proliferative liver lesions. The increased incidence of liver lesions observed in this study was considered to be related to exposure to indium phosphide. Although there was an increased incidence of rare neoplasms of the small intestine in male mice, this was not statistically significant and it was uncertain whether these neoplasms were a result of exposure to indium phosphide (National Toxicology Program, 2001).

3.1.2   Rat

In a study undertaken by the National Toxicology Program (2001), groups of 60 male and 60 female Fischer 344/N rats, 6 weeks of age, were exposed to particulate aerosols of indium phosphide (purity, > 99%; MMAD, 1.2 μm; GSD, 1.7–1.8 μm) at concentrations of 0, 0.03, 0.1, or 0.3 mg/m³ for 6 h per day on 5 days per week for 22 weeks (0.1 and 0.3 mg/m³ groups) or 105 weeks (0 and 0.03 mg/m³ groups). An interim sacrifice of 10 males and 10 females per group after 3 months showed increased lung weights, microcytic erythrocytosis, and lesions in the respiratory tract and lung-associated lymph nodes in animals exposed to 0.1 or 0.3 mg/m³. These changes were considered sufficiently severe to justify discontinuing exposure after 22 weeks and these animals were maintained on filtered air from termination of exposure at week 22 until the end of the study. No adverse effects on survival were observed in treated males or females compared with chamber controls (survival rates: 27/50 (control), 29/50 (low dose), 29/50 (mid dose) or 26/50 (high dose) in males and 34/50, 31/50, 36/50 or 34/50 in females, respectively; mean survival times: 667, 695, 678 or 688 days in males and 682, 671, 697 or 686 days in females, respectively). No adverse effects on mean body weight were observed in treated males or females compared with chamber controls. Incidences of neoplasms and non-neoplastic lesions are reported in Tables 4 and 5.

Table 4. Incidence of neoplasms and non-neoplastic lesions of the lung in rats in 2-year inhalation study of indium phosphide

Lesions observed	No. of rats exposed to indium phosphide at concentrations (mg/m^3) of			
	0 (chamber control)	0.03	0.1[a]	0.3[a]
Males				
Lung				
Total no. examined	50	50	50	50
Atypical hyperplasia	0	16^c (3.1)[b]	23^c (3.3)	39^c (3.8)
Chronic active inflammation	5 (1.2)	50^c (3.8)	50^c (3.4)	50^c (4.0)
Alveolar epithelium, metaplasia	0	45^c (3.1)	45^c (2.8)	48^c (3.2)
Foreign body	0	50^c (2.2)	50^c (1.9)	50^c (2.1)
Alveolus, proteinosis	0	50^c (3.7)	48^c (2.0)	47^c (3.4)
Interstitium, fibrosis	0	49^c (3.7)	50^c (3.5)	50^c (3.9)
Alveolar epithelium, hyperplasia	11 (1.5)	20 (2.4)	21^d (2.1)	31^c (2.6)
Squamous metaplasia	0	1 (2.0)	3 (3.0	4 (2.5)
Squamous cyst	0	1 (4.0)	3 (3.0)	2 (3.0)
Alveolar/bronchiolar adenoma, multiple	1	5	8^d	12^c
Alveolar/bronchiolar adenoma (includes multiple)	6	13	27^c	30^c
Alveolar/bronchiolar carcinoma, multiple	0	2	1	5^d
Alveolar/bronchiolar carcinoma (includes multiple)	1	10^c	8^d	16^c
Alveolar/bronchiolar adenoma or carcinoma	7/50	22/50^c	30/50^c	35/50^c
Squamous cell carcinoma	0/50	0/50	0/50	4/50
Females				
Lung				
Total no. examined	50	50	50	50
Atypical hyperplasia	0	8^c (2.8)	8^c (2.9)	39^c (3.8)
Chronic active inflammation	10 (1.0)	49^c (3.0)	50^c (2.6)	49^c (3.9)
Alveolar epithelium, metaplasia	0	46^c (3.3)	47^c (2.4)	48^c (3.8)
Foreign body	0	49^c (2.1)	50^c (1.8)	50^c (2.0)
Alveolus, proteinosis	0	49^c (3.7)	47^c (2.0)	50^c (3.8)
Interstitium, fibrosis	0	48^c (2.9)	50^c (2.6)	49^c (3.9)
Alveolar epithelium, hyperplasia	8 (1.5)	15 (2.1)	22^c (2.0)	16^d (1.8)
Squamous metaplasia	0	2 (1.5)	1 (2.0)	4 (2.5)
Squamous cyst	0	1 (4.0)	1 (4.0)	10^c (3.6)
Alveolar/bronchiolar adenoma, multiple	0	1	1	1
Alveolar/bronchiolar adenoma (includes multiple)	0	7^c	5^d	19^c
Alveolar/bronchiolar carcinoma, multiple	0	1	0	7^c

Table 4 (contd)

Lesions observed	No. of rats exposed to indium phosphide at concentrations (mg/m³) of			
	0 (chamber control)	0.03	0.1[a]	0.3[a]
Alveolar/bronchiolar carcinoma (includes multiple)	1	3	1	11[c]
Alveolar/bronchiolar adenoma or carcinoma	1/50	10/50[c]	6/50	26/50[c]

From National Toxicology Program (2001)
[a] Exposure stopped after 22 weeks.
[b] Average severity grade of lesions in affected animals: 1, minimal; 2, mild; 3, moderate; 4, marked
[c] Significantly different ($p \leq 0.01$) from the chamber control group by the Poly-3 test
[d] Significantly different ($p \leq 0.05$) from the chamber control group by the Poly-3 test

There was an increased incidence of lung neoplasms in male and female rats exposed to indium phosphide but no increased incidence of neoplasms in other tissues was observed. Proliferative lesions of the lung included alveolar/bronchiolar neoplasms and squamous-cell carcinomas as well as alveolar epithelial hyperplasia and atypical hyperplasia of alveolar epithelium. Alveolar/bronchiolar adenomas, typical of those observed spontaneously in Fischer 344/N rats, were generally distinct masses that often compressed surrounding tissue. Alveolar/bronchiolar carcinomas had similar cellular patterns but were generally larger and had one or more of the following histological features: heterogenous growth pattern, cellular pleomorphism and/or atypia, and local invasion or metastasis. A number of exposed males and females had multiple alveolar/bronchiolar neoplasms. It was not usually possible to determine microscopically if these represented intrapulmonary metastases of a malignant neoplasm or were multiple independent neoplasms. Included in the spectrum of lesions was a proliferation of alveolar/bronchiolar epithelium with a very prominent fibrous component not typically seen in alveolar/bronchiolar tumours of rodents. The smallest lesions were usually observed adjacent to areas of chronic inflammation. Small lesions with modest amounts of peripheral epithelial proliferation were diagnosed as atypical hyperplasia, while larger lesions with florid epithelial proliferation, marked cellular pleomorphism, and/or local invasion were diagnosed as alveolar/bronchiolar adenoma or carcinoma. While squamous epithelium is not normally observed within the lung, squamous metaplasia of alveolar/bronchiolar epithelium is a relatively common response to pulmonary injury and occurred in a few rats in each exposed group. Squamous metaplasia consisted of a small cluster of alveoli in which the normal epithelium was replaced by multiple layers of flattened squamous epithelial cells that occasionally formed keratin. Cystic squamous lesions also occurred and were rimmed by a band (varying in thickness from a few to many cell layers) of viable squamous epithelium with a large central core of keratin. Squamous-cell carcinomas were observed in four males exposed to 0.3 mg/m³ indium phosphide. These

Table 5. Incidence of neoplasms and non-neoplastic lesions of the adrenal medulla in rats in a 2-year inhalation study of indium phosphide

Lesions observed	No. of rats exposed to indium phosphide at concentrations (mg/m^3) of			
	0 (chamber control)	0.03	0.1[a]	0.3[a]
Males				
Adrenal medulla				
Number examined microscopically	50	50	49	50
Hyperplasia	26 (2.2)[b]	26 (2.4)	24 (2.4)	32 (2.3)
Benign pheochromocytoma, bilateral	0	6[c]	4	5[c]
Benign pheochromocytoma (includes bilateral)	10	22	16	23
Complex pheochromocytoma	0	1	0	0
Malignant pheochromocytoma	0	3	3	1
Benign, complex or malignant pheochromo-cytoma	10	26[d]	18[c]	24[d]
Females				
Adrenal medulla				
Number examined microscopically	50	48	50	40
Hyperplasia	6 (1.8)	13[c] (2.2)	9 (2.3)	15[c] (2.1)
Benign pheochromocytoma, bilateral	0	0	0	2
Benign pheochromocytoma (includes bilateral)	2	6	2	9
Malignant pheochromocytoma	0	0	0	1
Benign or malignant pheochromocytoma	2	6	2	9

From National Toxicology Program (2001)
[a] Exposure stopped after 22 weeks.
[b] Average severity grade of lesions in affected animals: 1, minimal; 2, mild; 3, moderate; 4, marked
[c] Significantly different ($p \leq 0.05$) from the chamber control group by the Poly-3 test
[d] Significantly different ($p \leq 0.01$) from the chamber control group by the Poly-3 test

neoplasms ranged from fairly well-differentiated squamous-cell carcinomas to poorly-differentiated and anaplastic ones.

There was an increased incidence of pheochromocytoma in male and female rats and an increased incidence of medullary hyperplasia in females. Focal hyperplasia and pheo-chromocytoma were considered to constitute a morphologic continuum in the adrenal medulla. There was also a marginal increase in neoplasms typical of those observed spontaneously in male and female Fischer 344/N rats. These included fibromas of the skin in males, mammary gland carcinomas in females, and mononuclear cell leukaemia in males and females. It was uncertain whether these neoplasms were a result of exposure to indium phosphide (National Toxicology Program, 2001).

3.1.3   *Comparison of findings from the rat and mouse inhalation studies*

The alveolar/bronchiolar adenomas found in rats exposed to indium phosphide (National Toxicology Program, 2001) closely resembled those found spontaneously in aged rats. Most alveolar/bronchiolar adenomas and carcinomas in mice exposed to indium phosphide also resembled those occurring spontaneously in B6C3F$_1$ mice (National Toxicology Program, 2001). However, some of the carcinomas were different from those occurring spontaneously in that they were very anaplastic with papillary and sclerosing patterns and often spread outside the lung into the mediastinum and distant metastases. A few appeared extensively throughout the lung and thus were diagnosed as multiple carcinomas. The neoplastic responses in the lungs of mice were even more significant than those in rats, because mice generally do not respond to particulate exposure by developing lung neoplasms, even at higher exposure concentrations.

In mice, exposure to indium phosphide also caused inflammatory and proliferative lesions of the mesothelium of the visceral and parietal pleura, another uncommon response to nonfibrous particulate exposure. Pleural fibrosis was a prominent component of the chronic inflammation and involved both visceral and parietal pleura with adhesions. Significantly, pulmonary interstitial fibrosis was uncommon in mice exposed to indium phosphide.

As a result of discontinuing exposure of the 0.1 and 0.3 mg/m^3 groups to indium phosphide at 21 or 22 weeks, only the groups receiving 0.03 mg/m^3 were exposed for 2 years. Therefore, typical concentration-related responses in neoplasms, based solely on external exposure concentration of particulate indium phosphide, were not expected. The amount of indium retained in the lung and that absorbed systemically must also be considered (see Table 6). The lung deposition and clearance model was used to estimate the total amount of indium deposited in the lungs of mice and rats after termination of exposure, the lung burdens at the end of the 2-year study, and the area under the lung-burden curves (AUC). For both species, the estimates at the end of 2 years indicated that the lung burdens in the groups exposed continuously to 0.03 mg/m^3 were greater than those of the other exposed groups (0.1 or 0.3 mg/m^3), with the lung burdens of the groups exposed to 0.1 mg/m^3 being the lowest. Because of the slow clearance of indium, the lung burdens in the groups exposed to 0.1 and 0.3 mg/m^3 were approximately 25% of the maximum levels in rats and 8% in mice, 83 to 84 weeks after exposure was stopped. The AUCs and the total amount of indium deposited per lung indicated that the groups exposed to 0.3 mg/m^3 received a greater amount of indium phosphide than the other two groups with the group exposed to 0.1 mg/m^3 being the lowest. Regardless of how the total 'dose' of indium to the lung was estimated, the group exposed to 0.1 mg/m^3 had less total exposure than the other two groups, implying that this group may be considered the 'low dose' in these studies. Therefore, lung-burden data should be considered when evaluating lung neoplasia incidence.

Table 6. Estimates of exposure of rats and mice to indium phosphide for 2 years based on a lung deposition and clearance model

Parameters of exposure	Exposure group		
	Rat/mouse	Rat[a]/mouse[b]	Rat[a]/mouse[b]
	0.03 mg/m^3	0.1 mg/m^3	0.3 mg/m^3
Lung burden at 2 years (μg In/lung)	65.1/6.2	10.2/0.5	31.9/2.3
Total amount deposited per lung (μg In/lung)	72/15	57/11	150/37
First-year AUC (μg In/lung × days of study)	6368/1001	11 502/1764	31 239/6078
Second-year AUC (μg In/lung × days of study)	18 244/2032	6275/486	18 532/1986
Total AUC (μg In/lung × days of study)	24 612/3000	17 777/2200	49 771/8000

AUC, area under the lung burden curve
From National Toxicology Program (2001)
[a] Exposure was discontinued and animals were maintained on filtered air from exposure termination at week 22 until the end of the study.
[b] Exposure was discontinued and animals were maintained on filtered air from exposure termination at week 21 until the end of the study.

3.2   Intratracheal instillation

Hamster

Tanaka and colleagues (1996) studied indium phosphide in hamsters. Groups of 30 male Syrian golden hamsters, 8 weeks of age, received intratracheal instillations of 0 or 0.5 mg phosphorus/animal indium phosphide (purity, ≥ 99.99 %; particle mean count diameter, 3.9 μm [GSD, 2.88 μm]) in phosphate buffer solution once a week for 15 weeks and were observed during their total life span (approximately 105 weeks). Survival after 15 instillations was 29/30 controls and 26/30 treated hamsters. There was no exposure-related mortality (survival time, 433 ± 170 days in exposed hamsters versus 443 ± 169 days in controls) and all exposed animals had died by 689 days (controls, 737 days). Histopathological examination of 23 exposed hamsters showed proteinosis-like lesions in 19/23, alveolar or bronchiolar cell hyperplasia in 9/23, squamous-cell metaplasia in 1/23 and particle deposition in 23/23 animals. There was no treatment-related increase in neoplasms of the lungs or other organs (liver, forestomach, pancreas or lymph nodes). [The Working Group concluded that because of the small number of animals, and because of the extent and duration of exposure by intratracheal instillation, this study may not have provided for adequate assessment of carcinogenic activity.]

4.   Other Data Relevant to an Evaluation of Carcinogenicity and its Mechanisms

4.1   Deposition, retention, clearance and metabolism

The absorption and distribution of indium is highly dependent on its chemical form. Indium phosphide has low solubility in synthetic simulated body fluids (Gamble solution) (Kabe *et al.*, 1996).

4.1.1   *Humans*

A study (Miyaki *et al.*, 2003) of concentrations of indium in blood, serum and urine of workers exposed (*n* = 107) or not exposed (*n* = 24) to water-insoluble indium-containing particulates in workplace air is described in detail in Section 1.3.2. In each of the three biological fluids, concentrations of indium were clearly higher in exposed workers than in unexposed workers.

4.1.2   *Experimental systems*

(*a*)   *Indium phosphide*

(i)   *Inhalation studies in rats and mice*

The deposition and clearance of indium phosphide have been studied by the National Toxicology Program (2001). Groups of 15 male Fischer 344 rats designated for tissue burden analyses and five male rats designated for post-exposure tissue burden analyses were exposed to particulate aerosols of indium phosphide at concentrations of 0, 1, 3, 10, 30, or 100 mg/m^3 for 6 h (plus 12 min build-up time) per day on 5 days per week for 14 weeks. Indium continued to accumulate in lung tissue, blood, serum and testes throughout the exposure period. At day 5, the concentrations of indium ranged from 13 to 500 µg/g lung and concentrations of up to 1 mg/g lung were measured after exposure to 100 mg/m^3 indium phosphide for 14 weeks.

Lung clearance half-lives during exposure were in the order of 47–104 days. At 14 days after exposure, the half-life increased to about 200 days. Blood and serum indium concentrations in all exposed animals were found to be similar at the end of exposure and at 112 days after exposure. Concentrations of indium in testis tissue continued to increase more than twofold after exposure ended in rats exposed to 10- and 30-mg/m^3 concentrations of indium phosphide. Indium concentrations reached 7.20 ± 2.4 µg/g testis 14 days after the end of exposure to 100 mg/m^3.

In a further study (National Toxicology Program, 2001), groups of 60 male and 60 female rats and mice were exposed to particulate aerosols of indium phosphide at concentrations of 0, 0.03, 0.1, or 0.3 mg/m^3 (MMAD ~1.2 μm), for 6 h (plus 12 min build-up time) per day on 5 days per week for 22 weeks (rats) and 21 weeks (mice) (0.1 and 0.3 mg/m^3 groups) or 105 weeks (0 and 0.03 mg/m^3 groups, rats and mice). Animals in the 0.1- and 0.3-mg/m^3 groups were maintained on filtered air from exposure termination at week 22 until the end of the study. In rats, the lung indium burden at 5 months was proportional to exposure. At 12 months, 34.3 ± 1.87 μg indium per lung was measured in the male rats of the 0.03-mg/m^3 exposure group. The estimated lung clearance was long (half-life, 2422 days) and the mean indium concentration in serum at 12 months was high (3.4 ± 0.2 ng/g) in the 0.03-mg/m^3 exposure group. Results for B6C3F$_1$ mice exposed to 0.03, 0.1 or 0.3 mg/m^3 were similar although there were quantitative differences in lung burden and kinetic parameters. The mean indium concentration in the lungs at 12 months was 4.87 ± 0.65 μg per lung for male mice in the low-exposure group (0.03 mg/m^3). Lung clearance half-lives of 144 and 163 days were estimated for mice in the 0.1- and 0.3-mg/m^3 exposure groups, respectively, compared with 262 and 291 days for rats exposed to the same concentrations.

Exposure of male rats for 5 days per week for 2 years to 0.03 mg/m^3 indium phosphide resulted in a mean indium concentration of 7.65 ± 0.36 μg/g lung tissue at 5 months, i.e. a fourfold lower concentration compared with that found at 14 weeks exposure to 1 mg/m^3 indium phosphide. Lung clearance half-lives for indium phosphide in male rats in the 2-year studies were estimated to be 2422, 262 and 291 days for 0.03-, 0.1- and 0.3-mg/m^3 exposure concentrations of indium phosphide, respectively. In male B6C3F$_1$ mice exposed to 0.03 mg/m^3 for 2 years, the mean indium concentration in the lung at 5 months was 8.52 ± 1.44 ng/g lung. Indium phosphide lung clearance half-lives were 230, 144 and 163 days for male mice exposed to 0.03, 0.1 and 0.3 mg/m^3 indium phosphide, respectively (National Toxicology Program, 2001).

Deposition and clearance during long-term exposure of rats and mice to indium phosphide appeared to follow zero-order (constant rate) kinetics. The burden of indium retained in the lung throughout the experiments was proportional to exposure concentration and duration. The studies indicated that elimination of indium was quite slow. For both species, estimates at the end of 2 years indicated that the lung burdens in the groups continuously exposed to 0.03 mg/m^3 were greater than those in the groups exposed to 0.1 or 0.3 mg/m^3 where exposure was terminated at 22 weeks. Because of the slow clearance of indium, the lung burdens in the groups exposed to 0.1 and 0.3 mg/m^3, 83 weeks after exposure was stopped, were approximately 35–50% and 16–28% of the maximum concentrations in rats and mice, respectively. These findings were also compatible with the results from the 14-week study in which concentrations in testes of rats exposed to 10 and 30 mg/m^3 indium phosphide continued to increase more than twofold after exposure ended (National Toxicology Program, 2001).

(ii) *Intratracheal administration in rats*

After an intratracheal instillation into male Fischer rats of 10 mg/kg bw particulate indium phosphide (1.73 ± 0.85-μm particles), Zheng *et al.* (1994) found minimal absorption, i.e. < 0.23% urinary excretion over a 10-day period. Retention at 96 h in the body (except in lung) was 0.36%; 73% of the administered dose was recovered in faeces, probably reflecting mucociliary transport followed by ingestion.

Uemura *et al.* (1997) exposed Fischer 344 rats to 0, 1, 10 and 100 mg/kg bw particulate indium phosphide (80% of the particles were < 0.8 μm in diameter) by intratracheal instillation. Indium, determined by use of AAS was detected at concentrations of 25 ng/g and 58 ng/g in liver and spleen, respectively, 1 day after instillation of 1 mg/kg bw indium phosphide. On day 7, the concentrations were 14 and 19 ng/g in these organs. Indium concentrations in serum increased significantly from day 1 to day 7 in animals that had received the highest dose. Toxic effects were obvious in the lungs but all rats survived. In this experiment, toxicity of indium phosphide was found to be much lower than that of more soluble compounds, such as indium chloride and indium nitrate (see e.g. Zheng *et al.*, 1994).

(iii)   *Intraperitoneal administration and gavage*

Kabe *et al.* (1996) studied male ICR mice after gavage and intraperitoneal injection of 0, 1000, 3000 and 5000 mg/kg bw indium phosphide suspended in 0.3 mL physiological saline and found minimal absorption after gavage with 2.4-μm particles but a dose-dependent increase in indium concentrations in serum after intraperitoneal administration. Mean indium concentrations were 1 and 4 μg/g in the liver and kidney, respectively, in mice given a single oral dose of 5000 mg/kg bw. Intraperitoneal administration resulted in accumulation of indium mainly in the lung (> 200 μg/g) and liver (about 300 μg/g) as measured by GF-AAS.

(b)   *Other indium compounds*

(i)   *Mice*

After intravenous injection of [113]In in mice, Stern *et al.* (1967) found that 50–60% of the injected radioactivity remained in the blood after 3 h. Castronovo and Wagner (1973) studied [114]In administered to mice as ionic indium chloride or as colloidal hydrated indium oxide and reported biphasic excretion patterns for both compounds, with half-life values of 1.9 and 69 days for indium chloride and 2 and 74 days for indium oxide. Ionic indium chloride concentrated primarily in the kidney while colloidal indium oxide was concentrated in the liver and reticuloendothelial system 4 days after a dose sufficient to cause the death of all animals.

(ii)   *Rats*

Smith *et al.* (1960) studied the metabolism of [114]InCl$_3$ in rats and found that more than half of the administered dose had been absorbed or excreted 4 days after intratracheal instillation, and intramuscular and subcutaneous injections. At 30 days after administration, 33–40% of the indium dose had been eliminated via faeces and urine independent of the route of administration.

Blazka *et al.* (1994) studied the distribution of indium trichloride after intratracheal instillation of 1.3 mg/kg bw in Fischer 344 rats. The rats were killed at different time-points up to 56 days after exposure and indium content of the lungs was determined. During the first 8 days after treatment, 87% of the indium was removed from the lung. Over the following 48 days less than 10% of the indium retained at 8 days was eliminated. It was concluded that indium chloride was capable of causing severe lung damage. [The Working Group noted the significant pulmonary retention for this soluble indium compound.]

4.2   Toxic effects

4.2.1   *Humans*

There are no published reports specific to the toxicity of indium phosphide in humans. A study by Raiciulescu *et al.* (1972) reported vascular shock in three of 770 patients injected with colloidal [113]In during liver scans.

4.2.2   *Experimental systems*

There is little information about the toxic effects in animals of indium phosphide either *in vivo* or *in vitro*. In general, the toxicity of indium compounds is dependent upon the form (solubility), the dose and the route of administration. When compared with the acute toxicity of other indium compounds, indium phosphide is less toxic (Venugopal & Luckey, 1978; National Toxicology Program, 2001).

(*a*)   *Indium phosphide*

Oda (1997) investigated the toxicity of indium phosphide particles (78% < 1 µm diameter) administered by intratracheal instillation of 0, 0.2, 6.0 and 62.0 µg/kg bw in male Fischer 344 rats that were subsequently observed for 8 days. Indium was not detected in the serum, liver, kidney, spleen, thymus or brain. A dose-related increase in SOD activity was observed in BALF on day 1 in all exposed groups, with no increase in inflammatory cells or total protein. LDH activity was increased on day 1 in the group that received the highest dose. On day 8, an increase in neutrophil and lymphocyte counts, LDH activity, and total protein, phospholipid and cholesterol concentrations was observed in BALF, together with desquamation of alveolar epithelial cells and the presence of amorphous exudate in the alveolar lumen as determined by histopathological examination, but only in rats that received the highest dose (62.0 µg/kg bw).

In another experiment from the same laboratory (Uemura *et al.*, 1997), male Fischer 344 rats (SPF grade) were exposed to intratracheal instillations of 0, 1, 10 or 100 mg/kg indium phosphide (mean diameter, 0.8 µm). The number of neutrophils in BALF increased considerably, in a dose-dependent manner, 1 and 7 days after indium phosphide administration. Indium phosphide particles were phagocytosed by macrophages and there was a large number of collapsed or broken macrophages at 7 days. LDH activity and the concentrations of total protein, total phospholipid and total cholesterol in BALF had increased in a dose-dependent manner 7 days after administration of indium phosphide. Histopathological examination of the lungs showed infiltration of macrophages and neutrophils, accompanied by broken macrophages, exfoliated alveolar cells and eosinophilic exudate. Indium phosphide particles were observed in the interstitium as well as in the lumen of the lung.

In a study conducted by the National Toxicology Program (2001) (described in Section 4.1.2), rats and mice were exposed to 0, 1, 3, 10, 30 or 100 mg/m³ of indium phosphide by inhalation 5 days per week for 14 weeks. Examination of the lungs at the end of the exposure period revealed pulmonary inflammation characterized by alveolar proteinosis, chronic inflammation, interstitial fibrosis and alveolar epithelial hyperplasia. In addition, microcytic erythrocytosis, consistent with bone-marrow hyperplasia and haematopoietic cell proliferation of the spleen, were observed in both rats and mice. Hepatocellular necrosis was indicated by the increased activities in serum of alanine aminotransferase and sorbitol dehydrogenase in all groups of male and female rats exposed to concentrations of 10 mg/m³ or greater. These findings were confirmed by histopathological examination of the liver in both sexes exposed to 100 mg/m³.

In further studies (National Toxicology Program, 2001; see also Section 4.1.2), groups of 60 male and 60 female B6C3F₁ mice and 60 male and 60 female Fischer 344/N rats, 6 weeks of age, were exposed to particulate aerosols of indium phosphide (purity, > 99%; MMAD, 1.2 μm; GSD, 1.7–1.8 μm) at concentrations of 0, 0.03, 0.1 or 0.3 mg/m³ for 6 h per day on 5 days per week for 22 weeks (rats) and 21 weeks (mice) (0.1 and 0.3 mg/m³) or 105 weeks (0 and 0.03 mg/m³). Exposure to indium phosphide caused dose-related increases in the incidence of proliferative and inflammatory lesions, especially in the lung, in both rats and mice (see Tables 2 and 3 in Section 3). In a subsequent evaluation of lung tissues collected during the 2-year National Toxicology Program study, Gottschling *et al.* (2001) used immunohistochemical techniques to show that concentrations of inducible nitric oxide synthase and cyclooxygenase-2 were elevated in inflammatory foci after 3 months of exposure to indium phosphide. In lungs of animals exposed for 2 years, inducible nitric oxide synthase, cyclooxygenase-2 and glutathione-*S*-transferase Pi were expressed and 8-OHdG was increased in non-neoplastic and neoplastic lesions. Glutathione-*S*-transferase Pi and 8-OHdG enhancement was observed in cells of carcinoma epithelium, atypical hyperplasia and squamous cysts. The results suggested that oxidative stress in pulmonary lesions may contribute to the carcinogenic process (Upham & Wagner, 2001).

(b)   Other indium compounds

In a study by Tanaka *et al.* (1996), male Syrian golden hamsters received indium arsenide or indium phosphide particles by intratracheal instillation of a dose containing 0.5 mg arsenic or phosphorus once a week for 15 weeks and were observed until the animals died [for about 105 weeks]. The cumulative gain in body weight was suppressed significantly in the indium arsenide-treated hamsters and not in the indium phosphide-treated group, compared with the control animals. Histopathological examination of the lungs showed that, in the animals treated with indium phosphide or indium arsenide, the incidence of proteinosis-like lesions, alveolar or bronchiolar cell hyperplasia, pneumonia, emphysema and metaplastic ossification, including infiltration of macrophages and lymphocytes into the alveolar space was significantly higher than that observed in controls. Particles of each compound were observed in the region of the alveolar septum and space as well as in the lymph nodes (Tanaka *et al.*, 1996).

A number of studies (Woods *et al.*, 1979; Fowler, 1986; Conner *et al.*, 1995) have shown that soluble indium administered as indium chloride, or indium arsenide particles, is a potent inducer of haeme oxygenase which is the rate-limiting enzyme in the haeme degradation pathway. Induction of this enzyme is used as a molecular marker of oxidative stress and, following acute administration of indium, is associated with marked decreases in cytochrome P450 and attendant mixed function oxidase activities in the liver of rats. Alterations in the activities of these mixed function oxidases may change cellular responsiveness to a number of known organic carcinogens found in semiconductor production facilities (Woods *et al.*, 1979; Fowler *et al.*, 1993).

Exposure to indium, indium arsenide and indium chloride has been shown to produce a number of effects on gene-expression patterns, including inhibition of expression of a number of stress proteins induced by arsenic (Fowler, 1986, 1988; Conner *et al.*, 1993). The marked inhibitory effects of indium on protein synthesis may play a role in altering the activities of DNA repair enzymes and the expression of proteins involved in regulating apoptosis: low doses of indium chloride induced apoptosis in rat thymocytes, whereas higher doses caused necrotic cell death (Bustamante *et al.*, 1997). These results provide another possible mechanism by which this element may contribute to the carcinogenic process, depending upon dose.

4.3   Reproductive and developmental effects

4.3.1   *Humans*

No data were available to the Working Group.

4.3.2   *Experimental systems*

Six studies in experimental animals have been published; indium nitrate $(In(NO_3)_3 \cdot 4.5H_2O)$ (Ferm & Carpenter, 1970) or indium trichloride $(InCl_3)$ (Chapin *et al.*,1995; Nakajima *et al.*, 1998, 1999, 2000) were used in five of these studies and given by oral gavage or intravenous injection. The overall results show that fetal development in rats is more affected than female or male reproductive capacity. Gross congenital malformations were observed in rat embryos. Mice were less susceptible to the teratogenicity of indium.

In the National Toxicology Program (2001) study, developmental toxicity was examined in Swiss (CD-1) mice and Sprague-Dawley rats exposed to 0, 1, 10 or 100 mg/m^3 indium phosphide by inhalation. Rats were exposed on gestation days 4–19 and mice were exposed on days 4 –17. In rats, exposure to indium phosphide by inhalation did not

induce maternal or fetal toxicity, malformations or effects on any developmental parameters. Exposure of mice to the highest dose resulted in early deaths and slightly reduced
body weight gain (not statistically significant); lung weights were significantly increased
in all mice exposed to indium phosphide. Renal haemorrhage was observed in some
fetuses in the group exposed to 100 mg/m³, but no significant teratogenicity or developmental effects could be attributed to exposure.

4.4   Genetic and related effects

No reports of genetic effects of indium phosphide in humans were found in the
literature.

In a study carried out by the National Toxicology Program (2001) (described in detail
in Section 3.1.1), no significant increases in the frequencies of micronucleated normochromatic erythrocytes were noted in the peripheral blood samples of male or female
B6C3F₁ mice exposed by inhalation to indium phosphide in concentrations up to
30 mg/m³ in a 14-week study. There was a significant increase in micronucleated polychromatic erythrocytes in male, but not in female mice exposed to 30 mg/m³. The percentage of polychromatic erythrocytes was not altered in males or females (National Toxicology Program, 2001).

In the 2-year inhalation study of indium phosphide (0.03 and 0.3 mg/m³) in male and
female B6C3F₁ mice (National Toxicology Program, 2001), β-catenin and *H-ras* mutations were assessed in hepatocellular adenomas and carcinomas. The frequency of *H-ras*
codon 61 mutations in the indium phosphide-induced hepatocellular neoplasms was
similar to that observed in controls. The frequency of β-catenin mutations was concentration-dependent: in the group exposed to 0.3 mg/m³ indium phosphide, 40% of the hepatocellular neoplasms showed β-catenin mutations compared with 10% in controls.

4.5   Mechanistic considerations

Inhalation of indium phosphide causes pulmonary inflammation associated with oxidative stress. The data of Gottschling *et al.* (2001) suggest that this inflammation may
progress to atypical hyperplasia and neoplasia in the lungs in rats.

It has been suggested that induction of apoptosis *in vitro* in rat thymocytes by indium
chloride at low concentrations occurs through alterations of the intracellular redox status,
or of intracellular homeostasis (Bustamante *et al.*, 1997). This apoptotic effect has been
shown to trigger repair-associated cell proliferation and may contribute to the risk for development of neoplasia.

Analysis of genetic alterations in indium phosphide-induced hepatocellular adenomas
and carcinomas revealed mutations in H-*ras* and β-catenin that were identical to those
found in human hepatocellular neoplasms (De la Coste *et al.*, 1998). This suggests a similar
pathway of carcinogenesis in both species.

5. Summary of Data Reported and Evaluation

5.1   Exposure data

Indium phosphide is used in the microelectronics industry because of its photovoltaic properties. It is produced as high-purity, single crystals cut into wafers and other shapes, which are used primarily for optoelectronic devices and in integrated circuits. Exposure to indium phosphide may occur in the microelectronics industry where workers are involved in the production of indium phosphide crystals, ingots and wafers, in grinding and sawing operations and in device fabrication.

5.2   Human carcinogenicity data

See Introduction to the Monographs on Gallium Arsenide and Indium Phosphide.

5.3   Animal carcinogenicity data

Indium phosphide was tested for carcinogenicity in a single study in mice and rats by inhalation exposure. Exposure to indium phosphide caused an increased incidence of alveolar/bronchiolar carcinomas in male mice and alveolar/bronchiolar adenomas and carcinomas in female mice and male and female rats. There was also a significant increase in the incidence of hepatocellular adenomas/carcinomas in exposed male and female mice and an increased incidence of benign and malignant pheochromocytomas of the adrenal gland in male and female rats. Other findings, which may have been exposure-related, were marginal increases in the incidences of adenomas/carcinomas of the small intestine in male mice, mononuclear-cell leukaemia in males and female rats, fibroma of the skin in male rats and carcinoma of the mammary gland in female rats. Indium phosphide was tested by intratracheal instillation in male hamsters and showed no carcinogenic response. However, due to the study design, it was not considered for evaluation.

5.4   Other relevant data

Indium phosphide has low solubility, and uptake from the gastrointestinal tract is low. Lung toxicity has been observed in long-term inhalation studies with indium phosphide. The lung tissue burden is high and elimination from the lung is very slow. In rats, concentrations of indium phosphide in blood, serum and testes could be followed for over 100 days after cessation of exposure by inhalation. The concentration of indium in the testes continued to increase, but the testicular tissue burden remained much lower than that in the lung. In various experimental systems using different routes of administration, accumulation of indium phosphide has also been demonstrated in liver, spleen and kidney. Indium is eliminated via urine and faeces.

Important toxic effects of intratracheally instilled indium phosphide particles are the induction of pulmonary inflammation, alveolar or bronchiolar hyperplasia, pneumonia and emphysema. Indium phosphide gave rise to enhanced activities of superoxide dismutase, nitric oxide synthase, cyclooxygenase and lactate dehydrogenase in bronchoalveolar lavage fluid, and to increased neutrophil and lymphocyte counts. At high doses, eosino-philic exudates and desquamation of alveolar epithelial cells were observed. Soluble indium was a potent inducer of haeme oxygenase, a marker of oxidative stress. Indium also showed inhibitory effects on protein synthesis and, at higher doses, on apoptosis.

No data were available on reproductive and developmental effects of indium phosphide in humans. Apart from slightly reduced pregnancy rates, no reproductive effects were observed in rats exposed to indium phosphide by inhalation. Mice exposed under com-parable conditions were much more sensitive, showing early fetal deaths and reduced body weight gain. There is no evidence that indium phosphide is teratogenic.

Micronucleus formation was observed in male, but not in female mice exposed to indium phosphide by inhalation. No other data on genetic and related effects as a result of exposure to indium phosphide were available. An association between oxidative stress and inflammation, possibly leading to lung neoplasia has been described in rats *in vivo*. Expo-sure of mice to indium phosphide by inhalation for 2 years was shown to cause an increase in β-catenin somatic mutations in liver neoplasms. Indium phosphide triggers apoptosis *in vitro*.

5.5   Evaluation

There is *inadequate evidence* in humans for the carcinogenicity of indium phosphide.

There is *sufficient evidence* in experimental animals for the carcinogenicity of indium phosphide.

Overall evaluation

Indium phosphide is *probably carcinogenic to humans (Group 2A)*.

In the absence of data on cancer in humans, the final evaluation for the carcino-genicity of indium phosphide was upgraded from 2B to 2A based on the following: extra-ordinarily high incidences of malignant neoplasms of the lung in male and female rats and mice; increased incidences of pheochromocytomas in male and female rats; and increased incidences of hepatocellular neoplasms in male and female mice. Of significance is the fact that these increased incidences of neoplasms occurred in rats and mice exposed to extremely low concentrations of indium phosphide (0.03–0.3 mg/m^3) and, even more significant, is the fact that these increased incidences occurred in mice and rats that were exposed for only 22 weeks (0.1 and 0.3 mg/m^3) and followed for 2 years.

6. References

ACGIH Worldwide® (2003) *Documentation of the TLVs® and BEIs® with Other Worldwide Occupational Exposure Values — CD-ROM — 2003*, Cincinnati, OH

Aucélio, R.Q., Smith, B.W. & Winefordner, J.D. (1998) Electrothermal atomization laser-excited atomic fluorescence spectroscopy for the determination of indium. *Appl. Spectr.*, **52**, 1457–1464

Beliles, R.P. (1994) The metals. In: Clayton G.D. & Clayton F.E., eds, *Patty's Industrial Hygiene and Toxicology*, 4th Ed., Vol. 2C, New York, John Wiley & Sons, pp. 2032–2038

Blazka, M.E. (1998) Indium. In: Zelikoff, J.T. & Thomas, P.T., eds, *Immunotoxicology of Environmental and Occupational Metals*, Philadelphia, PA, Taylor & Francis, pp. 93–110

Blazka, M.E., Dixon, D., Haskins, E. & Rosenthal, G.J. (1994) Pulmonary toxicity to intratracheally administered indium trichloride in Fischer 344 rats. *Fundam. appl. Toxicol.*, **22**, 231–239

Bliss, D.F. (2001) Recent highlights of bulk indium phosphide crystal growth. *AFRL Horizons*, **June** (http://afrlhorizons.com/Briefs/June01/SN0008.html; accessed 18.09.2003)

Busheina, I.S. & Headridge, J.B. (1982) Determination of indium by hydride generation and atomic–absorption spectrometry. *Talanta*, **29**, 519–520

Bustamante, J., Dock, L., Vahter, M., Fowler, B. & Orrenius, S. (1997) The semiconductor elements arsenic and indium induce apoptosis in rat thymocytes. *Toxicology*, **118**, 129–136

Castronovo, F.P., Jr & Wagner, H.N., Jr (1973) Comparative toxicity and pharmacodynamics of ionic indium chloride and hydrated indium oxide. *J. nucl. Med.*, **14**, 677–682

Chapin, R.E., Harris, M.W., Hunter, E.S., III, Davis, B.J., Collins, B.J. & Lockhart, A.C. (1995) The reproductive and developmental toxicity of indium in the Swiss mouse. *Fundam. appl. Toxicol.*, **27**, 140–148

Chemical Information Services (2003) *Directory of World Chemical Producers*, Dallas, TX (http://www.chemicalinfo.com; accessed 18.09.2003)

Conner, E.A., Yamauchi, H., Fowler, B.A. & Akkerman, M. (1993) Biological indicators for monitoring exposure/toxicity from III-V semiconductors. *J. Expos. anal. environ. Epidemiol.*, **3**, 431–440

Conner, E.A., Yamauchi, H. & Fowler, B.A. (1995) Alterations in the heme biosynthetic pathway from the III-V semiconductor metal, indium arsenide (InAs). *Chem.-biol. Interact.*, **96**, 273–285

CyOptics (2002) *Product Data Sheet: What is Indium Phosphide?*, Waltham, MA

De La Coste, A., Romagnolo, B., Billuart, P., Renard, C.-A., Buendia, M.-A., Soubrane, O., Fabre, M., Chelly, J., Beldjord, C., Kahn, A. & Perret, C. (1998) Somatic mutations of the β-catenin gene are frequent in mouse and human hepatocellular carcinomas. *Proc. natl Acad. Sci. USA*, **95**, 8847–8851

ESPI (1994) *Material Safety Data Sheet: Indium Phosphide*, Ashland, OR, Electronic Space Products International

Felix, N. (2003) Indium and indium compounds. In: *Ullmann's Encyclopedia of Industrial Chemistry*, 6th Ed., Vol. 17, Weinheim, Wiley-VCH Verlag GmbH & Co., pp. 681–691

Ferm, V.H. & Carpenter, S.J. (1970) Teratogenic and embryopathic effects of indium, gallium, and germanium. *Toxicol. appl. Pharmacol.*, **16**, 166–170

Fowler, B.A. (1986) Indium. In: Friberg, L., Nordberg, G.F. & Vouk, V.B., eds, *Handbook on the Toxicology of Metals*, Vol. II, New York, Elsevier Science, pp. 267–275

Fowler, B.A. (1988) Mechanisms of indium, thallium, and arsine gas toxicity: Relationships to biological indicators of cell injury. In: Clarkson, T.W., Friberg, L., Nordberg, G.F. & Sager, P.R., eds, *Biological Monitoring of Toxic Metals*, New York, Plenum Press, pp. 469–478

Fowler, B.A., Yamauchi, H., Conner, E.A. & Akkerman, M. (1993) Cancer risks for human exposure to the semiconductor metals. *Scand. J. Work Environ. Health*, **19** (Suppl. 1), 101–103

Gottschling, B.C., Maronpot, R.R., Hailey, J.R., Peddada, S., Moomaw, C.R., Klaunig, J.E. & Nyska, A. (2001) The role of oxidative stress in indium phosphide-induced lung carcinogenesis in rats. *Toxicol. Sci.*, **64**, 28–40

InPACT (2003) *The Indium Phosphide Substrates (InP) Specialist*, Moutiers (http://www.inpactsemicon.com; accessed 18.09.2003)

Jorgenson, J.D. (2002) *Minerals Yearbook: Indium*, Reston, VA, US Geological Survey (http://minerals.usgs.gov/minerals/pubs/commodity/indium/index.html; accessed 19.09.2003)

Jorgenson, J.D. (1997–2003) *Mineral Commodity Summaries: Indium*, Reston, VA, US Geological Survey (http://minerals.usgs.gov/minerals/pubs/commodity/indium/index.html; accessed 19.09.2003)

Kabe, I., Omae, K., Nakashima, H., Nomiyama, T., Uemura, T., Hosoda, K., Ishizuka, C., Yamazaki, K. & Sakurai, H. (1996) *In vitro* solubility and *in vivo* toxicity of indium phosphide. *J. occup. Health*, **38**, 6–12

Kucera, J., Havránek, V., Smolík, J., Schwarz, J., Vesely, V., Kugler, J., Sykorová, I. & Šantroch, J. (1999) INAA and PIXE of atmospheric and combustion aerosols. *Biol. Trace Elem. Res.*, **71–72**, 233–245

Landsberger, S., Hopke, P.K. & Cheng, M.D. (1992) Nanogram determination of indium using epithermal neutrons and its application in potential source contribution function of airborne particulate matter in the arctic aerosol. *Nucl. Sci. Eng.*, **110**, 79–83

Liao, Y. & Li, A. (1993) Indium hydride generation atomic absorption spectrometry with *in situ* preconcentration in a graphite furnace coated with palladium. *J. anal. Atom. Spectr.*, **8**, 633–636

Lide, D.R., ed. (2003) *CRC Handbook of Chemistry and Physics*, 84th Ed., Boca Raton, FL, CRC Press LLC, p. 4-61

Materials Database (2003) *Purdue University Materials Database* (http://yara.ecn.purdue.edu/~mslhub/MaterialsDBase/MATERIALS/InP/inp_intro.html; accessed 18.09.2003)

McCutcheon, B. (2001) *Canadian Minerals Yearbook: Indium*, Natural Resources Canada, pp. 27.1–27.4

Minamisawa, H., Murashima, K., Minamisawa, M., Arai, N. & Okutani, T. (2003) Determination of indium by graphite furnace atomic absorption spectrometry after coprecipitation with chitosan. *Anal. Sci.*, **19**, 401–404

Miyaki, K., Hosoda, K., Hirata, M., Tanaka, A., Nishiwaki, Y., Takebayashi, T., Inoue, N. & Omae, K. (2003) Biological monitoring of indium by means of graphite furnace atomic absorption spectrophotometry in workers exposed to particles of indium compounds. *J. occup. Health*, **45**, 228–230

Nakajima, M., Takahashi, H., Sasaki, M., Kobayashi, Y., Awano, T., Irie, D., Sakemi, K., Ohno, Y. & Usami, M. (1998) Developmental toxicity of indium chloride by intravenous or oral administration in rats. *Teratog. Carcinog. Mutag.*, **18**, 231–238

Nakajima, M., Sasaki, M., Kobayashi, M., Ohno, Y. & Usami, M. (1999) Developmental toxicity of indium in cultured rat embryos. *Teratog. Carcinog. Mutag.*, **19**, 205–209

Nakajima, M., Takahashi, H., Sasaki, M., Kobayashi, Y., Ohno, Y. & Usami, M. (2000) Comparative developmental toxicity study of indium in rats and mice. *Teratog. Carcinog. Mutag.*, **20**, 219–227

National Institute for Occupational Safety and Health (1985) *Hazard Assessment of the Electronic Component Manufacturing Industry*, US Department of Health and Human Services, Cincinnati, OH

National Toxicology Program (2001) *Toxicology and Carcinogenesis Studies of Indium Phosphide (CAS No. 22398-80-7) in F3344/N rats and B6C3F₁ mice (Inhalation Studies)* (Technical Report Series No. 499; NIH Publication No. 01-4433), Research Triangle Park, NC

Oda, K. (1997) Toxicity of a low level of indium phosphide (InP) in rats after intratracheal instillation. *Ind. Health*, **35**, 61–68

Raiciulescu, N., Niculescu-Zinca, D. & Stoichita-Papilan, M. (1972) Anaphylactoid reactions induced in In 113m and Au 198 radiopharmaceuticals used for liver scanning. *Rev. Roum. Med. Intern.*, **9**, 55–60

Reade Advanced Materials (1997) *Indium Phosphide*, Providence, RI (http://www.reade.com/ Products/Phosphides/indium_phosphides.html; accessed 18.09.2003)

Scansetti, G. (1992) Exposure to metals that have recently come into use. *Sci. total Environ.*, **120**, 85–91

Slattery, J.A. (1995) Indium and indium compounds. In: Kroschwitz, J.I. & Howe-Grant, M., eds, *Kirk-Othmer Encyclopedia of Chemical Technology*, 4th Ed., Vol. 14, New York, John Wiley & Sons, pp. 155–160

Slattery, J.A. (1999) Indium and indium compounds. In: Kroschwitz, J., ed., *Kirk-Othmer Concise Encyclopedia of Chemical Technology*, 4th Ed., New York, John Wiley & Sons, pp. 1113–1114

Smith, G.A., Thomas, R.G. & Scott, J.K. (1960) The metabolism of indium after administration of a single dose to the rat by intratracheal subcutaneous intramuscular and oral injection. *Health Physics*, **4**, 101–108

Stern, H.S., Goodwin, D.A., Scheffel, U. & Kramer, H.N. (1967) In [113]-M for blood-pool and brain scanning. *Nucleonics*, **25**, 62–65

Suva (2003) *Grenzwerte am Arbeitsplatz 2003*, Luzern [Swiss OELs]

Szweda, R. (2003) *Indium Phosphide Powers Ahead in the Chip Business*, IOP Publishing Ltd (http://wireless.iop.org/articles/; accessed 18.09.2003)

Tanaka, A., Hisanaga, A., Hirata, M., Omura, M., Makita, Y., Inoue, N. & Ishinishi, N. (1996) Chronic toxicity of indium arsenide and indium phosphide to the lungs of hamsters. *Fukuoka Acta med.*, **87**, 108–115

Työsuojelusäädöksiä (2002) *HTP arvot 2002*, Sosiaali-ja terveysministeriön, Tempere [Finnish OELs]

Uehara, N., Jinno, K., Hashimoto, M & Shijo, Y. (1997) Selective fluorometric determination of indium(III) by high-performance liquid chromatography with 2-methyl-8-quinolinol based on a ligand-exchange reaction of silanol groups. *J. Chromatogr.*, **A789**, 395–401

Uemura, T., Oda, K., Omae, K., Takebayashi, T., Nomiyama, T., Ishizuka, C., Hosoda, K., Sakurai, H., Yamazaki, K. & Kabe, I. (1997) Effects of intratracheally administered indium phosphide on male Fisher 344 rats. *J. occup. Health*, **39**, 205–210

Upham, B.L. & Wagner, J.G. (2001) Toxicological highlight: Toxicant-induced oxidative stress in cancer. *Toxicol. Sci.*, **64**, 1–3

Venugopal, B. & Luckey, T.D. (1978) *Metal Toxicity in Mammals. Vol. 2. Chemical Toxicity of Metals and Metalloids*, New York, Plenum Press

Woods, J.S., Carver, G.T. & Fowler, B.A. (1979) Altered regulation of hepatic heme metabolism by indium chloride. *Toxicol. appl. Pharmacol.*, **49**, 455–461

Zheng, W., Winter, S.M., Kattnig, M.J., Carter, D.E. & Sipes, I.G. (1994) Tissue distribution and elimination of indium in male Fischer 344 rats following oral and intratracheal administration of indium phosphide. *J. Toxicol. environ. Health*, **43**, 483–494

VANADIUM PENTOXIDE

VANADIUM PENTOXIDE

1. Exposure Data

1.1 Chemical and physical data

1.1.1 *Nomenclature*

The nomenclature of selected vanadium compounds is given in Table 1.
Chem. Abstr. Serv. Reg. No.: 1314-62-1
Deleted CAS Reg. No.: 12503-98-9; 56870-07-6; 87854-55-5; 87854-56-6; 166165-37-3; 172928-47-1; 184892-22-6; 200577-85-1; 203812-34-4; 251927-12-5; 410546-90-6
Chem. Abstr. Serv. Name: Vanadium oxide (V_2O_5)
IUPAC Systematic Name: Vanadium oxide
Synonyms: CI 77938; divanadium pentaoxide; pentaoxodivanadium; vanadic acid anhydride; vanadin (V) oxide (see also Table 1)

1.1.2 *Empirical formula and relative molecular mass*

V_2O_5 Relative molecular mass: 181.88

1.1.3 *Chemical and physical properties of the pure substance*

- (*a*) *Description*: Yellow to rust-brown orthorhombic crystals (O'Neil, 2001; Lide, 2003); yellow-orange powder or dark-gray flakes (Bauer *et al.*, 2003; National Institute for Occupational Safety and Health, 2005)
- (*b*) *Boiling-point*: 1800 °C, decomposes (Lide, 2003)
- (*c*) *Melting-point*: 670 °C (Lide, 2003); 690 °C (O'Neil, 2001)
- (*d*) *Density*: 3.36 (O'Neil, 2001; Lide, 2003)
- (*e*) *Solubility*: Slightly soluble in water (0.1–0.8 g/100 cm^3); soluble in concentrated acids and alkalis; insoluble in ethanol (Woolery, 1997; O'Neil, 2001)
- (*f*) *Stability*: Reacts with chlorine or hydrochloric acid to form vanadium oxytrichloride; absorbs moisture from the air (ESPI, 1994).

Table 1. Nomenclature of selected vanadium compounds

Molecular formula	Name used in Monograph [Registry number]	Synonyms
NH_4VO_3	Ammonium metavanadate [7803-55-6]	Ammonium monovanadate Ammonium trioxovanadate Ammonium trioxovanadate(1-) Ammonium vanadate Ammonium vanadate(V) Ammonium vanadium oxide Ammonium vanadium trioxide Vanadate (VO_3^-), ammonium Vanadic acid, ammonium salt Vanadic acid (HVO_3), ammonium salt
Na_3VO_4	Sodium orthovanadate [13721-39-6]	Sodium pervanadate Sodium tetraoxovanadate(3-) Sodium vanadate Sodium vanadate(V) (Na_3VO_4) Sodium vanadium oxide (Na_3VO_4) (9CI) Trisodium orthovanadate Trisodium tetraoxovanadate Trisodium vanadate Vanadic acid (H_3VO_4), trisodium salt (8CI) Vanadic(II) acid, trisodium salt
VO^{2+}	Vanadyl [20644-97-7]	Oxovanadium(2+) Oxovanadium(IV) ion Vanadium monoxide(2+) Vanadium oxide (VO), ion(2+) Vanadium oxide (VO^{2+}) Vanadyl(II) Vanadyl ion(2+) (8CI, 9CI)
VO_3^-	Vanadate [13981-20-9]	Metavanadate Metavanadate(1-) Trioxovanadate(1-) Vanadate (VO_3^-) Vanadate, ion −
$NaVO_4$	Sodium peroxyvanadate [15593-26-7]	Hydrogen peroxide, vanadium complex Peroxyvanadic acid ($HVO_2(O_2)$), sodium salt
Unspecified	Sodium vanadium oxide [11105-06-9]	Peroxyvanadic acid, sodium salt Sodium peroxyvanadate Sodium vanadate Vanadic acid, sodium salt

Table 1 (contd)

Molecular formula	Name used in Monograph [Registry number]	Synonyms
$VO(SO_4)$	Vanadyl sulfate [27774-13-6]	Oxo(sulfato)vanadium Oxovanadium(IV) sulfate Vanadic sulfate Vanadium oxide sulfate Vanadium(IV) oxide sulfate Vanadium oxosulfate Vanadium, oxosulfato- (8CI) Vanadium, oxo(sulfato(2-)-O)- Vanadium, oxo(sulfato(2-)-κO)- (9CI) Vanadium oxysulfate Vanadium sulfate
Unknown	Ferrovanadium [12604-58-9]	Ferrovanadium alloy Ferrovanadium dust
V_2O_3	Vanadium trioxide [1314-34-7]	Divanadium trioxide Vanadic oxide Vanadium oxide (V_2O_3) (8CI, 9CI) Vanadium(3+) oxide Vanadium sesquioxide Vanadium trioxide
V_2O_5	Vanadium pentoxide [1314-62-1]	Divanadium pentoxide Pentaoxodivanadium Vanadia Vanadic anhydride Vanadium oxide Vanadium oxide (V_2O_5) (8CI, 9CI) Vanadium(V) oxide Vanadium pentoxide
VCl_3	Vanadium trichloride [7718-98-1]	Vanadium chloride (VCl_3) (8CI, 9CI) Vanadium(3+) chloride Vanadium(III) chloride Vanadium trichloride

From STN International (2003); National Library of Medicine (2003)

1.1.4   *Technical products and impurities*

Vanadium pentoxide is commercially available in the USA in purities between 95% and 99.6%, with typical granulations between 10 mesh [~ 1600 μm] and 325 mesh [~ 35 μm] × down (Reade Advanced Materials, 1997; Strategic Minerals Corp., 2003). Vanadium pentoxide is also commercially available as a flake with the following specifications: purity, 98–99%; silicon, < 0.15–0.25%; iron, < 0.20–0.40%; and phosphorus, < 0.03–0.05%; and as a powder with the following specifications: purity, 98%; silicon dioxide, < 0.5%; iron, 0.3%; and arsenic, < 0.02% (American Elements, 2003).

Vanadium pentoxide is commercially available in Germany as granules and powder with a minimum purity of 99.6% (GfE mbH, 2003), and in the Russian Federation as a powder with the following specifications: purity, 98.6–99.3%; iron, < 0.05–0.15%; silicon, < 0.05–0.10%; manganese, < 0.04–0.10%; chromium, < 0.02–0.07%; sulfur, < 0.005–0.010%; phosphorus, < 0.01%; chlorine, < 0.01–0.02%; alkali metals (sodium and potassium), < 0.1–0.3%; and arsenic, < 0.003–0.010% (AVISMA titanium-magnesium Works, 2001).

Vanadium pentoxide is also commercially available in South Africa as granular and R-grade powders with a minimum purity of 99.5% and grain sizes of > 45 μm and < 150 μm, respectively (Highveld Steel & Vanadium Corporation Ltd, 2003).

1.1.5   *Analysis*

Occupational exposure to vanadium pentoxide is determined by measuring total vanadium in the workplace air or by biological monitoring.

(a)   *Monitoring workplace and ambient air*

Respirable fractions (< 0.8 μm) of airborne vanadium pentoxide are collected by drawing air in a stationary or personal sampler through a membrane filter made of polycarbonate, cellulose esters and/or teflon. The filter containing the collected air particulates can be analysed for vanadium using several methods. In destructive methods, the filter is digested in a mixture of concentrated mineral acids (hydrochloric acid, nitric acid, sulfuric acid, perchloric acid) and the vanadium concentration in the digest determined by GF–AAS (Gylseth *et al.*, 1979; Kiviluoto *et al.*, 1979) or ICP–AES (Kawai *et al.*, 1989). Non-destructive determination of the vanadium content on a filter can be performed using INAA (Kucera *et al.*, 1998).

Similar methods can be used for the measurement of vanadium in ambient air.

X-ray powder diffraction allows quantification of vanadium pentoxide, vanadium trioxide and ammonium metavanadate separately on the same sample of airborne dust (Carsey, 1985; National Institute for Occupational Safety and Health, 1994).

(b)   *Biological monitoring*

(i)   *Tissues suitable for biomonitoring of exposure*

Vanadium concentrations in urine, blood or serum have been suggested as suitable indicators of occupational exposure to vanadium pentoxide (Gylseth *et al.*, 1979; Kiviluoto *et al.*, 1979, 1981; Pyy *et al.*, 1984; Kawai *et al.*, 1989; Kucera *et al.*, 1998). The concentration of vanadium in urine appears to be the best indicator of recent exposure, since it rises within a few hours after the onset of exposure and decreases within a few hours after cessation of exposure (Kucera *et al.*, 1998). Table 2 presents data of vanadium concentrations in urine from workers exposed to vanadium.

Detailed information on the kinetics of vanadium in human blood after exposure is still lacking. Kucera *et al.* (1998) regarded vanadium concentrations in blood as the most suitable indicator of the long-term body burden (see Section 4.1.1). However, in a study of vanadium pentoxide exposure in rats, blood concentrations showed only marginal increases. This seems to indicate that there was limited absorption of vanadium (National Toxicology Program, 2002).

(ii)   *Precautions during sampling and sample handling*

Biological samples are prone to contamination from metallic parts of collection devices, storage containers, some chemicals and reagents; as a result, contamination-free sampling, sample handling and storage of blood and urine samples prior to analysis are of crucial importance (Minoia *et al.*, 1992; Sabbioni *et al.*, 1996). There is also a great risk of contamination during preconcentration, especially when nitric acid is used (Blotcky *et al.*, 1989).

(iii)   *Analytical methods*

Several reviews are available on analytical methods used for the determination of vanadium concentrations in biological materials (Seiler, 1995) and on the evaluation of normal vanadium concentrations in human blood, serum, plasma and urine (Versieck & Cornelis, 1980; Sabbioni *et al.*, 1996; Kucera & Sabbioni, 1998). Determination of vanadium concentrations in blood and/or its components and in urine is a challenging analytical task because the concentrations in these body fluids are usually very low (below the μg/L level). A detection limit of < 10 ng/L is therefore required and only a few analytical techniques are capable of this task, namely GF–AAS, isotope dilution mass spectrometry (IDMS), ICP–MS and NAA. Furthermore, sufficient experience in applying well-elaborated analytical procedures is of crucial importance for accurate determination of vanadium concentrations in blood, serum and urine.

Direct determination of vanadium concentrations in urine or diluted serum by GF–AAS is not feasible because the method is not sufficiently sensitive and because the possibility of matrix interferences; however, GF–AAS with a preconcentration procedure has been applied successfully (Ishida *et al.*, 1989; Tsukamoto *et al.*, 1990).

IDMS has good potential for the determination of low concentrations of vanadium. This technique has been applied for the determination of vanadium concentrations in human

Table 2. Vanadium concentrations in workplace air and urine from workers occupationally exposed to vanadium

Industrial process	No. of subjects	Vanadium in air mean ± SD or range of means in mg/m³	Vanadium in urine mean ± SD (range) in μg/L[b]	Reference
Ferrovanadium production	16	NK[c]	152 (44–360) nmol/mmol creatinine	Gylseth et al. (1979)
Smelting, packing and filtering of vanadium pentoxide	8	0.19 ± 0.24	73 ± 50 nmol/mmol creatinine	Kiviluoto et al. (1981)
Vanadium pentoxide processing	2	NK	13.9	Pyy et al. (1984)
Boiler cleaning	4	2.3–18.6 (0.1–6.4)[a]	(2–10.5)	White et al. (1987)
Vanadium pentoxide staining	2	[< 0.04–0.13]	(< 7–124)	Kawai et al. (1989)
Boiler cleaning	21	NK	0.7 (0.1–2.1)	Arbouine & Smith (1991)
Vanadium alloy production	5	NK	3.6 (0.5–8.8)	Arbouine & Smith (1991)
Removal of ashes in oil-fired power station	11	NK	2.2–27.4	Pistelli et al. (1991)
Boiler cleaning	10 (– RPE)[d] 10 (+ RPE)	NK	92 (20–270) 38 ± 26	Todaro et al. (1991)
Boiler cleaning	30	0.04–88.7	(0.1–322)	Smith et al. (1992)
Maintenance in oil-fired boiler	NK	0.28	57.1 ± 15.4 μg/g creatinine	Barisione et al. (1993)
Vanadium pentoxide production	58	Up to 5	28.3 (3–762)	Kucera et al. (1994)
Waste incineration workers	43	NK	0.66 ± 0.53 (< 0.01–2)	Wrbitzky et al. (1995)

Table 2 (contd)

Industrial process	No. of subjects	Vanadium in air mean ± SD or range of means in mg/m³	Vanadium in urine mean ± SD (range) in µg/L[b]	Reference
Boilermakers	20	0.02 (0.002–0.032)	1.53 ± 0.53 mg/g creatinine	Hauser et al. (1998)

Updated from WHO (2001)
[a] Time-weighted average (TWA)
[b] Unless stated otherwise
[c] NK, not known
[d] RPE, respiratory protective equipment

serum in only one study (Fassett & Kingston, 1985); however, the high mean value obtained (2.6 ± 0.3 mg/L) suggested the possibility of contamination (Sabbioni et al., 1996; Kucera & Sabbioni, 1998).

ICP–MS cannot be used for the determination of low concentrations of vanadium because of spectral and non-spectral interferences, unless high-resolution ICP–MS is used (Moens et al., 1994; Moens & Dams, 1995).

The problems of various interferences encountered with the above methods are mostly avoided by using NAA (Byrne, 1993). However, interfering radionuclides such as ^{24}Na or ^{38}Cl must be removed, preferably by post-irradiation radiochemical separation, so-called radiochemical NAA (RNAA). Also, because of the short half-life of the analytical radionuclide ^{52}V ($T_{1/2}$, 3.75 min), sample decomposition by irradiation and vanadium separation must be completed within 6–12 min (Byrne & Kosta, 1978a; Sabbioni et al., 1996). This technique has been mastered by only a few research groups (Byrne & Kosta, 1978b; Cornelis et al., 1980, 1981; Byrne & Versieck, 1990; Heydorn, 1990; Byrne & Kucera, 1991a,b; Kucera et al., 1992, 1994). If dry ashing is carried out prior to irradiation, the separation time can be shortened by a few minutes and a lower detection limit can be achieved (Byrne & Kucera, 1991a,b). Various procedures of pre-irradiation separation have been employed to circumvent the necessity for speedy operations with radioactive samples; however, high values were obtained, indicating that contamination and problems with blank samples could not be excluded (Heydorn, 1990). The only exception to date is an analysis performed by NAA in a clean Class 100 laboratory (Greenberg et al., 1990), which yielded a vanadium concentration in serum similar to that determined by RNAA.

(iv) *Reference values in occupationally non-exposed populations*

The values for blood and serum vanadium concentrations obtained by RNAA (Byrne & Kosta, 1978a; Cornelis et al., 1980, 1981; Byrne & Versieck, 1990; Heydorn, 1990; Byrne & Kucera, 1991a,b; Kucera et al., 1992, 1994), by NAA with pre-irradiation separation (Greenberg *et al.*, 1990), by GF–AAS with preconcentration (Ishida *et al.*, 1989; Tsukamoto *et al.*, 1990) and by high-resolution ICP–MS (Moens *et al.*, 1994) suggest that the true normal vanadium concentration in blood and serum of occupationally non-exposed populations is in the range of 0.02–0.1 µg/L. The accuracy of the results obtained by RNAA was confirmed by concomitant analysis of a variety of biological reference materials and comparison of the values obtained with certified or literature values. For the Second Generation Biological Reference Material (freeze-dried human serum), vanadium concentrations of 0.67 ± 0.05 µg/kg (dry mass) and 0.66 ± 0.10 µg/kg (dry mass) obtained by RNAA in two separate studies (Byrne & Versieck, 1990; Byrne & Kucera, 1991a) were consistent with the mean of 0.83 ± 0.09 µg/kg (dry mass) obtained by high-resolution ICP–MS (Moens *et al.*, 1994). These values correspond to serum concentrations of 0.060–0.075 µg/L, which are in the range of the normal vanadium concentrations in blood and/or serum suggested above. [The concentration in µg/kg dry mass can be converted into a concentration in µg/L by dividing by a factor of 11 (Versieck *et al.*, 1988).]

Vanadium concentrations in urine of occupationally non-exposed populations determined by RNAA (Kucera *et al.*, 1994) and by GF–AAS with preconcentration (Buchet *et al.*, 1982; Buratti *et al.*, 1985; Ishida *et al.*, 1989; Minoia *et al.*, 1990) have been shown consistently to have mean values ranging from 0.2 to 0.8 µg/L.

1.2   Production and use

1.2.1   *Production*

Although vanadium is widely dispersed and relatively abundant in the earth's crust, deposits of ore-grade minable vanadium are rare (see Section 1.3.1). The bulk of vanadium production is derived as a by-product or coproduct in processing iron, titanium, phosphorus and uranium ores. Vanadium is most commonly recovered from these ores in the form of pentoxide, but sometimes as sodium and ammonium vanadates.

Only about a dozen vanadium compounds are commercially significant; of these, vanadium pentoxide is dominant (Woolery, 1997; Nriagu, 1998; O'Neil, 2001; Atomix, 2003).

Vanadium was discovered twice. In 1801, Andres Manuel del Rio named it erythronium, but then decided he had merely found an impure form of chromium. Independently, Nils Gabriel Sefstrom found vanadium in 1830, and named it after the Scandinavian goddess of beauty and youth — the metal's compounds provide beautiful colours in solution. Henry Enfield Roscoe first isolated the metal in 1867, from vanadium dichloride. It was not until 1925 that relatively pure vanadium was obtained — by reducing vanadium pentoxide with calcium metal (Atomix, 2003).

According to the US Geological Survey (2002), nearly all the world's supply of vanadium comes from primary sources. Seven countries (China, Hungary, Japan, Kazakhstan, the Russian Federation, South Africa and the USA) recover vanadium from ores, concentrates, slag or petroleum residues. In five of the seven countries, the mining and processing of magnetite-bearing ores was reported to be an important source of vanadium production. Japan and the USA are believed to be the only countries to recover significant quantities of vanadium from petroleum residues. World demand for vanadium fluctuates in response to changes in steel production. It is anticipated to increase due to the demands for stronger and lighter steels and new applications, such as the vanadium battery (Magyar, 2002).

Raw materials processed into vanadium compounds include the titanomagnetite ores and their concentrates, which are sometimes processed directly, vanadium slags derived from ores, oil combustion residues, residues from the hydrometallization process and spent catalysts (secondary raw materials) (Hilliard, 1994; Bauer *et al.*, 2003). Primary industrial compounds produced directly from these raw materials are principally 98% (by weight) fused pentoxide, air-dried (technical-grade) pentoxide and technical-grade ammonium metavanadate (Woolery, 1997).

The titanomagnetite ore in lump form, containing approximately 1.5–1.7% vanadium pentoxide, is first reduced by coal at approximately 1000 °C in directly-heated rotary kilns. A further reduction is then performed in an electric furnace to obtain a pig iron which contains approximately 1.4% vanadium pentoxide. The molten pig iron is oxidized in a shaking ladle, causing the vanadium to be transferred to the slag in the form of a water-soluble trivalent iron spinel. A typical vanadium slag has the following approximate composition: 14% vanadium (equivalent to 25% vanadium pentoxide), 9% metallic iron, 32% total iron, 7% silica, 3.5% manganese, 3.5% titanium, 2.5% magnesium, 2.0% aluminium and 1.5% calcium. This is the world's principal raw material for vanadium production (Hilliard, 1994; Bauer *et al.*, 2003).

The main process used today to produce vanadium pentoxide from vanadium slags is alkaline roasting. The same process, with minor differences, can also be used for processing titanomagnetite ores and vanadium-containing residues. The slag is first ground to < 100 μm, and the iron granules are removed. Alkali metal salts are added, and the material is roasted with oxidation at 700–850 °C in multiple-hearth furnaces or rotary kilns to form water-soluble pentavalent sodium orthovanadate. The roasted product is leached with water, and ammonium polyvanadate or sparingly-soluble ammonium metavanadate are precipitated in crystalline form from the alkaline sodium orthovanadate solution by adding sulfuric or hydrochloric acid and ammonium salts at elevated temperature. These compounds are converted to high-purity, alkali-free vanadium pentoxide by roasting. The usual commercial 'flake' form of vanadium pentoxide is obtained from the solidified melt (Hilliard, 1994; Bauer *et al.*, 2003).

Hydrometallurgical methods or a combination of pyrometallurgical and hydrometallurgical processes are used to produce vanadium oxides and salts from other raw materials. In the combined processes, thermal treatment is followed by alkaline or, more rarely, acid processing (Hilliard, 1994; Bauer *et al.*, 2003).

Uranium production from carnotite and other vanadium-bearing ores also yields significant amounts of vanadium pentoxide (Atomix, 2003).

Total world production of vanadium pentoxide in 1996 was approximately 131 million pounds [59 500 tonnes] (Woolery, 1997). Based on vanadium pentoxide production capacity in 1994 from all sources, it has been estimated that the world's production of vanadium was split as follows: South Africa, 43%; USA, 17%; the Russian Federation, 15%; China, 13%; Venezuela, 4%; Chile, 4%; and others, 4% (Perron, 1994). In 2001, vanadium production capacity was estimated as follows: South Africa, 44%; the Russian Federation, 21%; Australia, 10%; USA, 8%; China, 8%; New Zealand, 4%; Kazakhstan, 2%; Japan, 1%; and others, 4% (Perron, 2001).

Available information indicates that vanadium pentoxide is produced by 12 companies in China, seven companies in the USA, six companies in India, five companies in Japan, four companies in the Russian Federation, two companies each in Germany and Taiwan, China, and one company each in Austria, Brazil, France, Kazakhstan, South Africa and Spain (Chemical Information Services, 2003).

1.2.2   Use

The major use of vanadium pentoxide is in the production of metal alloys. Iron–vanadium and aluminium–vanadium master alloys (e.g. for automotive steels, jet engines and airframes) are produced preferably from vanadium pentoxide fused flakes because of the low loss on ignition, low sulfur and dust contents, and high density of the molten oxide compared with powder.

Vanadium pentoxide is also used as an oxidation catalyst in heterogeneous and homogeneous catalytic processes for the production of sulfuric acid from sulfur dioxide, phthalic anhydride from naphthalene or *ortho*-xylene, maleic anhydride from benzene or *n*-butane/butene, adipic acid from cyclohexanol/cyclohexanone, acrylic acid from propane and acetaldehyde from alcohol. Minor amounts are used in the production of oxalic acid from cellulose and of anthraquinone from anthracene. Vanadium pentoxide has not found any significant uses in microelectronics but does have some applications in cathodes in primary and secondary (rechargeable) lithium batteries and in red phosphors for high-pressure mercury lamps and television screens. Vanadium pentoxide is used in the industries of enamelling, electrics and electronics, metallurgy, glass, catalysts, petrochemistry, and paint and ceramics. It is also used as a corrosion inhibitor in industrial processes for the production of hydrogen from hydrocarbons, as a coating for welding electrodes, as ultraviolet absorbent in glass, as depolariser, for glazes, for yellow and blue pigments, as a photographic developer, and in colloidal solution for anti-static layers on photographic material. It is also used as starting material for the production of carbides, nitrides, carbonitrides, silicides, halides, vanadates and vanadium salts (Woolery, 1997; O'Neil, 2001; ACGIH Worldwide®, 2003; Bauer et al., 2003).

1.3   Occurrence and exposure

1.3.1   *Natural occurrence*

Vanadium is widely but sparsely distributed in the earth's crust at an average concentration of 150 mg/kg and is found in about 80 different mineral ores, mainly in phosphate rock and iron ores. The concentration of vanadium measured in soil appears to be closely related to that of the parent rock from which it is formed and a range of 3–300 mg/kg has been recorded, with shales and clays exhibiting the highest concentrations (200 mg/kg and 300 mg/kg, respectively) (Byerrum *et al.*, 1974; Waters, 1977; WHO, 1988; Nriagu, 1998).

Vanadium is also found in fossil fuels (oil, coal, shale). It is present in almost all coals, in concentrations ranging from extremely low to 10 g/kg. It is found in crude oil and residual fuel oil, but not in distillate fuel oils. Venezuelan crude oils are thought to have the highest vanadium content, reaching 1400 mg/kg. Flue-gas deposits from oil-fired furnaces have been found to contain up to 50% vanadium pentoxide. In crude oil, residual fuel oil and asphaltenes, the most common form of vanadium is the +4 oxidation state (Byerrum *et al.*, 1974; Lagerkvist *et al.*, 1986; WHO, 1988; Nriagu, 1998).

1.3.2   *Occupational exposure*

Exposure to vanadium pentoxide in the workplace occurs primarily during the processing and refining of vanadium-rich ores and slags, during production of vanadium and vanadium-containing products, during combustion of fossil fuels (especially oil), during the handling of catalysts in the chemical industry, and during the cleaning of oil-fuelled boilers and furnaces (Plunkett, 1987). Data on vanadium concentrations in workplace air and the urine of workers exposed to vanadium in various industries are summarized in Table 2.

The processing of metals containing vanadium includes chemical treatment and high-temperature operations. However, only moderate concentrations of vanadium have been recorded in air in the breathing zone of workers engaged in these operations: $0.006–0.08$ mg/m^3 during the addition of vanadium to furnaces, $0.004–0.02$ mg/m^3 during tapping, $0.008–0.015$ mg/m^3 during oxyacetylene cutting and $0.002–0.006$ mg/m^3 during arc-welding (WHO, 1988).

In the main work areas of vanadium pentoxide production facilities where vanadium slag is processed, Roshchin (1968) recorded vanadium concentrations in dust of $20–55$ mg/m^3 (reported to be mainly vanadium trioxide) and < 0.17 mg/m^3 vanadium pentoxide (cited by WHO, 1988). In another study in a vanadium pentoxide production plant, Kucera *et al.* (1998) recorded the highest concentration of total air particulates of 271 mg/m^3 at a pelletizer, with a corresponding vanadium concentration of 0.5 mg/m^3; the highest concentrations of vanadium were detected in air at a vibratory conveyer and reached 4.9 mg/m^3. Similarly high concentrations of vanadium (4.7 mg/m^3) were reported in air in the breathing zone of workers in the steel industry (Kiviluoto *et al.*, 1979).

Breaking, loading and unloading, crushing and grinding, and magnetic separation of vanadium slag (about 120 g/kg vanadium pentoxide) causes formation of thick dust, with vanadium concentrations of 30–120 mg/m³. About 70–72 % of the particles were reported to have a diameter of < 2 μm and 86–96% a diameter of < 5 μm. When the slag is roasted, free vanadium pentoxide is discharged and concentrations of vanadium in the vicinity of the furnace have been found to range from 0.04 to 1.56 mg/m³. During leaching and precipitation, vanadium concentrations in the air can exceed 0.5 mg/m³. Smelting and granulation of technical-grade vanadium pentoxide are accompanied by the formation of a vanadium-containing aerosol. During the loading of smelting furnaces, vanadium pentoxide concentrations in the surrounding air have been found to range from 0.15 to 0.80 mg/m³; during smelting and granulation, from 0.7 to 11.7 mg/m³; during the crushing, unloading and packaging of pure vanadium pentoxide, dusts are formed in the facilities and concentrations of 2.2–49 mg/m³ vanadium pentoxide in air have been recorded (Roshchin, 1968; cited by WHO, 1988).

In the production of ferrovanadium alloys, a continuous discharge of vanadium pentoxide occurs during the smelting process. Vanadium pentoxide concentrations in air were reported to be 0.1–2.6 mg/m³ in the work area of smelters and helpers, 2–124.6 mg/m³ during charging of vanadium pentoxide in furnace, 0.07–9.43 mg/m³ in the crane driver's cabin during smelting, 0.97–12.6 mg/m³ during cutting up of ferrovanadium and 7.5–30 mg/m³ during furnace maintenance (Roshchin, 1968; cited by WHO, 1988).

When ductile vanadium is produced by the aluminothermic process (based on the reduction of pure vanadium pentoxide with aluminium), a condensation aerosol of vanadium pentoxide is released, with 98% of the particles having a diameter of < 5 μm and 82% a diameter of < 2 μm. Vanadium pentoxide concentrations recorded in the surrounding air were 19–25.1 mg/m³ during the preparation of the charge mixture, 64–240 mg/m³ during placing of the burden inside the smelting chambers and 0.2–0.6 mg/m³ in smelting operator's workplace (Roshchin, 1968; cited by WHO, 1988).

Usutani et al. (1979) measured vanadium pentoxide concentrations in air in a vanadium refinery. The highest concentrations (> 1 mg/m³) were detected in samples collected during removal of vanadium pentoxide flakes from the slag (cited by WHO, 1988).

In facilities producing aluminium from bauxite, concentrations of vanadium pentoxide up to 2.3 mg/m³ have been recorded in workplace air during tapping, packing and loading (Roshchin, 1968; cited by WHO, 1988).

Workers may be exposed to vanadium pentoxide in air during the handling of catalysts in chemical manufacturing plants. Exposure depends on the type of operations being carried out. During the removal and replacement of the catalyst, exposure to 0.01–0.67 mg/m³ have been reported. Sieving of the catalyst can lead to higher exposures, and concentrations between 0.01 and 1.9 mg/m³ (total inhalable vanadium) have been observed. Air-fed respiratory protective equipment is normally worn during catalyst removal and replacement and sieving (WHO, 2001).

Concentrations of vanadium pentoxide in the air during vanadium catalyst production have been reported as 1–7 mg/m³ during grinding and unloading of vanadium pentoxide,

3.2–7.5 mg/m^3 during loading into the bin and 0.1–1 mg/m^3 during sifting and packing granules of contact substance (Roshchin, 1968; cited by WHO, 1988).

Hery *et al.* (1992) assessed exposures to chemical pollutants during the handling (loading and unloading of reactors, sieving of catalysts) of inorganic catalysts, including vanadium pentoxide. Concentrations of vanadium pentoxide in air were reported to be 0.08–0.9 mg/m^3 during unloading, 1.1–230 mg/m^3 during screening and 600–1200 mg/m^3 during loading.

Hery *et al.* (1994) assessed exposures during the manufacture and reprocessing of inorganic catalysts, including vanadium pentoxide. In one of four 1-h air samples taken in a reprocessing plant during the oven-cleaning operation, a vanadium pentoxide concentration of 2.2 mg/m^3 was measured.

Fuel oil combustion results in the formation of vanadium-containing dust, and large amounts of dust result from operations connected with removal of ash encrustations when cleaning boilers and the blades of gas turbines. Dust concentrations in the air inside the boilers have been reported to range from 20 to 400 mg/m^3, the most common range being 50–100 mg/m^3, with the dust containing 5–17% vanadium pentoxide (Roshchin, 1968; cited by WHO, 1988).

Occupational exposure to vanadium occurs during the cleaning of oil-fired boilers and furnaces in oil-fired heating and power plants and ships, although workers probably spend less than 20% of their time cleaning oil-fired boilers. Vanadium concentrations in air (total inhalable fraction) as high as 20 mg/m^3 were recorded when these tasks were performed, but typically were lower than 0.1 mg/m^3. The lowest results were obtained where wet cleaning methods were used. Respiratory protective equipment was usually worn during boiler cleaning operations (WHO, 2001).

Williams (1952) published air sampling data on boiler-cleaning operations in the British power industry. A vanadium concentration of 40.2 mg/m^3 was recorded in air in the superheater chamber, while the concentration was 58.6 mg/m^3 in the combustion chamber; 93.6% of the dust particles had a diameter of 0.15–1 µm (cited by WHO, 1988).

Kuzelova *et al.* (1977) reported dust concentrations during boiler-cleaning operations of about 136–36 000 mg/m^3 in the workplace air, in which vanadium concentrations ranged from 1.7 to 18.4 mg/m^3 (cited by WHO, 1988).

Barisione *et al.* (1993) assessed the acute exposure to vanadium pentoxide in maintenance personnel working inside an oil-fired boiler at an electric power station in Italy. The vanadium pentoxide concentration in the air in the work room was 0.28 mg/m^3, which exceeded exposure standards. The concentration of vanadium in the urine of the arc welders did not correlate with vanadium pentoxide concentration in the air (see Table 2).

In 26 boilermakers overhauling an oil-powered boiler in the USA, Hauser *et al.* (1995a) investigated exposure to air particulates with an aerodynamic diameter of ≤ 10 µm (PM$_{10}$) and respirable vanadium-containing dust for up to 15 work days. The peak PM$_{10}$ concentration (1- to 10-h TWA) ranged from 1.48 to 7.30 mg/m^3; the peak vanadium concentration ranged from 2.2 to 32.2 µg/m^3, with a mean (SD) of 20.2 (11.4) µg/m^3. In a later study, the authors determined vanadium concentrations in the urine of a subgroup of workers (Hauser *et al.*, 1998; see Table 2).

In another study of boilermakers overhauling an oil-fired boiler in the USA, lower exposures to PM_{10} particulates and to respirable vanadium-containing dust were reported (median, 0.6 mg/m^3 and 12.7 µg/m^3, respectively) (Woodin *et al.*, 1999).

The National Institute of Occupational Safety and Health in the USA conducted surveys on exposure to vanadium pentoxide in the industry. The National Occupational Hazard Survey, conducted in 1972–74, estimated that 2562 workers in 333 plants were potentially exposed to vanadium pentoxide in 1970. The largest number of workers exposed worked in the stone, clay and glass products industries, and the second largest group was involved with electric, gas and sanitary services (National Institute for Occupational Safety and Health, 1976). The National Occupational Exposure Survey, conducted in 1980–83, reported that approximately 5319 workers in 151 plants were potentially exposed to vanadium in 1980. Among them, 84% were exposed specifically to vanadium pentoxide. The largest number of workers were exposed in the chemical and allied products industry (National Institute for Occupational Safety and Health, 1984).

Workers in the manufacture of vanadium-containing pigments for the ceramics industry may be exposed to vanadium compounds. Exposure is controlled by the use of local exhaust ventilation, and data indicate that vanadium concentrations in air are normally below 0.2 mg/m^3 (total inhalable fraction) (WHO, 2001).

Other reports of occupational exposures to vanadium have been reviewed (Zenz, 1994).

1.3.3   *Environmental exposure*

(*a*)   *Air*

(i)   *Natural sources*

Natural sources of atmospheric vanadium include continental dust, marine aerosols (sea salt sprays) and volcanic emissions. The quantities entering the atmosphere from each of these sources are uncertain; however, continental dust is believed to account for the largest portion of naturally-emitted atmospheric vanadium; contributions from volcanic emissions are believed to be small (Zoller *et al.*, 1973; Byerrum *et al.*, 1974). Atmospheric emissions of vanadium from natural sources had been estimated at 70 000 to 80 000 tonnes per year. However, more recent estimates report much lower values (1.6–54.2 tonnes per year) and suggest that fluxes from natural sources were overestimated by earlier workers (Mamane & Pirrone, 1998; Niriagu & Pirrone, 1998).

Concentrations of vanadium in the atmosphere in unpopulated areas such as Antarctica have been found to range from 0.0006 to 0.0024 ng/m^3 (Zoller *et al.*, 1974). Measurements taken over the eastern Pacific Ocean averaged 0.17 ng/m^3 (range of means, ≤ 0.02– 0.8 ng/m^3) (Hoffman *et al.*, 1969). Measurements over rural north-western Canada and Puerto Rico were one order of magnitude higher (0.2–1.9 ng/m^3) (Martens *et al.*, 1973; Zoller *et al.*, 1973).

(ii)   *Anthropogenic sources*

Estimates of global anthropogenic emissions of vanadium into the atmosphere over the last decade range from 70 000 tonnes to 210 000 tonnes per year (Hope, 1994; Mamane & Pirrone, 1998; Nriagu & Pirrone, 1998).

The major point sources are metallurgical works (30 kg vanadium/tonne vanadium produced), and coal and residual oil burning (0.2–2 kg vanadium/1000 tonnes and 30–300 kg/10^6 L burnt, respectively) (Zoller *et al.*, 1973; Lagerkvist *et al.*, 1986). Crude oils have an average vanadium content of 50 mg/kg (see above). [*Residual fuel oils* (heavy fuel oils) are petroleum refining residues remaining after distillation or cracking, and blends of these residues with distillates. They are used primarily in industrial burners and boilers as sources of heat and power (IARC, 1989). During refining and distillation, the vanadium remains in the residual oil because of its low volatility, and as a result becomes more concentrated than in the original crude.] During combustion, most of the vanadium in residual oils is released into the atmosphere in the form of vanadium pentoxide as part of fly ash particulates. Vanadium concentrations in coal fly ash range from 0.1 to 1 mg/g, and in residual oil from 10 to 50 mg/g (Mamane & Pirrone, 1998).

Vanadium was found in 87% of all air samples taken in the vicinity of large metallurgical plants at concentrations in the range of 0.98–1.49 µg/m^3, and in 11% of the samples exceeded 2 µg/m^3 (Pazhynich, 1967). At a steel plant in the USA in 1967, concentrations of vanadium in ambient air ranged from 40 to 107 ng/m^3 and averaged 72 ng/m^3 (WHO, 1988). Concentrations as high as 1000 ng/m^3 vanadium pentoxide were found in air by Pazhynich (1967) in the former Soviet Union at a site 1500 m from areas of extensive metallurgical activity unconnected with vanadium production. In the same country, near a plant producing technical vanadium pentoxide, 24-h mean concentrations of vanadium pentoxide of 4–12, 1–6, and 1–4 µg/m^3 in air were recorded at distances of 500, 1000 and 2000 m from the source, respectively (WHO, 1988).

According to the US Toxic Release Inventory (TRI, 1987–2001), the amount of vanadium released into the atmosphere from manufacturing and processing facilities in the USA fluctuated between 5–9 tonnes between 1987 and 1997 and had dramatically increased to over 100 tonnes by 2001. However, this estimate is believed to be limited because the largest anthropogenic releases of vanadium to the atmosphere are attributed to the combustion of residual fuel oils and coal, which are probably not included.

Vanadium-containing particulates emitted from anthropogenic sources into the atmosphere are simple or complex oxides (Byerrum *et al.*, 1974) or may be associated with sulfates (Mamane & Pirrone, 1998). Generally, lower oxides formed during combustion of coal and residual fuel oils, such as vanadium trioxide, undergo further oxidation to the pentoxide form before leaving the stacks (Environmental Protection Agency, 1985).

Concentrations of vanadium measured in ambient air vary widely between rural and urban locations; in general, these are higher in urban than in rural areas. Earlier reports suggested concentrations of 1–40 ng/m^3 (van Zinderen Bakker & Jaworski, 1980) or 0.2–75 ng/m^3 (Environmental Protection Agency, 1977) in air in rural sites, although the annual average was below 1 ng/m^3. This was attributed to the local burning of fuel oils with a high vanadium content. Recent data from rural areas show concentrations ranging from 0.3 to about 5 ng/m³, with annual averages frequently below 1 ng/m³, which can be regarded as the natural background concentration in rural areas (Mamane & Pirrone, 1998).

Annual average concentrations of vanadium in air in large cities may often be in the range of 50–200 ng/m³, although concentrations exceeding 200–300 ng/m³ have been recorded, and the maximum 24-h average may exceed 2000 ng/m³ (WHO, 1988). In the USA, cities can be divided into two groups based on the concentrations of vanadium present in their ambient air. The first group consists of cities widely distributed throughout the USA and characterized by vanadium concentrations in ambient air that range from 3 to 22 ng/m³, with an average of 11 ng/m³. Cities in the second group, primarily located in the north-eastern USA, have mean concentrations of vanadium that range from 150 to 1400 ng/m³, with an average of about 600 ng/m³. The difference is attributed to the use of large quantities of residual fuel oil in cities in the second group for the generation of heat and electricity, particularly during winter months (Zoller et al., 1973; WHO, 2000). Vanadium concentrations in ambient urban air vary extensively with the season. However, there are indications that vanadium concentrations in urban locations in 1998 were lower than those reported in the 1960s and 1970s (Mamane & Pirrone, 1998).

Hence, the general population may be exposed to airborne vanadium through inhalation, particularly in areas where use of residual fuel oils for energy production is high (Zoller et al., 1973). For instance, assuming vanadium concentrations in air of approximately 50 ng/m³, Byrne and Kosta (1978b) estimated a daily intake of 1 μg vanadium by inhalation.

(b)   Water

Vanadium dissolved in water is present almost exclusively in the pentavalent form. Its concentration ranges from approximately 0.1 to 220 μg/L in fresh water and from 0.3 to 29 μg/L in seawater. The highest concentrations in fresh waters were recorded in the vicinity of metallurgical plants or downstream of large cities (WHO, 1988; Bauer et al., 2003). Anthropogenic sources account for only a small percentage of the dissolved vanadium reaching the oceans (Hope, 1994).

(c)   Food

Vanadium intake from food has been reasonably well established, based on the analysis of dietary items (Myron et al., 1977; Byrne & Kosta, 1978b; Minoia et al., 1994) and total diets (Myron et al., 1978; Byrne & Kucera, 1991a). Considering consumption of about 500 g (dry mass) total diet, daily dietary vanadium intake in the general population has been estimated at 10–30 μg per person per day, although it can reach 70 μg per day in some countries (Byrne & Kucera, 1991a).

An increased daily intake of vanadium may result from the consumption of some wild-growing mushrooms (Byrne & Kosta, 1978b) and some beverages (Minoia et al., 1994), especially beer. Contamination of the marine environment with oil in the Gulf War resulted in increased concentrations of vanadium in certain seafood (WHO, 2001).

Considering the poor absorption of vanadium from the gastrointestinal tract, dietary habits can be expected to have only a minor influence on vanadium concentrations in body fluids (WHO, 1988; Sabbioni *et al.*, 1996) (see Section 4.1).

1.4   Regulations and guidelines

Occupational exposure limits and guidelines for vanadium pentoxide in workplace air are presented in Table 3.

ACGIH Worldwide® (2003) recommends a semi-quantitative BEI for vanadium in urine of 50 μg/g creatinine. ACGIH recommends monitoring vanadium in urine collected at the end of the last shift of the work week as an indicator of recent exposure to vanadium pentoxide. Germany recommends a biological tolerance value for occupational exposure for vanadium in urine of 70 μg/g creatinine. Germany also recommends monitoring vanadium in urine collected at the end of the exposure, for example at the end of the shift or, for long-term exposures, after several shifts (Deutsche Forschungsgemeinschaft, 2002).

2.   Studies of Cancer in Humans

No data were available to the Working Group.

3.   Studies of Cancer in Experimental Animals

3.1   Inhalation exposure

3.1.1   *Mouse*

In a study undertaken by the National Toxicology Program (2002), groups of 50 male and 50 female B6C3F$_1$ mice, 6–7 weeks of age, were exposed to vanadium pentoxide particulate (light orange, crystalline solid; purity, ≈ 99%; MMAD, 1.2–1.3 μm; GSD, 1.9 μm) at concentrations of 0, 1, 2 or 4 mg/m^3 by inhalation for 6 h per day on 5 days per week for 104 weeks. Survival was significantly decreased in males exposed to 4 mg/m^3 compared with chamber controls (survival rates: 39/50 (control), 33/50 (low concentration), 36/50 (mid concentration) or 27/50 (high concentration) in males and 38/50, 32/50 30/50 or 32/50 in females, respectively; mean survival times, 710, 692, 704 or 668 days in males and 692, 655, 653 or 688 days in females, respectively). Mean body weights were decreased in females exposed to ≥ 1 mg/m^3 and in males exposed to ≥ 2 mg/m^3. Exposure to vanadium pentoxide caused an increase in the incidence of alveolar/ bronchiolar neoplasms, but did not cause an increased incidence of neoplasms in other tissues. The incidence of neoplasms and non-neoplastic lesions of the respiratory system

Table 3. Occupational exposure limits and guidelines for vanadium (as V_2O_5 unless otherwise specified)

Country or region	Concentration (mg/m³)	Classification[a]	Interpretation[b]
Australia	0.05 (respirable dust and fume)		TWA
Belgium	0.5		TWA
Canada			
Alberta	0.05 (respirable dust and fume)		TWA
	0.15 (respirable dust and fume)		STEL
Quebec	0.05 (respirable dust and fume)		TWA
China	0.05 (dust and fume, as V)		TWA
	0.15 (dust and fume, as V)		STEL
Finland	0.05 (dust, as V)		TWA
	0.5 (fume, as V)		TWA
France	0.05 (respirable dust and fume)		TWA
Germany	0.05 (respirable fraction)		TWA (MAC)
	0.05 (respirable fraction)		STEL
Hong Kong SAR	0.05 (respirable dust and fume)	A4	TWA
Ireland	0.04 (respirable dust, as V)		TWA
	0.05 (fume, as V)		TWA
	0.5 (total inhalable dust, as V)		TWA
Japan	0.1 (fume)		TWA
	0.5 (dust)		TWA (JSOH)
Malaysia	0.05		TWA
Mexico	0.5 (dust and fume)	A4	TWA
Netherlands	0.01		TWA
	0.03		STEL
New Zealand	0.05 (respirable dust and fume)		TWA
Poland	0.05 (dust and fume)		TWA
	0.1 (fume); 0.5 (dust)		STEL
Russian Federation	0.1 (fume)		MAC
	0.5 (dust)		NG
South Africa	0.05 (respirable dust and fume)		TWA (DOL-RL)
	0.5 (total inhalable dust)		TWA
Spain	0.05 (respirable dust and fume)		TWA
Sweden	0.2 (total dust, as V)		TWA
	0.05 (respirable dust, as V)		Ceiling
Switzerland	0.05		TWA
	0.05		STEL
United Kingdom	0.05		TWA (MEL)

Table 3 (contd)

Country or region	Concentration (mg/m^3)	Classification[a]	Interpretation[b]
USA[c]			
ACGIH	0.05 (respirable dust and fume)	A4	TWA (TLV)
NIOSH	0.05 (total dust and fume, as V)		Ceiling (REL)
OSHA	0.1 (fume); 0.5 (respirable dust)		Ceiling (PEL)

From Sokolov (1981); INRS (1999); Työsuojelusäädöksiä (2002); ACGIH Worldwide® (2003); Suva (2003)

[a] A4, not classifiable as a human carcinogen; the absence of any classification does not necessarily mean that vanadium pentoxide has been evaluated by individual organizations as non-carcinogenic to humans.

[b] TWA, time-weighted average; STEL, short-term exposure limit; MAC, maximum allowed concentration; JSOH, Japanese Society for Occupational Health; NG, not given; DOL-RL, Department of Labour-Recommended Limit; MEL, maximum exposure limit; TLV, threshold limit value; REL, recommended exposure limit; PEL, permissible exposure limit.

are reported in Table 4. Alveolar/bronchiolar adenomas were typical of those that occur spontaneously in mice. Carcinomas had one or more of the following histological features; heterogeneous growth pattern, cellular pleomorphism and/or atypia, and local invasion or metastasis. A number of exposed males and females had multiple alveolar/bronchiolar neoplasms. This last finding is an uncommon response in mice and, in some cases, it was difficult to distinguish between multiplicity and metastases from other lung neoplasms. Mice are generally not considered to respond to particulate exposure by the development of lung neoplasms, even at high concentrations. There was a significantly-increased incidence of alveolar epithelial hyperplasia and bronchiolar epithelial hyperplasia in the lungs of exposed male and female mice. The hyperplasia was essentially a diffuse change with proliferation of epithelium in the distal terminal bronchioles and the immediately associated alveolar ducts and alveoli. The hyperplasia of the alveolar epithelium was pronounced and increased in severity with increasing exposure concentration, while the hyperplasia of the distal bronchioles was minimal to mild. Histiocytic infiltration occurred primarily within alveoli in close proximity to alveolar/bronchiolar neoplasms, particularly carcinomas (National Toxicology Program, 2002; Ress *et al.*, 2003).

3.1.2 *Rat*

In a study undertaken by the National Toxicology Program (2002), groups of 50 male and 50 female Fischer 344/N rats, 6–7 weeks of age, were exposed to vanadium pentoxide particulate (light orange, crystalline solid; purity, ≈ 99%; MMAD, 1.2–1.3 μm; GSD, 1.9 μm) at concentrations of 0, 0.5, 1 or 2 mg/m^3 by inhalation for 6 h per day on 5 days per week for 104 weeks. No adverse effects on survival were observed in treated males or females compared with chamber controls (survival rates: 20/50 (control), 29/50 (low

Table 4. Incidence of neoplasms and non-neoplastic lesions of the respiratory system and bronchial lymph nodes in mice in a 2-year inhalation study of vanadium pentoxide

	No. of mice exposed to vanadium pentoxide at concentrations (mg/m³) of			
	0 (chamber control)	1	2	4
Males				
Lung				
Total no. examined	50	50	50	50
No. with:				
Alveolar epithelium, hyperplasia	3 (3.0)[a]	41[b] (2.2)	49[b] (3.3)	50[b] (3.9)
Bronchiole epithelium, hyperplasia	0	15[b] (1.0)	37[b] (1.1)	46[b] (1.7)
Inflammation, chronic	6 (1.5)	42[b] (1.5)	45[b] (1.6)	47[b] (2.0)
Alveolus, infiltration cellular, histiocyte	10 (2.4)	36[b] (2.4)	45[b] (2.6)	49[b] (3.0)
Interstitial fibrosis	1 (1.0)	6 (1.7)	9[b] (1.2)	12[b] (1.7)
Alveolar/bronchiolar adenoma, multiple	1	1	11[b]	5
Alveolar/bronchiolar adenoma (includes multiple)	13	16	26[b]	15
Alveolar/bronchiolar carcinoma, multiple	1	10[b]	16[b]	13[b]
Alveolar/bronchiolar carcinoma (includes multiple)	12	29[b]	30[b]	35[b]
Alveolar/bronchiolar adenoma or carcinoma	22	42[b]	43[b]	43[b]
Larynx				
Total no. examined	49	50	48	50
No. with:				
Respiratory epithelium, epiglottis, metaplasia, squamous	2 (1.0)	45[b] (1.0)	41[b] (1.0)	41[b] (1.0)
Nose				
Total no. examined	50	50	50	50
No. witth:				
Inflammation, suppurative	16 (1.3)	11 (1.4)	32[b] (1.2)	23[c] (1.3)
Olfactory epithelium, atrophy	6 (1.0)	7 (1.6)	9 (1.3)	12 (1.2)
Olfactory epithelium, degeneration, hyaline	1 (1.0)	7[c] (1.0)	23[b] (1.1)	30[b] (1.2)
Respiratory epithelium, degeneration, hyaline	8 (1.1)	22[b] (1.0)	38[b] (1.2)	41[b] (1.4)
Respiratory epithelium, metaplasia, squamous	0	6[c] (1.2)	6[c] (1.3)	2 (1.5)
Lymph node, bronchial				
Total no. examined	40	38	36	40
No. with:				
Hyperplasia	7 (2.1)	7 (2.4)	12 (2.1)	13 (2.2)

Table 4 (contd)

	No. of mice exposed to vanadium pentoxide at concentrations (mg/m³) of			
	0 (chamber control)	1	2	4
Females				
Lung				
Total no. examined	50	50	50	50
No. with:				
Alveolar epithelium, hyperplasia	0	31[b] (1.6)	38[b] (2.0)	50[b] (3.3)
Bronchiole epithelium, hyperplasia	0	12[b] (1.0)	34[b] (1.0)	48[b] (1.5)
Inflammation, chronic	4 (1.0)	37[b] (1.3)	39[b] (1.8)	49[b] (2.0)
Alveolus, infiltration cellular, histiocyte	0	34[b] (2.4)	35[b] (2.4)	45[b] (2.7)
Interstitial fibrosis	0	1 (2.0)	4[c] (2.5)	8[b] (1.5)
Alveolar/bronchiolar adenoma, multiple	0	3	5[c]	6[c]
Alveolar/bronchiolar adenoma (includes multiple)	1	17[b]	23[b]	19[b]
Alveolar/bronchiolar carcinoma, multiple	0	9[b]	5[c]	5[c]
Alveolar/bronchiolar carcinoma (includes multiple)	0	23[b]	18[b]	22[b]
Alveolar/bronchiolar adenoma or carcinoma	1	32[b]	35[b]	32[b]
Larynx				
Total no. examined	50	50	50	50
No. with:				
Respiratory epithelium, epiglottis, metaplasia, squamous	0	39[b] (1.0)	45[b] (1.0)	44[b] (1.1)
Nose				
Total no. examined	50	50	50	50
No. with:				
Inflammation, suppurative	19 (1.1)	14 (1.2)	32[b] (1.2)	30[b] (1.3)
Olfactory epithelium, atrophy	2 (1.5)	8[c] (1.3)	5 (1.0)	14[b] (1.3)
Olfactory epithelium, degeneration, hyaline	11 (1.2)	23[b] (1.0)	34[b] (1.2)	48[b] (1.3)
Respiratory epithelium, degeneration, hyaline	35 (1.3)	39 (1.5)	46[b] (1.7)	50[b] (1.8)
Respiratory epithelium, metaplasia, squamous	0	3 (1.3)	7[b] (1.1)	8[b] (1.1)
Respiratory epithelium, necrosis	0	0	1 (2.0)	7[b] (1.4)
Lymph node, bronchial				
Total no. examined	39	40	45	41
No. with:				
Hyperplasia	3 (2.0)	13[b] (1.8)	14[b] (2.3)	20[b] (2.3)

From National Toxicology Program (2002)

[a] Average severity grade of lesions in affected animals: 1, minimal; 2, mild; 3, moderate; 4, marked

[b] Significantly different ($p \leq 0.01$) from the chamber control group by the Poly-3 test

[c] Significantly different ($p \leq 0.05$) from the chamber control group by the Poly-3 test

concentration), 26/50 (mid concentration) or 27/50 (high concentration) in males and 33/50, 24/50, 29/50 or 30/50 in females, respectively; mean survival times: 668, 680, 692 or 671 days in males and 688, 678, 679 or 683 days in females, respectively). Mean body weights were slightly decreased in females exposed to 2.0 mg/m^3 throughout the study compared with chamber controls. Although there was a marginally increased incidence of alveolar/bronchiolar neoplasms in female rats, the increase was not statistically significant, did not occur in a concentration-related fashion and was in the historical control range. Thus, it was uncertain whether the increased incidence observed was exposure-related. Exposure to vanadium pentoxide caused an increase in the incidence of alveolar/bronchiolar neoplasms in male rats. Although not statistically significant, the incidence of alveolar/bronchiolar adenoma in males exposed to 0.5 mg/m^3 and of alveolar/bronchiolar carcinoma and alveolar/bronchiolar adenoma or carcinoma (combined) in males exposed to 0.5 and 2 mg/m^3 exceeded the historical ranges in controls (all routes) given NTP-2000 diet and inhalation controls given NIH-07 diet. This response was considered to be related to exposure to vanadium pentoxide. However, exposure to vanadium pentoxide did not cause increased incidence of neoplasms in other tissues. The incidence of neoplasms and non-neoplastic lesions of the respiratory system in male rats is reported in Table 5. Alveolar bronchiolar adenomas, typical of those occurring spontaneously, were generally distinct masses that compressed surrounding tissue. Component epithelial cells were generally uniform in appearance and were arranged in acinar and/or irregular papillary structures and occasionally in a solid cellular pattern. Alveolar/bronchiolar carcinomas had similar cellular patterns but were generally larger and had one or more of the following histological features; heterogeneous growth pattern, cellular pleomorphism and/or atypia, and local invasion or metastasis. Three male rats exposed to 0.5 mg/m^3, one male rat exposed to 1 mg/m^3 and three male rats exposed to 2 mg/m^3 developed alveolar/bronchiolar carcinomas, one of which metastasized. There were no primary lung carcinomas in the chamber control rats. Alveolar/bronchiolar adenomas and especially carcinomas with metastases from the site of origin are uncommon in rats (Hahn, 1993). Exposure to vanadium pentoxide caused a spectrum of inflammatory and proliferative lesions in the lungs that were similar in male and female rats. There was a significantly-increased incidence of alveolar epithelial hyperplasia in the lungs of males exposed to 0.5 mg/m^3 or greater and females exposed to 1 or 2 mg/m^3. Squamous metaplasia of the alveolar epithelium occurred in 21/50 male and 6/50 female rats exposed to 2.0 mg/m^3 vanadium pentoxide. Squamous epithelium is not a normal component of the lung parenchyma. It is a more resilient epithelium and its occurrence in the lung generally represents a response to injury (National Toxicology Program, 2002; Ress *et al.*, 2003).

3.1.3   *Comparison of findings from the rat and mouse inhalation studies*

A wide range of proliferative lesions in the lungs were observed in rats and mice exposed to vanadium pentoxide for 2 years. The incidence of hyperplasia of the alveolar and bronchiolar epithelium was increased in exposed rats and mice. Although given

Table 5. Incidence of neoplasms and non-neoplastic lesions of the respiratory system and bronchial lymph nodes in male rats in a 2-year inhalation study of vanadium pentoxide

	No. of rats exposed to vanadium pentoxide at concentrations (mg/m^3) of			
	0 (chamber control)	0.5	1	2
Lung				
Total no. examined	50	49	48	50
No. with:				
Alveolar epithelium, hyperplasia	7 (2.3)[a]	24[b] (2.0)	34[b] (2.0)	49[b] (3.3)
Bronchiole epithelium, hyperplasia	3 (2.3)	17[b] (2.2)	31[b] (1.8)	49[b] (3.3)
Alveolar epithelium, metaplasia, squamous	1 (1.0)	0	0	21[b] (3.6)
Bronchiole epithelium, metaplasia, squamous	0	0	0	7[b] (3.7)
Inflammation, chronic active	5 (1.6)	8 (1.8)	24[b] (1.3)	42[b] (2.4)
Interstitial fibrosis	7 (1.4)	7 (2.0)	16[c] (1.6)	38[b] (2.1)
Alveolus, infiltration cellular, histiocyte	22 (1.3)	40[b] (2.0)	45[b] (2.3)	50[b] (3.3)
Alveolus, pigmentation	1 (2.0)	0	2 (1.5)	28[b] (2.1)
Alveolar/bronchiolar adenoma, multiple	0	2	0	0
Alveolar/bronchiolar adenoma (includes multiple)	4	8	5	6
Alveolar/bronchiolar carcinoma, multiple	0	1	0	0
Alveolar/bronchiolar carcinoma (includes multiple)	0	3	1	3
Alveolar/bronchiolar adenoma or carcinoma	4	10	6	9
Larynx				
Total no. examined	49	50	50	49
No. with:				
Inflammation, chronic	3 (1.0)	20[b] (1.1)	17[b] (1.5)	28[b] (1.6)
Respiratory epithelium, epiglottis, degeneration	0	22[b] (1.1)	23[b] (1.1)	33[b] (1.5)
Respiratory epithelium, epiglottis, hyperplasia	0	18[b] (1.5)	34[b] (1.5)	32[b] (1.9)
Respiratory epithelium, epiglottis, metaplasia, squamous	0	9[b] (1.7)	16[b] (1.8)	19[b] (2.1)
Nose				
Total no. examined	49	50	49	48
No. with:				
Goblet cell, respiratory epithelium, hyperplasia	4 (1.8)	15[b] (1.8)	12[c] (2.0)	17[b] (2.1)

From National Toxicology Program (2002)

[a] Average severity grade of lesions in affected animals: 1, minimal; 2, mild; 3, moderate; 4, marked

[b] Significantly different ($p \leq 0.01$) from the chamber control group by the Poly-3 test

[c] Significantly different ($p \leq 0.05$) from the chamber control group by the Poly-3 test

distinct diagnoses, the lesions were considered to be one pathogenic process. The authors concluded that this hyperplastic change was striking and appeared more prominent than had been observed in other National Toxicology Program inhalation studies. Although the exact pathogenesis was not determined in this study, the hyperplasia of the alveolar and bronchiolar epithelium was consistent with bronchiolization, a process in which bronchiolar epithelium proliferates and migrates down into alveolar ducts and adjacent alveoli. Although there was clearly proliferation, it was thought primarily to represent a metaplastic change. Whether this represented a precursor lesion for development of pulmonary neoplasms is not known. The lung tumour response in rats and mice following exposure to vanadium pentoxide was not concentration-related; there was a flat dose response. Several dose metrics and lung-burden data were used to aid in interpretation of lung pathology in exposed rats and mice. In the case of all dose metrics, rats received more vanadium than mice. In mice, the total 'dose' was similar in the groups exposed to 1 mg/m^3 and 2 mg/m^3 and this may help explain the flat dose response in the lung neoplasms in male and female mice. The total dose does not explain the differences in neoplasms in rats compared with mice. However, when the total dose is corrected for body weight, mice received a three- to five-fold higher dose of vanadium than rats at comparable exposure concentrations of 1 and 2 mg/m^3. Therefore, on a body weight basis, mice received considerably more vanadium than rats, and this may help explain the differences in responses between the species (National Toxicology Program, 2002; Ress *et al.*, 2003).

4. Other Data Relevant to an Evaluation of Carcinogenicity and its Mechanisms

4.1   Deposition, retention, clearance and metabolism

Vanadium pentoxide (V_2O_5) is a poorly soluble oxide which, in water or body fluids, releases some vanadium ions which may speciate either in cationic (VO_2^+) or anionic (HVO_4^{2-}) forms [at physiological pH: $H_2VO_4^-$].

Toya *et al.* (2001) showed that vanadium pentoxide powder (geometric mean diameter, 0.31 μm) was eight times more soluble in an artificial biological fluid (Gamble's solution) than in water.

Elimination from the lung, and distribution to and elimination from tissues, is partly a function of solubility. Sodium vanadate is more soluble than vanadium pentoxide and is consequently cleared more rapidly from the lung (Sharma *et al.*, 1987).

Vanadium (V) is reduced to vanadium (IV) in humans and other mammals. It is considered to be an essential element in chickens, rats and probably humans (Nielsen, 1991; French & Jones, 1993; Crans *et al.*, 1998; Hamel, 1998; National Toxicology Program, 2002). The main source of vanadium intake for the general human population is food (see also Section 1.3.5).

4.1.1 *Humans*

Zenz and Berg (1967) studied responses in nine human volunteers exposed to 0.2 mg/m^3 vanadium pentoxide (particle size, 98% < 5 μm) for 8 h in a controlled environmental chamber. The highest concentration of vanadium was found in the urine (0.13 mg/L [2.6 μM/L]) 3 days after exposure; none of the volunteers had detectable concentrations 1 week after exposure.

Pistelli *et al.* (1991) studied 11 vanadium pentoxide-exposed workers 40–60 h after they had removed ashes from boilers of an oil-fired power station. Seven of the workers were smokers compared with eight of 14 controls. Vanadium concentrations in urine were determined by AAS and ranged between 1.4 and 27 μg/L in the exposed group. Four of the controls had detectable concentrations of vanadium in the urine (range, 0.5–1.0 μg/L).

Hauser *et al.* (1998) determined concentrations of vanadium by means of GF-AAS in the urine of workers overhauling an oil-fired boiler where concentrations of vanadium pentoxide in the air ranged from 0.36 to 32.2 μg/m^3 (mean, 19.1 μg/m^3). On the first day of work on the overhaul, the mean vanadium concentrations in urine were 0.87 mg/g creatinine before a shift and 1.53 mg/g creatinine after a shift. However, the vanadium concentrations in the start-of-shift urine samples on the last Monday of the study were not significantly different from the start-of-shift concentrations on the previous Saturday, a time interval of about 38 h between the end of exposure and sample collection. Spearman rank correlation between start-of-shift concentration of vanadium in urine and concentration of vanadium in workplace dust during the previous day was not strong (r = 0.35) due to incomplete and insufficient information on respirator usage as noted by the authors. These data support a rapid initial clearance of inhaled vanadium occurring on the first day of work followed by a slower clearance phase that was not complete 38 h after the end of exposure (Hauser *et al.*, 1998).

Kucera *et al.* (1998) analysed vanadium in biological samples from workers engaged in the production of vanadium pentoxide by a hydrometallurgical process and occupationally non-exposed controls. Average exposure time was 9.2 years (range, 0.5–33 years). Concentrations of vanadium in workplace air samples were high (range, 0.017–4.8 mg/m^3). Concentrations of vanadium in the blood of a subsample of workers was 12.1 ± 3.52 μg/L (geometric mean ± GSD) compared with 0.055 ± 1.41 μg/L among the non-exposed controls. Vanadium concentrations in morning urine were 29.2 ± 3.33 μg/L in exposed workers and 0.203 ± 1.61 μg/L for the non-exposed. The finding of high concentrations in morning urine is compatible with the fact that long-term exposure results in vanadium accumulation in the bone from which it can be released slowly.

Vanadium pentoxide was found to be rapidly absorbed following inhalation exposure, but poorly through dermal contact or when ingested as ammonium vanadyl tartrate (Dimond *et al.*, 1963; Gylseth *et al.*, 1979; Kiviluoto *et al.*, 1981; Ryan *et al.*, 1999). When given orally, 0.1–1% is absorbed from the gut, although absorption of more soluble vanadium compounds is greater. About 60% of absorbed vanadium is excreted in the urine within 24 h (McKee, 1998). Based on samples from autopsies, vanadium was found to be distributed to the lungs and the intestine. It was not detected in heart, aorta, brain, kidney, ovary or testes, although detection methods were reported to be insensitive (Schroeder *et al.*, 1963; Ryan *et al.*, 1999).

Using AAS, Fortoul *et al.* (2002) analysed vanadium concentrations in lung tissue samples from autopsies of Mexico city residents in the 1960s and 1990s (n = 39 and 48, respectively). Vanadium concentrations were 1.04 ± 0.05 µg/g in lung samples from the 1960s and 1.36 ± 0.08 µg/g in samples from the 1990s, indicating an increase in ambient exposure to vanadium.

4.1.2   *Experimental systems*

(*a*)   *In-vivo studies*

Absorption of vanadium compounds after oral administration is known to be strongly affected by such dietary components as type of carbohydrate, fibre protein concentration, other trace elements, chelating agents and electrolytes (Nielsen, 1987). Associated pathology or physiological state may also affect vanadium absorption and hence may render a consistent determination of a lethal dose (e.g. LD_{50}) by the oral route very difficult (Thompson *et al.*, 1998).

In general, the absorption, distribution and elimination of vanadium pentoxide and other vanadium compounds are similar. There are, however, variations depending on the solubility of the administered compound, the route of exposure and the form of vanadium administered (National Toxicology Program, 2002).

(i)   *Inhalation studies*

Mice

In a National Toxicology Program tissue burden study (2002), male and female $B6C3F_1$ mice were exposed to 1, 2, or 4 mg/m³ vanadium pentoxide by inhalation for 104 weeks (for details, see Section 3.1.1). Tissue burden analyses were performed on days 1, 5, 12, 26, 54, 171, 362 and 535 after the start of treatment. Lung weights increased throughout the study, most markedly in the group exposed to the highest concentration. The mean lung weights of the two lower-dose groups were similar. Lung vanadium burden increased roughly in proportion to the exposure concentration, with strong indications of linear toxicokinetics. As with the rats (see below), lung burdens in the mice did not reach a steady state in the groups exposed to 2 and 4 mg/m³; they peaked near day 54 (at 5.9 and 11.3 µg, respectively), and then declined until day 535. In the low-dose group (1 mg/m³), the lung burden reached a steady state around day 26 at a level of 3 µg vanadium. The same toxicokinetic model could be applied to both mice and rats (see below), with an initial deposition rate increasing with increasing exposure concentration, and a decline in deposition rate over the course of the study. In the group exposed to 4 mg/m³, the deposition rate decreased from 0.62 to 0.27 µg/day between day 1 and day 535 and in the group exposed to 2 mg/m³ it decreased from 0.41 to 0.22 µg/day. However, in the group exposed to the lowest dose there was a minimal decline in deposition rate between days 1 and 535 (0.31 to 0.26 µg/day). Lung clearance half-lives in mice were 6, 11 and 14 days for the 1, 2 and 4 mg/m^3 exposure groups, respectively. Total vanadium lung doses were estimated to have been 153, 162 and 225 µg, respectively, while normalized lung doses were 153, 80.9 and 56.2 µg vanadium per mg vanadium pentoxide per m^3 exposure. On day 535, mice had retained approximately 2–3% of the total estimated lung doses (National Toxicology Program, 2002).

In an inhalation model described by Sánchez *et al.* (2003; abstract only), male CD-1 mice were exposed to an aerosol of 0.02 M vanadium pentoxide for 2 h twice a week for 4 weeks. Concentrations of vanadium (determined by AAS) in lung, liver, kidney, testes and brain increased after the first week of inhalation in all the organs examined and remained at almost the same values at the end of the fourth week. The organ with the highest concentrations of vanadium was the liver followed by the kidney. The lowest concentrations were found in testes. However, at the fourth week, a decrease in concentrations of vanadium was observed in the kidney.

Rats

In a study undertaken by the National Toxicology Program (2002), blood and lung concentrations, lung clearance half-life of vanadium, and the onset and extent of vanadium pentoxide-induced lung injury were determined in female Fischer 344 rats exposed to 0, 1 or 2 mg/m^3 vanadium pentoxide for 16 days. Lung weights of exposed rats were significantly greater than those of control animals on days 0, 1 and 4 post-exposure but were similar on day 8 post-exposure. There was little difference in lung weights between exposed groups. AUC analysis showed that lung burdens were proportional to exposure concentration throughout the recovery period. The results suggested linear toxicokinetics. Lung clearance half-lives during the 8-day recovery period were similar among exposed groups (range, 4.42–4.96 days). Concentrations of vanadium in blood were similar among exposed groups, but several orders of magnitude lower than the concentrations in lung tissue, and showed only marginal increases with increasing exposure doses.

In the 2-year inhalation study (National Toxicology Program, 2002), tissue burden analyses were performed on female Fischer 344 rats on days 1, 5, 12, 26, 54, 173, 360 and 540 after the start of exposure to 0.5, 1 or 2 mg/m^3 vanadium pentoxide. Lung weights increased throughout the study, with similar increases in the two lower-dose groups. When lung burden data were integrated over all time points, they did appear to be approximately proportional to exposure concentrations. During the two years, lung burdens in the two higher-dose groups (1 and 2 mg/m^3) did not reach a steady state, but showed an increase until day 173 followed by a decline until day 542. In contrast, the lung burden in the group exposed to 0.5 mg/m^3 increased with time and reached a steady state at 173 days. The data fitted a model in which the rate of deposition of vanadium in the lung decreased with time, while the initial deposition rates increased with the exposure concentration. Between days 1 and 542, the calculated deposition rate decreased from 0.41 to 0.25 µg/day in the 1-mg/m^3 exposure group and from 0.68 to 0.48 µg/day in the 2-mg/m^3 exposure group. There was no such change in deposition rate in the group exposed to the lowest dose (approximately 0.22 µg/day). These results are likely to be explained by altered pulmonary function in the higher-dose groups, resulting in lung clearance rates that were lower than in the low-dose group. Lung clearance half-lives were 37, 59 and 61 days for the high, medium and low exposure groups, respectively, i.e. much longer than in the 16-day study (see above). Apparently, vanadium is cleared more rapidly from the lungs of rats exposed to vanadium pentoxide for short periods of time or at low concentrations repeatedly for longer periods. From the deposition curves over the 542 days of the study, the estimated total vanadium lung doses were 130, 175 and 308 µg for the 0.5-, 1- and 2-mg/m^3 exposure groups, respectively. Normalized lung doses (µg vanadium/mg vanadium pentoxide per m^3) were not constant but decreased with increasing exposure, i.e., 260, 175 and 154 µg per mg/m^3 for low, medium and high dose groups, respectively. This decrease was due to the reduced deposition of vanadium with increasing exposure concentration. Rats retained approximately 10–15% of the estimated lung dose on day 542. Concentrations of vanadium in blood were much lower than in lung and were only marginally higher in exposed rats than in controls. Vanadium concentrations in blood of exposed animals peaked on days 26 or 54, then declined throughout the rest of the study. Because the changes were small, it was difficult to distinguish between decreased absorption from the lung, resulting from reduced deposition, and increased elimination from the blood (National Toxicology Program, 2002).

Kyono *et al.* (1999) showed that the health status of the lung influences the deposition and retention of vanadium. In an experimental model for nickel-induced bronchiolitis in rats, bronchiolitic rats and control animals were exposed to vanadium pentoxide (2.2 mg/m^3; MMAD, 1.1 µm) for 5 h. The vanadium content in the lungs of controls was higher (about 100%) than in bronchiolitic rats after 1 day of exposure, but 2 days later the retention was 20% in controls and 80% in bronchiolitic rats. Elimination of vanadium was found to be much slower in bronchiolitic rats.

(ii)   *Intratracheal instillation*

Several studies have shown that after intratracheal instillation of vanadium pentoxide in rats there was generally a rapid initial clearance of up to 50% during the first hour, a second phase with a half-life of about 2 days and a third phase during which vanadium remained in the lung for up to 63 days (Oberg *et al.*, 1978; Conklin *et al.*, 1982; Rhoads & Sanders, 1985).

(iii)   *Oral administration*

Administration of vanadium pentoxide by gavage resulted in absorption of 2.6% of the dose through the gastrointestinal tract 3 days after the treatment (Conklin *et al.*, 1982). Distribution was mainly to bone, liver, muscle, kidney, spleen and blood. Chronic treatment with inorganic vanadium salts or organic vanadium has been shown to result in significant accumulation in the bone, spleen and kidney (Mongold *et al.*, 1990; Thompson & McNeil, 1993; Yuen *et al.*, 1993).

Studies with non-diabetic and streptozotocin-diabetic rats given vanadyl sulfate in their drinking-water (0.5–1.5 mg/mL) for 1 year showed concentrations of vanadium to be in the following order [of distribution]: bone > kidney > testis > liver > pancreas > plasma > brain. Vanadium was found to be retained in these organs 16 weeks after cessation of treatment while the concentrations in plasma were below the limits of detection at this time (Dai *et al.*, 1994).

(b)   Cellular studies

Edel and Sabbioni (1988, 1989) showed accumulation of vanadium in hepatocytes and kidney cells (in the nucleus, cytosol and mitochondria) in rats exposed to vanadium as radioactive [48]V (V) pentavanadate ions and [48]V (IV) tetravalent ions by intratracheal instillation, oral administration or intravenous injection.

Cell cultures (human Chang liver cells, bovine kidney cells), incubated in medium supplemented with vanadium in the form of vanadate, have been shown to accumulate this element in the nucleus and mitochondria (Bracken *et al.*, 1985; Stern *et al.*, 1993; Sit *et al.*, 1996). In BALB/3T3 C1A31-1-1 cells incubated in the presence of sodium vanadate and vanadyl sulfate, the cellular retention of both compounds was similar. After exposure to a non-toxic dose (1 μM for 48 and 72 h), nearly all vanadium was present in the cytosol, but at a toxic dose (10 μM for 48 and 72 h), 20% of the vanadium was found in cellular organelles (Sabbioni *et al.*, 1991).

4.2   Toxic effects

4.2.1   Humans

In humans, acute vanadium poisoning can manifest itself in a number of symptoms including eye irritation and tremors of the hands (Lewis, 1959). In addition, a greenish colouration of the tongue has been observed in humans exposed to high concentrations of vanadium pentoxide and is probably due to the formation of trivalent and tetravalent vanadium complexes (Wyers, 1946). The green colour disappears within 2–3 days of cessation of exposure (Lewis, 1959).

(a)   Studies with volunteers

Zenz and Berg (1967) studied the effects of vanadium pentoxide in nine male volunteers exposed in an inhalation chamber to concentrations of vanadium pentoxide of 0.1, 0.25, 0.5 or 1.0 mg/m^3 (particle size, 98% < 5 μm) for 8 h, with follow-up periods of 11–19 months. Acute respiratory irritation was reported, which subsided within 4 days after exposure (see also Section 4.1.1).

No skin irritation was reported in 100 human volunteers after skin patch testing with 1, 2 and 10% vanadium pentoxide in petrolatum (Motolese *et al.*, 1993).

(b)   Studies of workers exposed to vanadium

There is an extensive published literature concerning the development of 'boiler-makers bronchitis' in persons cleaning boilers in which fuel oils containing high concentrations of vanadium were used (Hudson, 1964; Levy *et al.*, 1984). The clinical picture is characterized by dyspnoea which is largely reversible. Levy *et al.* (1984) studied 100 workers exposed to vanadium pentoxide ($0.05–5.3$ mg/m^3) during the conversion of a utility company power plant and found severe respiratory tract irritation in 74 individuals. Expiratory flow rates and forced vital capacity were decreased in about 50% of a subsample (35 individuals) of the workers studied.

Eye irritation has been reported in workers exposed to vanadium (Lewis, 1959; Zenz *et al.*, 1962; Lees, 1980; Musk & Tees, 1982). Skin patch testing in workforces produced two isolated reactions (but none in unexposed volunteers; see Section 4.2.1). The underlying reason for the skin responses in these workers is unclear (Motolese *et al.*, 1993).

Lewis (1959) investigated 24 men exposed to vanadium pentoxide for at least 6 months from two different centres, and age-matched with 45 control subjects from the same areas. Exposure to vanadium pentoxide was between 0.02 and 0.92 mg/m^3. In the exposed group, 62.5% complained of eye, nose, and throat irritation (6.6% in control), 83.4% had a cough (33.3% in control), 41.5% produced sputum (13.3% in control), and 16.6% complained of wheezing (0% in control). Physical findings included wheezes, rales, or rhonchi in 20.8% (0% in controls), hyperaemia of the pharynx and nasal mucosa in 41.5% (4.4% in controls), and 'green tongue' in 37.5% (0% in controls).

Zenz *et al.* (1962) reported on 18 workers exposed to varying concentrations of vanadium pentoxide dust (mean particle size, < 5 µm) in excess of 0.5 mg/m^3 during a pelletizing process. Three of the men most heavily exposed developed symptoms, including sore throat and dry cough. Examination of each on the third work day revealed markedly inflamed throats and signs of intense persistent coughing, but no evidence of wheezing. The three men also reported 'burning eyes' and physical examination revealed slight conjunctivitis. Upon resumption of work after a 3-day exposure-free period, the symptoms returned within 0.5–4 h, with greater intensity than before, despite the use of respiratory protective equipment. After the process had been operating for 2 weeks, all 18 workers, including those primarily assigned to office and laboratory duties, developed symptoms and signs to varying degrees, including nasopharyngitis, hacking cough, and wheezing. This study confirms that vanadium pentoxide exposure can produce irritation of the eye and respiratory tract.

Lees (1980) reported signs of respiratory irritation (cough, respiratory wheeze, sore throat, rhinitis, and nosebleed) and eye irritation in a group of 17 boiler cleaners. As there was no control group and it was unclear whether there was exposure to compounds other than vanadium, no conclusions can be drawn regarding the cause or significance of these symptoms. However, the findings are compatible with those of other studies on inhalation of vanadium pentoxide.

Huang *et al.* (1989) conducted a clinical and radiological investigation of 76 workers who had worked in a ferrovanadium factory for 2–28 years. In the exposed group, out of 71 workers examined, 89% had a cough (10% in controls), expectoration was seen in 74% (15% in controls), 38% were short of breath (0% in controls), and 44% had respiratory harshness or dry sibilant rale (0% in controls). In 66 of the exposed group examined, hyposmia or anosmia was reported in 23% (5% in controls), congested nasal mucosa in 80% (13% in controls), erosion or ulceration of the nasal septum in 9% (0% in controls), and perforation of the nasal septum in one subject (1.5%) (0 in controls). Chest radiographs of all 76 exposed subjects revealed 68% with increased, coarsened, and contorted broncho-vascular shadowing (23% in controls). [While exposure to vanadium compounds may have contributed to the clinical findings and symptoms reported, no firm conclusion can be drawn from this study in this regard, as mixed exposures are likely to have occurred.]

A prospective study (Hauser *et al.*, 1995a) of pulmonary function in 26 boiler workers exposed to fuel oil ash showed decreased FEV_1 (forced expiratory volume in 1 s) values which were associated with PM_{10} exposure but not with vanadium exposure. There was no post-exposure change in non-specific airway responsiveness. Hauser *et al.* (1995b) used nasal lavage analysis to study upper airway responses in 37 utility workers exposed to fuel oil ash. Responses were examined in relation to vanadium concentrations and PM_{10} particles using personal samplers. A significant increase in polymorphonuclear cells in nasal lavage was observed in samples from nonsmokers but not in smokers, suggesting that exposure to vanadium dust is associated with upper airway inflammation. In both non-smokers and smokers, a dose–response relationship between adjusted polymorphonuclear cell count and either PM_{10} or respirable vanadium dust exposure could not be found.

Woodin *et al.* (1998) studied the effects of vanadium exposure/PM_{10} concentrations in 18 boilermakers engaged in a utility boiler conversion; 11 utility workers acted as controls. The nasal lavage technique was used at various time points and interleukins (IL-6, IL-8), eosinophilic cationic protein (ECP) and myeloperoxidase (MPO) were investi-gated as biomarkers. Increases were observed in IL-8 and MPO concentrations but not IL-6 and ECP concentrations, in the exposed workers. The authors concluded that the changes observed in the upper airways were related to increased PM_{10} and upper airway vanadium concentrations. Subsequent studies in workers exposed to vanadium-rich fuel-oil ash (Woodin *et al.*, 2000) also demonstrated lower (72% versus 27% for controls) and upper (67% versus 36% for controls) airway symptoms.

While the majority of the above studies have noted reversibility of these acute pulmo-nary effects, asthma [now possibly labelled 'reactive airways disfunction syndrome'] has been reported to develop as a sequela to high, acute exposure to vanadium in some exposed workers (Musk & Tees, 1982).

(c)   *Environmental exposure*

A single epidemiological study has been conducted (Lener *et al.*, 1998) assessing indi-vidual exposure in the general population to dusts generated by a plant processing vana-dium-rich slag. It was estimated that an area with a radius of 3 km was exposed to the dust from the plant in Mnisek in the Czech Republic. The population in this area at the time of the study was 4850. The two-year study concentrated on three groups of 10–12-year-old schoolchildren: 15 children (11 boys, four girls) from the localities of Cisovice and Lisnice (Group A), the area potentially most affected by the emission of vanadium; 28 children (14 boys, 14 girls) from the locality of Mnisek (Group B), an area of medium exposure; and 32 children (17 boys, 15 girls) from the locality of Stechovive (Group C), a control area not affected by any emission from vanadium production. Vanadium concentrations in venous blood, hair and fingernail clippings were determined. The mean vanadium concentration in blood was 0.10 ± 0.07 µg/L in the exposed Group A (Group B data not given) and 0.05 ± 0.05 µg/L in the control group. In hair, the concentrations were 96 ± 42 µg/kg and 181 ± 114 µg/kg in the exposed groups A and B, respectively, compared with 69 ± 50 µg/kg in controls. Concentrations in fingernails were 189 ± 41 µg/kg and 186 ± 38 µg/kg in the exposed groups A and B, respectively, compared with 109 ± 68 µg/kg in the controls. Vanadium concentrations in blood, hair and fingernails were elevated in children living close to the plant. In group B, those with parent(s) working at the plant had higher vanadium concentrations in hair than those whose parent(s) did not, suggesting a secondary exposure in the home from dust transferred on working clothes.

Health status of the children in the study was assessed based on haematological parameters, specific immunity, cellular immunity and cytogenetic analysis. Children from the exposed groups A and B had lower red blood cell counts and lower concentrations of serum and salivary secretory IgA than control group, and a seasonal decrease in IgG. Marked differences between exposed and control groups were seen in natural cell-mediated immunity, with significantly higher mitotic activity of T-lymphocytes in children living in the immediate vicinity of the plant. A higher incidence of viral and bacterial infections was registered in children from the exposed area. However, the study could not control for confounding by exposures to compounds other than vanadium. Cytogenetic analysis revealed no genotoxic effects (see Section 4.4.1). The overall conclusion was that long-term exposure to vanadium had no negative impact on health; the differences observed were within the range of normal values in all cases (Lener *et al.*, 1998).

4.2.2   *Experimental systems*

(*a*)   *In-vivo studies*

(i)   *General toxicity*

The acute toxicity of vanadium is low when given orally, moderate when inhaled and high when injected. As a rule, the toxicity of vanadium increases as its valency increases, with vanadium (V), as in vanadium pentoxide, being the most toxic form (Lagerkvist *et al.*, 1986; WHO, 1988; National Toxicology Program, 2002).

Studies in animals have shown that equivalent doses of vanadium pentoxide are better tolerated by small animals, including rats and mice, than by larger animals, such as rabbits and horses (Hudson, 1964). The LD_{50} of vanadium pentoxide is highly species-dependent (Table 6). Differences in diet and route of vanadium administration may contribute to these discrepancies.

Table 6. Acute toxicity values for vanadium pentoxide in experimental animals

Species	Route of administration	Dose or concentration/ exposure	Parameter[a]	Reference
Mouse	Oral	23 mg/kg bw	LD_{50}	Lewis (2000)
	Subcutaneous	10 mg/kg bw	LD_{50}	Lewis (2000)
	Subcutaneous	87.5–117.5 mg/kg bw	LD	Hudson (1964)
	Subcutaneous	102 mg/kg bw	LD_{100}	Venugopal & Luckey (1978)
Rat	Oral	10 mg/kg bw	LD_{50}	Lewis (2000)
	Inhalation	70 mg/m³/2 h	LC_{LO}	Lewis (2000)
	Subcutaneous	14 mg/kg	LD_{50}	Lewis (2000)
	Intraperitoneal	12 mg/kg bw	LD_{50}	Lewis (2000)
Guinea-pig	Subcutaneous	20–28 mg/kg bw	LD	Hudson (1964)
Rabbit	Intravenous	1–2 mg/kg bw	LD	Hudson (1964)
	Intravenous	10 mg/kg	LD_{LO}	Lewis (2000)
	Inhalation	205 mg/m³/7 h	LC_{100}	Sjöberg (1950)
	Subcutaneous	20 mg/kg	LD_{LO}	Lewis (2000)
Cat	Inhalation	500 mg/m³/23 min	LC_{LO}	Lewis (2000)

[a] LD_{100}: dose which is lethal to 100% of the animals; LD_{50}, dose which is lethal to 50% of the animals; LC_{100}, concentration in air which is lethal to 100% of the animals; LC_{LO}, lethal concentration low: the lowest concentration in air which is lethal to animals; LD, lethal dose

Ammonium metavanadate given to six weanling pigs at a dose of 200 mg/kg of feed (200 ppm) for 10 weeks was found to suppress growth and increase mortality (Van Vleet *et al.*, 1981). In contrast, ammonium metavanadate was not markedly toxic when 200 mg/kg of feed (200 ppm) (approximately equivalent to 6.6 mg/kg bw) or less were fed to growing lambs for 84 days (Hansard *et al.*, 1978).

(ii)   *Respiratory effects*

Inhalation exposure

Male CD-1 mice exposed by inhalation to vanadium pentoxide (0.01-M and 0.02-M solution as aerosol, for 1 h) developed an increased mitochondrial matrix density and distorted nuclear morphology in non-ciliated bronchiolar Clara cells (Sánchez *et al.*, 2001; abstract only).

In rats and mice exposed to vanadium pentoxide at concentrations up to 16 mg/m³ for 3 months, inflammation and epithelial hyperplasia were observed in the nose and lung of rats and in the lung of mice at exposures ≥ 2 mg/m³. Non-neoplastic lesions in the nose and lung of rats were noted at all doses, and rats exposed to ≥ 4 mg/m^3 developed fibrosis (National Toxicology Program, 2002).

In addition, decreases in heart rate and in diastolic, systolic and mean blood pressure were seen in male and female F344/N rats exposed to 16 mg/m^3. These effects were not attributed to a direct cardiotoxic action of vanadium pentoxide but were considered to reflect the poor condition of the animals coupled with an effect of the anaesthesia (used to facilitate implantation of electrodes for electrocardiogram measurements). The overall pulmonary changes indicated the presence of restrictive lung disease in both sexes exposed to vanadium pentoxide concentrations of ≥ 4 mg/m^3, while an obstructive lung disease may have been present in the group exposed to 16 mg/m^3 (National Toxicology Program, 2002).

In a two-year study, F344/N rats and B6C3F$_1$ mice (50 animals per sex and per species) were exposed to vanadium pentoxide at concentrations of 0, 0.5, 1 or 2 (rats only), 1, 2 or 4 (mice only) mg/m^3, by inhalation for 2 years. Non-neoplastic proliferative and inflammatory lesions of the respiratory tract were observed in both species at increasing frequency with increased exposure concentration (see Tables 3.1.1 and 3.1.2, Section 3) (National Toxicology Program, 2002; Ress et al. 2003). The main differences observed between acute (3 months) and chronic (2 years) effects of exposure to vanadium pentoxide were the development by 2 years of chronic inflammation of the bronchi, septic bronchopneumonia, interstitial infiltration and proliferation, and emphysema (National Toxicology Program, 2002).

When rabbits were exposed to vanadium pentoxide by inhalation (8–18 mg/m^3, 2 h per day, 9–12 months) and rats to vanadium pentoxide condensation aerosol (3–5 mg/m^3, 2 h per day every 2 days, 3 months) or vanadium pentoxide dust (10–40 mg/m^3, 4 months), similar respiratory effects (sneezing, nasal discharge, dyspnoea and tachypnea) were produced in both species, which in some cases included attacks of bronchial asthma and a haemorrhagic inflammatory process (Roshchin, 1967b, 1968, cited by WHO, 1988).

In studies carried out by Sjöberg (1950), rabbits exposed to vanadium pentoxide dust (205 mg/m^3) developed tracheitis, pulmonary oedema and bronchopneumonia and died within 7 h. In another experiment, repeated inhalation of vanadium pentoxide (20–40 mg/m^3, 1 h per day, for several months) by rabbits produced chronic rhinitis and tracheitis, emphysema, patches of lung atelectasis and bronchopneumonia.

When adult male cynomolgus monkeys were exposed by inhalation to 0.5 or 5.0 mg/m^3 vanadium pentoxide dust aerosol for 1 week, significant air flow limitation was produced only at the 5.0 mg/m^3 dose in both central and peripheral airways, without changes in parenchymal function. However, analysis of BALF showed a significant increase in the absolute number and relative percentage of polymorphonuclear leukocytes, indicating that vanadium pentoxide induced pulmonary inflammatory effects (Knecht et al., 1985). In a study conducted to evaluate changes in pulmonary reactivity resulting from repeated vanadium pentoxide inhalation through the use of provocation challenges, and after different subchronic exposure regimens, one group of monkeys ($n = 8$) was exposed by inhalation (6 h per day, 5 days per week, for 26 weeks) to 0.1 mg/m^3 vanadium pentoxide on Mondays, Wednesdays and Fridays, with a twice-weekly peak exposure of

1.1 mg/m^3 on Tuesdays and Thursdays, and another group ($n = 8$) was exposed to a constant daily concentration of 0.5 mg/m^3; a control group ($n = 8$) received filtered, conditioned air. Pre-exposure challenges with vanadium pentoxide induced airway obstruction with a significant influx of inflammatory cells into the lung in both subchronic exposure groups. Inhalation of vanadium pentoxide with intermittent high exposure concentrations did not produce an increase in pulmonary reactivity to vanadium pentoxide, and cytological, immunological and skin test results indicated the absence of allergic sensitization (Knecht *et al.*, 1992).

Intratracheal exposure

Zychlinski *et al.* (1991) investigated the toxic effects of vanadium pentoxide in rats exposed intratracheally to 0.56 mg vanadium pentoxide/kg bw once a month for 12 months. Body weight gain of exposed animals slowed following the 10th treatment when compared with control animals. Lung weights were significantly greater than in controls, but other organ weights were unchanged. The glucose concentrations in blood of treated animals were slightly decreased whereas total cholesterol concentrations were reduced markedly. In parallel to this in-vivo study, in-vitro experiments with isolated untreated rat lung microsomes and mitochondria in the presence of reduced nicotinamide adenine dinucleotide phosphate (NADPH) were performed to investigate the mechanism of the chronic toxic effects of vanadium. The results showed that vanadium(V) undergoes one-electron redox cycling (enzymatic reduction) in rat lung biomembranes and that non-enzymatic reoxidation of vanadium(IV) initiates lipid peroxidation under aerobic conditions. It was postulated that free-radical redox cycling of vanadium may be responsible for the observed pulmonary toxicity.

When female CD rats were instilled intratracheally with 42 or 420 µg/kg bw vanadium pentoxide and followed from 1 h to 10 days, pulmonary inflammation was induced in a dose-dependent manner, but neutrophil influx was not detected until 24 h after exposure. Expression of mRNA for two cytokines, macrophage inflammatory protein-2 (MIP-2) and KC protein was also detected in the bronchoalveolar macrophages (Pierce *et al.*, 1996).

Bonner *et al.* (2000) reported that two weeks after a single intratracheal instillation of 1 mg/kg bw vanadium pentoxide, male Sprague-Dawley rats developed constrictive airway pathology including airway smooth muscle cell thickening, mucous cell metaplasia and fibrosis.

Evaluating the effects of a single intratracheal dose of residual oil fly ash in rats, Dreher *et al.* (1997), Kodavanti *et al.* (1998) and Silbajoris *et al.* (2000) concluded that vanadium compounds were the major toxic component inducing pulmonary injury, activation of alveolar macrophages and inflammatory changes. In addition, Silbajoris *et al.* (2000) described the induction of some mitogen-activated protein (MAP) kinases in the alveolar epithelium of the animals.

Rice *et al.* (1999) instilled Sprague-Dawley rats intratracheally with 1 mg/kg bw vanadium pentoxide and found proliferation of myofibroblasts, indicating pulmonary fibrosis. Toya *et al.* (2001), using the same model, found that intratracheal instillation with 0.88, 3.0 or 13.0 mg/kg bw vanadium pentoxide for 4 weeks induced pathological lung lesions that developed dose-dependently, and were characterized by exudative inflammation, injury of alveolar macrophages, and swelling and mucous degeneration of the broncho-bronchiolar epithelium.

(iii)   *Hepatic effects*

In mice exposed to vanadium pentoxide (0.02 M inhaled for 30 min), fatty degeneration, extramedullary haematopoietic activity and neutrophilic infiltration around the central veins were detected in the liver (Acevedo-Nava *et al.*, 2001; abstract only).

In rats and rabbits, fatty changes with necrosis in the liver and a drastic reduction in liver tissue respiration have been observed as a result of long-term exposure to vanadium pentoxide by inhalation (10–70 mg/m^3, 2 h per day, 9–12 months) (Roshchin, 1968, cited by Lagerkvist, 1986). Livers and kidneys of rats treated with vanadium(V) showed an electron paramagnetic resonance signal characteristic of vanadium(IV) (Johnson *et al.*, 1974).

The bioenergetic functions of liver mitochondria have been studied *in vivo* and *in vitro* following acute and chronic exposure of rats to vanadium pentoxide via the respiratory tract or exposure of isolated rat liver mitochondria to various vanadium pentoxide concentrations. *In vivo*, the mitochondrial respiration with glutamate (as nicotinamide adenine dinucleotide (NAD)-linked substrate) or succinate (as flavine adenine dinucleotide (FAD)-linked substrate) was inhibited significantly when compared with control animals. No inhibition was found with ascorbate as cytochrome c-linked substrate. The same effects were observed *in vitro*. These combined effects provide evidence that vanadium(V) acts as an inhibitor of respiration in rat liver mitochondria. It was postulated that significant amounts of vanadium(V) accumulated in the intermembrane space of liver mitochondria of exposed rats. The enzymatic process of detoxification, by reduction of vanadium(V) in the tissue, may be insufficient to prevent the deleterious action of this compound on liver mitochondria (Zychlinski & Byczkowski 1990).

(iv)   *Renal effects*

Glomerular hyperaemia and necrosis of convoluted tubules in the kidney were observed in some early studies of acute toxicity of vanadium compounds in various mammalian species (Hudson, 1964; Pazhynich, 1966; WHO, 1988).

Intraperitoneal administration of sodium orthovanadate to rats resulted in inhibition of tubular reabsorption of sodium and hypokalaemic distal renal tubular acidosis with increased urinary pH (Bräunlich *et al.*, 1989; Dafnis *et al.*, 1992). Vanadium, in the form of ammonium metavanadate injected subcutaneously into rats, was found to be toxic to the kidney at doses of 0.6 and 0.9 mg/kg bw per day for 16 days. Histological changes were observed, including necrosis, cell proliferation and fibrosis. Vanadium was shown to be more toxic for the kidneys in rats when given by a parenteral route (Al-Bayati *et al.*, 1989).

Chronic treatment of rats with vanadyl sulfate has been shown to result in significant accumulation of the element in the kidneys (Mongold *et al.*, 1990; Thompson & McNeill, 1993); however, most is probably bound to small peptides or macromolecules in the form of vanadyl and thus is not available as vanadate, a more potent inhibitor of Na^+/K^+-ATPases (Cantley *et al.*, 1977; Rehder, 1991; Thompson *et al.*, 1998).

(v)   *Nervous system effects*

Neurophysiological effects have been reported following acute exposure (by oral administration and subcutaneous injection) of dogs and rabbits to vanadium oxides and salts (vanadium trioxide, vanadium pentoxide, vanadium trichloride and ammonium meta-vanadate). These effects included disturbances of the central nervous system, such as impaired conditioned reflexes and neuromuscular excitability (Roshchin, 1967a). The animals behaved passively, refusing to eat, and lost weight. In cases of severe poisoning, diarrhoea, paralysis of the hind limbs and respiratory failure were followed by death (Hudson, 1964; Roshchin, 1967b, 1968).

In a study reported by Seljankina (1961 cited by Lagerkvist *et al.*, 1986 and WHO, 1988), solutions of vanadium pentoxide were administered orally to rats and mice at doses of 0.005–1 mg/kg bw per day for periods ranging from 21 days at the higher concentrations to 6 months at the lower concentrations. A dose of 0.05 mg/kg bw was found to be the threshold dose for functional disturbances in conditioned reflex activity in both mice and rats. Repeated exposure to aqueous solutions (0.05–0.5 mg/kg bw per day, for 80 days) of vanadium pentoxide impaired conditioned reflex mechanisms in rats.

In male CD-1 mice exposed by inhalation to 0.02 M vanadium pentoxide 2 h twice a week for 4 weeks, Golgi staining revealed a drastic reduction in dendritic spines in the striatum compared with controls, showing that the inhalation of vanadium causes severe neuronal damage in the corpus striatum (Montiel-Flores *et al.*, 2003; abstract only). Using the same inhalation model, after 12 weeks of exposure, a decrease in dendritic spines of granule cells of the olfactory bulb was observed (Mondragón *et al.*, 2003; abstract only). In addition, ultrastructural modifications in nuclear morphology of these cells were evident, Golgi apparatus was dilated and an increase in lipofucsin granules was observed, as well as necrosis of some cells (Colin-Barenque *et al.*, 2003; abstract only). In the cerebellum, necrosis and apoptosis of the Purkinje and granule cell layers were seen (Meza *et al.*, 2003; abstract only).

(vi)   *Cardiovascular system effects*

Perivascular swelling, as well as fatty changes in the myocardium, were observed by Roshchin (1968, cited by WHO, 1988) following chronic exposure of rats and rabbits to vanadium pentoxide ($10–70$ mg/m^3, 2 h per day, 9–12 months) by inhalation.

(vii)   *Skeletal alterations*

The effect of vanadium pentoxide on bone metabolism has been investigated in weanling rats. Vanadium pentoxide (10.0–200.0 µmol/kg bw [1.8–36.4 mg/kg bw]) was administered orally for 3 days. Low doses (10–100 µmol/kg bw [1.8–18 mg/kg bw]) caused increases in alkaline phosphatase activity and DNA content in the femoral diaphysis, indicating that vanadium pentoxide may play a role in the enhancement of bone formation *in vivo*. However, high doses (over 150 μmol/kg bw [27 mg/kg bw] had toxic inhibitory effects (Yamaguchi *et al.*, 1989).

(viii) *Immunological effects*

In the National Toxicology Program study (2002), a localized inflammatory response was seen in the lungs of male F344/N rats and female B6C3F$_1$ mice exposed by inhalation to 4, 8, or 16 mg/m^3 vanadium pentoxide in a 16-day study. Increases in cell numbers, protein, neutrophils and lysozymes in BALF were observed but the number of macro-phages in lavage fluids of male rats and female mice exposed to 8 or 16 mg/m^3 was decreased. No effects were seen on systemic immunity in rats and mice.

When weanling and adult ICR mice were given 6 mg/kg bw vanadium pentoxide by gavage (5 days per week for 6 weeks), an increase in the number of leukocytes and plaque-forming cells, as well as enhanced phytohaemagglutinin responsiveness, increased spleen weight and depression of phagocytosis were observed in treated mice. In Wistar rats given vanadium pentoxide in drinking-water (1 or 100 mg/L for 6 months), the higher dose resulted in increased spleen weight and concanavalin-A responsiveness; a depression of phagocytosis was found in a dose-dependent manner. These results suggest activation of T- and B-cell immune responses (Mravcová *et al.*, 1993).

(ix) *Biochemical effects*

Chakraborty *et al.* (1977) gave male albino rats vanadium pentoxide orally at a dose of 3 mg/kg bw five times a week for the first week and 4 mg/kg bw for a further 2 weeks and found that it induced histological and enzymatic alterations including inhibition of biosynthesis, enhanced catabolism and increased use of L-ascorbic acid in the liver and kidney tissues of the rats.

(*b*) *In-vitro studies*

(i) *Organ culture*

Garcia *et al.* (1981) found that treatment with vanadium pentoxide ($10^{-5}–10^{-2}$ M [1.82–1820 μg/mL]) produced dose-dependent contractions of the rat vas deferens organ cultures *in vitro*; a response that could be associated with the inhibition of Na$^+$/K$^+$-ATPase activity.

Schiff and Graham (1984) used organ cultures of hamster trachea to study the in-vitro effects of vanadium pentoxide (0.1, 1, 10 or 100 μg/mL) and oil-fired fly ash (10, 50, 100 or 250 μg/mL) on mucociliary respiratory epithelium following exposure for 1 h per day for 9 consecutive days. Vanadium pentoxide was found to decrease ciliary activity and produce ciliostasis in tracheal ring explants. The degree of change depended on the concentration and length of exposure; early morphological alterations consisted of vacuolization of both nuclei and cytoplasm of tracheal epithelium cells.

Preincubation of rat kidney brush border membrane vesicles with 1 mM [182 μg/mL] vanadium pentoxide for 8 h significantly inhibited citrate uptake in a time-dependent manner. This effect was attributed to a direct interaction of vanadium with the sodium citrate cotransporter. The results suggest that vanadium pentoxide has nephrotoxic potential (Sato *et al.*, 2002).

(ii)   *Cell culture*

In cultures of bovine alveolar macrophages, Fisher *et al.* (1986) found that vanadium pentoxide was the most cytotoxic compound when compared with other metals or metalloids (zinc oxide, nickel sulfide, manganese oxide, sodium arsenite, sodium selenite) tested. Vanadium caused a reduction in phagocytosis by macrophages to 50% of control values after incubation for 20 h at a concentration of 0.3 μg/mL, but this concentration was also associated with a substantial (59%) loss of macrophage viability. The authors concluded that their results confirmed those of previous studies (Waters *et al.*, 1974) which demonstrated that vanadium is a unique macrophage toxicant.

Vanadium(V) and related compounds are known to exert potent toxic effects on a wide variety of biological systems. One of the pathways of vanadium(V) toxicity is thought to be mediated by oxygen-derived free radicals (Zychlinski *et al.*, 1991; Shi *et al.*, 1997; Ding *et al.*, 1999).

Parfett and Pilon (1995) evaluated the effects of promoters such as vanadium compounds on oxidative stress-regulated gene expression and promotion of morphological transformation in C3H/10T1/2 cells. Promoters which elevate intracellular oxidant levels can be distinguished by a spectrum of induced gene expression which includes the oxidant-responsive murine proliferin gene family. Proliferin transcription was found to be induced 20-fold by 5 μM [0.9 μg/mL] vanadium pentoxide. Another pentavalent vanadium, ammonium metavanadate (5 μM [0.6 μg/mL]), added as promoter in two-stage morphological transformation assays, amplified yields of Type II and Type III foci in monolayers of 20-methylcholanthrene-initiated C3H/10T1/2 cells. These results suggest that pentavalent vanadium compounds could promote morphological transformation in these cells by creating a cellular state of oxidative stress, which induces the expression of proliferin. Proliferation of MCF-7 cells was found to be stimulated after 4-day treatments with 0.5–2 μM vanadium(V); the effect reached a plateau at 1 μM vanadium, declined at 3 μM and disappeared at 5 μM (Auricchio *et al.*, 1995; 1996).

To determine the effect of vanadium pentoxide on the release of two major immunoregulatory cytokines, mouse macrophage-like WEHI-3 cells were treated *in vitro* (Cohen *et al.*, 1993). Vanadium pentoxide decreased the release of IL-1 and TNFα stimulated with lipopolysaccharide endotoxin. Spontaneous release of the IL-1/TNF-regulating prostanoid prostaglandin E_2 (PGE$_2$) was significantly increased by the highest concentration of ammonium metavanadate tested, although lipopolysaccharide endotoxin-stimulated PGE$_2$ production was unaffected. These results showed that pentavalent vanadium could alter the host's immunocompetence. In another study with WEHI-3 cells treated with 100 μM or 100 nM vanadium pentoxide or ammonium metavanadate, the capacity of macrophage-like cells to bind and respond to interferon γ was altered (Cohen *et al.*, 1996).

When mice and rat hepatocytes or human Hep G2 cells were treated *in vitro* with vanadium pentoxide (1, 10 or 100 μM), gene expression (after 2-h treatment) and secretion of IL-8, MIP-2 chemokines and TNFα (after 18-h treatment) were increased. The induction of IL-8 and MIP-2 secretion was inhibited by antioxidants such as tetra-methylthiourea and *N*-acetylcysteine, showing that the events responsible for this gene expression involve cellular redox changes (Dong *et al.*, 1998). Vanadium pentoxide caused a several-fold increase in heparin-binding epidermal growth factor-like growth factor (HB-EGF) mRNA expression and protein in normal human bronchial epithelial cells and increased the release of HB-EGF mitogenic activity of these cells (Zhang *et al.*, 2001a).

Wang and Bonner (2000) showed that vanadium pentoxide activated extracellular signal-regulated kinases 1 and 2 (ERK-1/2) in rat pulmonary myofibroblasts. This acti-vation was an oxidant-dependent event and required components of an epidermal growth factor-receptor signalling cascade.

Ingram *et al.* (2003) showed that vanadium pentoxide stimulated HB-EGF mRNA expression and hydrogen peroxide production by human lung fibroblasts. Both vanadium pentoxide and hydrogen peroxide activated ERK-1/2 and p38 MAP kinases. Inhibitors of these two kinase-pathways significantly reduced both vanadium and H_2O_2-induced HB-EGF expression. These data indicate that vanadium upregulates HB-EGF via ERK and p38 MAP kinases.

Evidence suggests that some forms of vanadium (sodium metavanadate, peroxovana-date and pervanadate) or vanadium-containing particles from environmental and occupa-tional sources can trigger or potentiate apoptosis. The pentavalent form of vanadium has been shown to cause apoptosis in a JB6 P^+ mouse epidermal cell line (Cl 41) and in lym-phoid cell lines, but may be anti-apoptotic in others such as malignant glioma cells (Hehner *et al.*, 1999; Chin *et al.*, 1999; Huang *et al.*, 2000; Chen *et al.*, 2001).

Rivedal *et al.* (1990) found that vanadium pentoxide exposure for 5 days promoted the induction of morphological transformation of hamster embryo cells pre-exposed to a low concentration of benzo[*a*]pyrene for 3 days. However, when vanadium pentoxide (0.25, 0.50 or 0.75 μg/mL) was tested in the Syrian hamster embryo (SHE) assay, the results were nega-tive after a 24-h exposure, but significant morphological transformation was produced after a 7-day exposure. This pattern of response (24-h SHE negative/7-day SHE positive) has been seen with other chemicals (i.e., 12-*O*-tetradecanoylphorbol 13-acetate, butylbenzyl phthalate, methapyrilene) that have tumour promotion-like characteristics (Kerckaert *et al.*, 1996a,b).

(iii)   *Cell-free systems*

In cell-free systems, vanadium(V) caused the oxidation of thiols, including GSH and cysteine, and induced the formation of thiyl radicals (Shi *et al.*, 1990; Byczkowski & Kulkarni, 1998). It has been shown that depletion of GSH not only decreases the antioxi-dant defence in the cytosol, but also prevents regeneration of a vital lipid-soluble antioxi-dant, α-tocopherol, thereby increasing the vulnerability of phospholipid-rich biomem-branes to oxidative stress and lipid peroxidation (Byczkowski & Kulkarni, 1998).

Vanadium can inhibit a variety of enzymes such as heart adenyl cyclase and protein kinase, ribonucleases, phosphatases, and several adenosinetriphosphatases (ATPases), but it can stimulate a number of others. The enzymes inhibited include phosphoenzyme iontransport ATPases, acid and alkaline phosphatases, $Na^++K^+ATPase$, $H^++K^+ATPase$, phosphotyrosyl protein phosphatase, dynein (contractile protein ATPase associated with microtubules of cilia and flagella), myosin ATPase, phosphofructokinase, adenylate kinase and cholinesterase (Nechay, 1984; WHO, 1988).

Vanadium(V) appears to undergo a redox cycling when the inner mitochondrial membrane permeability barrier to vanadate polyanions is broken. It has been proposed that vanadium(V) stimulates the oxidation of NAD(P)H by biological membranes and amplifies the initial generation of $O_2^{-\bullet}$ produced by membrane-associated NAD(P)H oxidase. This stimulatory effect is due to interaction of vanadium(V) with $O_2^{-\bullet}$ but not with the membrane-associated enzymes (Liochev & Fridovich, 1988).

Using ESR spin trapping, Shi and Dalal (1992) demonstrated that rat liver microsomes/NADH, in the absence of exogenous H_2O_2, generated hydroxyl ($^\bullet$OH) radicals from the reduction of vanadium(V) via a Fenton-like mechanism. This radical generation may play a role in vanadium(V)-induced cellular injury.

4.3   Reproductive and developmental effects

4.3.1   *Humans*

No data were available to the Working Group.

4.3.2   *Experimental systems*

(*a*)   *In-vivo studies*

Several studies describe the reprotoxic (male or female reproductive capability) and developmental (teratological) effects of vanadium pentoxide (Lagerkvist *et al.*, 1986; Domingo, 1994; Leonard & Gerber, 1994; Domingo, 1996; Leonard & Gerber, 1998; National Toxicology Program, 2002).

(i)   *Toxicokinetics in pregnant animals*

Li *et al.* (1991) treated non-pregnant and pregnant Wistar rats with 5 mg/kg vanadium pentoxide intraperitoneally and reported the tissue distribution of this compound. Non-pregnant rats had significant concentrations of vanadium in kidney, ovary, uterus and liver, suggesting that female genital organs are important target organs in the distribution of vanadium. Treatment of pregnant rats gave similar results, including the presence of vanadium in the placenta. The authors suggested that vanadium could pass the blood–placenta barrier.

Zhang *et al.* (1991a) analysed the passage of vanadium across the placenta into the embryo/fetus of pregnant Wistar rats at different times after different dose regimens: 4 h after treatment with a single intraperitoneal injection of vanadium pentoxide (5 mg/kg bw) on day 12 of gestation; 1, 4, 24 or 48 h after a single treatment (5 mg/kg bw) on days 16–18 of gestation; or 120 h after the final treatment with 0.33, 1 or 3 mg/kg bw given daily on days 6–15 of gestation. The concentrations of vanadium in maternal blood, placenta and fetus were elevated after these different treatments in comparison with those of the respective untreated groups. The vanadium concentration in fetuses increased with increasing doses, suggesting that the embryo/fetus accumulated vanadium (Zhang *et al.*, 1991a).

(ii)   *Effects on reproductive organs and fertility*

Male CD-1 mice were treated intraperitoneally with 8.5 mg/kg bw vanadium pentoxide once every 3 days for 60 days. Groups of five animals were killed every 10 days after the beginning of treatment. Twenty-four hours after the last injection, the males were mated with untreated females. A decrease in fertility rate, implantations, live fetuses and fetal weight, and an increase in the number of resorptions/dam was observed. In males, sperm count and motility were impaired as treatment advanced and the presence of abnormal sperm was observed on days 50 and 60 of treatment (Altamirano-Lozano & Alvarez-Barrera, 1996; Altamirano-Lozano *et al.*, 1996).

In a National Toxicology Program study (2002), reduced epididymal sperm motility was observed in B6C3F$_1$ mice exposed to vanadium pentoxide by inhalation (8- and 16 mg/m^3 dose groups) for 3 months. There were no effects on estrous cycle parameters in females. No effects were seen on reproductive parameters in male and female F344/N rats exposed by inhalation to 4, 8 or 16 mg/m^3 vanadium pentoxide (National Toxicology Program, 2002).

To evaluate the effect of vanadium pentoxide on the newborn rats, Altamirano *et al.* (1991) injected 12.5 mg/kg bw vanadium pentoxide intraperitoneally into male and female prepubertal CII-ZV rats every 2 days (from birth to 21 days), and into female rats from day 21 to the day of the first vaginal estrus. No changes in vaginal opening nor in the estrous cycle were observed in either prepubertal or adult female rats; however, the ovulation rate was reduced in the treated adult females. No differences were observed in the weights of ovaries, uterus, adrenal gland or pituitary gland, compared to those of untreated rats; the weights of thymus, liver, kidneys and submandibular glands of newborn treated females were similar to those of controls. However, when treatment began at 21 days of age, an increase in the weight of thymus, submandibular glands and liver was observed. In male prepubertal rats, an increase was observed in the weight of seminal vesicles, thymus and submandibular glands but not of testis and prostate of animals treated with vanadium from birth to 21 days. The results indicate that, as observed with other metals, the toxicological effects of vanadium pentoxide differ in males and females, with toxicity in prepubertal rats being higher in males than in females.

(iii)   *Developmental effects*

To evaluate the effects of vanadium pentoxide on the embryonic and fetal development of mice, Wide (1984) injected pregnant albino NMRI mice via the tail veins with

1.5 mM/animal [273 µg/animal ~ 10 mg/kg bw] vanadium pentoxide on day 3 or day 8 of gestation. All animals were killed 2 days before parturition (17th day of pregnancy) and fetuses were dissected and examined. Treatment with vanadium pentoxide on day 8 of gestation did not induce teratogenic effects but reduced fetal skeletal ossification.

In a study of the developmental toxicity of vanadium pentoxide, Zhang *et al.* (1991b) injected pregnant female NIH mice intraperitoneally with 5 mg/kg bw vanadium pentoxide per day on different days of gestation (days 1–5, 6–15, 7, 8, 9, 10, 11 or 14–17 of pregnancy). No effects on pre-implantation were found, nor malformations nor premature birth. However, an increased frequency of resorptions or fetal death was observed in animals treated on days 7, 6–15, and 14–17 of gestation. Delayed skeletal ossification was noted in mice treated on days 6–15, 8, 10 and 14–17 of gestation. The authors suggested that vanadium pentoxide acted as a weak developmental toxicant but not a teratogen.

To evaluate the teratogenic effects of vanadium pentoxide, female CD-1 mice were injected intraperitoneally once daily on days 6–15 of gestation with 8.5 mg/kg bw. Vanadium did not cause significant adverse effects on the number of live and dead fetuses (including resorptions) nor on fetal implants; however, a decrease in fetal weight and a delay in skeletal ossification were observed. Limb shortening was the most frequent alteration. No maternal toxicity was detected (Altamirano-Lozano *et al.*, 1993).

In female Wistar rats exposed to 0.33, 1 or 3 mg/kg bw vanadium pentoxide from days 6–15 of gestation, the highest dose was toxic. Increased fetal mortality and external or skeletal malformations with delay in ossification were also observed (Zhang *et al.*, 1993a). Similar results were found in one further study in Wistar rats (Zhang *et al.*, 1993b).

(b)   *In-vitro studies*

Li *et al.* (1995) investigated the toxicological effects of vanadium pentoxide (0.125, 0.25, 0.5, 2 or 3 mM) in rat Leydig cells *in vitro* and found no obvious relationship between testosterone secretion and the concentration of vanadium. The authors concluded that Leydig cells are not a target for vanadium pentoxide. This is in agreement with results of in-vivo studies previously reported by Altamirano *et al.* (1991) who had shown that the weight of the testis and prostate were not increased after vanadium treatment of rats (see Section 4.3.2(ii)).

Altamirano-Lozano *et al.* (1997, 1998a) tested the reprotoxic effects of various metal compounds on boar spermatozoa *in vitro*. Sperm were exposed to vanadium pentoxide (5.5, 16.5, 27.5, 55, 110 or 220 µM) and motility was analysed 0, 1, 2, 3, 4, 5 and 6 h after treatment. A dose- and time-dependent reduction in sperm motility was observed, in accordance with results obtained *in vivo* in mice by the same group (Altamirano-Lozano *et al.*, 1996).

4.4   Genetic and related effects

4.4.1   *Humans*

Lener *et al.* (1998) studied children exposed to vanadium in air in an area close to a plant processing vanadium-rich slag (see Section 4.2.3). Group A comprised 15 children from the area potentially most affected by vanadium emissions; Group B, 28 children from an area of medium exposure; and Group C, 32 children was the control group. No significant induction of chromosomal aberrations was found in the lymphocytes of exposed children (1.2 ± 1.2 in Group A; 1.3 ± 1.1 in Group B) compared with the control group (0.95 ± 0.97). Sister chromatid exchange was analysed in exposed children (4.6 ± 1.0 in Group A; 4.6 ± 0.87 in Group B) but no data were available from controls. However, the authors concluded that these results revealed no genotoxic effects of vanadium exposure.

Only one in-vivo study of the genotoxic action of vanadium pentoxide in adult humans has been reported. Ivancsigts *et al.* (2002) studied the effect of occupational exposure to vanadium pentoxide by measuring DNA strand breaks using the single-cell gel electrophoresis assay 'Comet Assay', formation of 8-hydroxy-2′-deoxyguanosine, and the frequency of sister chromatid exchange in whole blood or lymphocytes of 49 male workers in a vanadium-processing factory. Although there was significant vanadium uptake (mean vanadium concentration in serum, 5.38 µg/mL), no increase in cytogenetic end-points nor in oxidative DNA damage was observed in the cells from these workers.

4.4.2   *Experimental systems*

(*a*)   *Biochemical assays*

Effects of vanadium compounds on DNA-metabolizing enzymes have been reported by Sabbioni *et al.* (1983). Vanadate(V) ions (10^{-7}–10^{-3} M) inhibited calf thymus terminal deoxynucleotidyl transferase (with an apparent Ki of 2.5 µM) and the catalytic activity of mammalian DNA polymerase α (at I_{50} of 60 µM), while bacterial DNA polymerase-I was inhibited when the concentration was increased to about 0.5 mM.

(*b*)   *Mutagenicity* (see Table 7)

(i)   *In-vitro studies*

The mutagenicity of vanadium compounds has been reviewed (Graedel *et al.*, 1986; Léonard & Gerber, 1994; Altamirano-Lozano *et al.*, 1998b; Léonard & Gerber, 1998; National Toxicology Program, 2002).

The majority of the results of mutagenic activity of vanadium have been shown in *Escherichia coli* and *Salmonella typhimurium* (Hansen & Stern, 1984; Graedel *et al.*, 1986; Leonard & Gerber, 1994); there is one study only with exogenous metabolic activation (National Toxicology Program, 2002).

Early studies demonstrated that vanadium pentoxide was more genotoxic in recombination-repair-deficient (rec⁻) strains of *Bacillus subtilis* than in the wild-type rec⁺

Table 7. Genetic and related effects of vanadium pentoxide

Test system	Result[a] Without exogenous metabolic system	Result[a] With exogenous metabolic system	Dose[b] (LED/HID)	Reference
Escherichia coli, spot test B/r WP2try⁻, WP2hcr⁻try⁻	−	NT	0.5 M	Kanematsu *et al.* (1980)
Escherichia coli; WP₂, WP₂uvrA, CM₈₉₁, reversion assay	+	NT	1200 µg/plate	Si *et al.* (1982)[c]
Escherichia coli, ND160 and MR102, frameshift mutation	−		1200 µg/plate	Si *et al.* (1988)[c]
Bacillus subtilis, M45 recombination-repair-deficient (rec⁻)	+	NT	0.5 M	Kanematsu & Kada (1978); Kada *et al.* (1980); Kanematsu *et al.* (1980)
Bacillus subtilis H17 (rec⁺) and M45 (rec⁻) recombination-repair-deficient	+	NT	100 000	Sun (1996)
Salmonella typhimurium, TA100, TA1535, TA1537, TA1538, (his⁻)	−	NT	0.5 M	Kanematsu *et al.* (1980)
Salmonella typhimurium, TA100, TA98, TA102, TA1535 reverse mutation	−	−	333 µg/plate	National Toxicology Program (2002)
Salmonella typhimurium, TA97, TA98, TA100, TA102 reverse mutation	−	NT	200 µg/plate	Zen *et al.* (1988)[c]
Gene mutation, 6-thioguanine resistant mutation, Chinese hamster lung fibroblast cell line (V79) *in vitro*	−		4	Zhong *et al.* (1994)
Sister chromatid exchanges, Chinese hamster lung fibroblast cell line (V79) *in vitro*	−		4	Zhong *et al.* (1994)
Micronucleus formation in binucleated cells, cytochalasin-B assay, Chinese hamster lung fibroblast cell line (V79) *in vitro*	+	NT	1	Zhong *et al.* (1994)
Numerical chromosomal aberrations, endoreduplication, Chinese hamster lung fibroblast cell line (V79) *in vitro*	+		1	Zhong *et al.* (1994)

Table 7 (contd)

Test system	Result[a]		Dose[b] (LED/HID)	Reference
	Without exogenous metabolic system	With exogenous metabolic system		
Numerical chromosomal aberrations, aneuploidy, kinetochore staining of micronuclei in binucleated cells, Chinese hamster lung fibroblast cells line (V79) in vitro	+		1	Zhong et al. (1994)
DNA strand breaks, alkaline 'Comet Assay', human lymphocytes in vitro	+	NT	0.3 µM	Rojas et al. (1996a, b)
Inhibition of double-strand DNA breaks repair, alkaline and neutral 'Comet Assay', human fibroblasts in vitro	+		UV (4.8 kJ/m^2) + V$_2$O$_5$ 0.5 µM Bleomycin (1 µg/mL) + V$_2$O$_5$ 0.5 µM	Ivancsists et al. (2002)
Sister chromatid exchanges, human lymphocytes in vitro	−	NT	47 M	Sun et al. (1989)[c]
Sister chromatid exchanges, human lymphocytes in vitro	−	NT	6	Roldán & Altamirano (1990)
Sister chromatid exchanges, human lymphocytes in vitro	+	NT	4[d]	Roldán-Reyes et al. (1997)
Structural chromosomal aberrations, human lymphocytes in vitro	−	NT	6	Roldán & Altamirano (1990)
Numerical chromosomal aberrations, polyploidy, human lymphocytes in vitro	+	NT	2	Roldán & Altamirano (1990)
Aneuploidy, FISH centromeric probes, human lymphocytes in vitro	+	NT	0.001 µM	Ramírez et al. (1997)
Inhibition of microtubule polymerisation, immunostaining, human lymphocytes in vitro	+	NT	0.1 µM	Ramírez et al. (1997)
Chromosomes associated and satellite association, human lymphocyte in vitro	+	NT	4	Roldán & Altamirano (1990)

Table 7 (contd)

Test system	Result[a] Without exogenous metabolic system	With exogenous metabolic system	Dose[b] (LED/HID)	Reference
DNA strand breaks, alkaline 'Comet Assay', in several organs of CD-1 mice in vivo	+		5.75 ip	Altamirano-Lozano et al. (1996, 1999)
DNA synthesis, inhibition assay, mice testes, spleen, liver and lymphocytes in vivo	–		58.4 po	Zen et al. (1988)[c]
Sister chromatid exchanges, CD-1 mice, bone marrow, in vivo	-		23 ip	Altamirano-Lozano et al. (1993); Altamirano-Lozano & Alvarez-Barrera (1996)
Micronucleus formation, 615 and Kunming albino mice, bone marrow, in vivo	+		0.17 ip	Si et al. (1982)[c]
Micronucleus formation, 615 and Kunming albino mice, bone marrow, in vivo	+		0.25 sc	Si et al. (1982)[c]
Micronucleus formation, 615 and Kunming albino mice, bone marrow, in vivo	+		0.5 mg/m^3, inhal.	Si et al. (1982)[c]
Micronucleus formation, Kunming albino mice, bone marrow, in vivo	–		11.3 po	Sun et al. (1989)[c]
Micronucleus formation, Kunming albino pregnant mice, fetal liver, maternal bone marrow, maternal spleen, in vivo	+		0.2–5 ip[e]	Liu et al. (1992)[c]
Micronucleus formation, B6C3F1 mice, peripheral blood erythrocytes, in vivo	–		16 mg/m^3, inhal.	National Toxicology Program (2002)
Structural chromosomal aberrations, CD-1 mice, bone marrow, in vivo	–		23 ip	Altamirano-Lozano & Alvarez-Barrera (1996)

Table 7 (contd)

Test system	Result[a] Without exogenous metabolic system	Result[a] With exogenous metabolic system	Dose[b] (LED/HID)	Reference
Structural chromosomal aberrations, albino rat, bone marrow cells, in vivo	?		4 po	Giri et al. (1979)
Dominant lethal mutations, CD-1 mice in vivo	+		8.5 ip	Altamirano-Lozano et al. (1996)
Dominant lethal mutations, CD-1 mice in vivo	−		4 sc	Si et al. (1982)[c]

FISH, fluorescence in-situ hybridization

[a] +, positive; −, negative; (+), weak positive; NT, not tested;?, inconclusive

[b] LED, lowest effective dose; HID, highest ineffective dose; in-vitro tests, µg/mL, except where stated otherwise; in-vivo tests, mg/kg bw per day; po, orally, by gavage; sc, subcutaneously; ip, intraperitoneally; inhal., by inhalation

[c] Cited in Sun (1996)

[d] Combined with 20 µg of caffeine

[e] LED not given

(Kanematsu & Kada, 1978; Kanematsu *et al.*, 1980). However, vanadium pentoxide was not mutagenic in several strains of *E. coli* or *S. typhimurium*. But Si *et al.* (1982) (cited by Sun *et al.*, 1996) demonstrated that vanadium pentoxide induced reverse mutations in *E. coli* WP2, WP2uvrA and CM-981, but not frameshift mutations in strains ND-160 or MR102. This compound showed negative results in *S. typhimurium* strains TA100, TA1535, TA1537, TA1538, TA97, and TA98.

Bis(cyclopentadienyl)vanadium chloride (1 to 33 µg/plate) was mutagenic or weakly mutagenic in strains TA97 and TA100 without exogenous metabolic activation system, but not mutagenic in strains TA1535 and TA98 with or without metabolic activation (Zeiger *et al.*, 1992).

In another series of studies, vanadium pentoxide (0.33 to 333.00 µg/plate) was not mutagenic in *S. typhimurium* strains TA97, TA98, TA100, TA102 or TA1535, with or without induced rat or hamster liver S9 enzymes (National Toxicology Program, 2002).

No increase in the frequency of micronucleated normochromatic erythrocytes was seen in peripheral blood samples from male or female B6C3F$_1$ mice exposed to vanadium pentoxide by inhalation in concentrations up to 16 mg/m^3 for 3 months. Furthermore, no effect was seen in the ratio of polychromatic erythrocytes/normochromatic erythrocytes in peripheral blood, indicating a lack of toxicity to the bone marrow by vanadium pentoxide (National Toxicology Program, 2002).

[The Working Group was aware of positive results on induction of mitotic recombination by vanadium pentoxide in *Drosophila*; the data were reported in BSc and MSc theses].

In Chinese hamster lung fibroblast cell lines, vanadium pentoxide induced endoreduplication and micronuclei which were shown to be kinetochore-positive, but did not induce gene mutation nor sister chromatid exchange.

In human lymphocytes cultured *in vitro*, positive genotoxic effects of vanadium pentoxide were demonstrated for the induction of DNA damage with the alkaline 'Comet Assay' (two studies from the same laboratory), sister chromatid exchange when the compound was given in combination with caffeine (one study out of three), chromosomes associated, satellite associations and polyploidy with Hoechst staining (a single study), aneuploidy with fluorescence in-situ hybridization staining and inhibition of microtubule polymerization with immunostaining (a single study).

Vanadium pentoxide was shown to inhibit repair of double-strand breaks induced in human fibroblasts by UV radiation or bleomycin in both the neutral and alkaline comet assays.

(ii)   *In-vivo studies*

In CD-1 mice, induction of DNA damage by vanadium pentoxide administered intraperitoneally was demonstrated with the alkaline 'Comet Assay' in several organs. In the same mouse strain, a lack of sister chromatid exchange and chromosomal aberrations was reported in bone marrow; however, dominant lethal effects were observed after intraperitoneal injection of vanadium pentoxide (8.5 mg/kg bw).

In 615 and Kunming albino mice, micronuclei were induced in bone marrow by vanadium pentoxide administered by inhalation, by subcutaneous injection or by intraperitoneal injection. The results were negative following oral administration. Micronuclei
were also seen in fetal liver after intraperitoneal injection of vanadium pentoxide into
pregnant mice. No induction of dominant lethals was observed.

A single in-vivo study of the induction of chromosomal aberrations in albino rats was
inconclusive (number of animals not reported).

(c)   Genetic changes in vanadium pentoxide-induced tumours

In a National Toxicology Program study (2002), male and female B6C3F$_1$ mice were
exposed by inhalation to 1, 2, or 4 mg/m^3 vanadium pentoxide for 2 years (see
Section 3.1.1). The lung carcinomas that developed as a result of this exposure showed a
high frequency of K-*Ras* mutation, loss of heterozygosity in the region of the K-*Ras* gene
on chromosome 6 and activation of MAP kinase (Zhang *et al.*, 2001b; Devereux *et al.*, 2002;
National Toxicology Program, 2002). The authors concluded that these genetic alterations
played an important role in vanadium pentoxide-induced lung carcinogenesis. On the other
hand, there was no evidence of overexpression of mutant p53 suggesting no evidence of a
role for altered p53 function in the lung carcinomas due to exposure to vanadium pentoxide
(Devereux *et al.*, 2002; National Toxicology Program, 2002).

4.5   Mechanistic considerations

Vanadium pentoxide is considered to induce oxidative damage leading to DNA alkali-
labile sites and DNA strand breakage.

Inhibition of microtubule polymerization may explain the aneugenic effects of vanadium pentoxide. Whether these spindle disturbances are related to oxidative damage or to
direct interaction with vanadium cations is unclear. Indirect effects of vanadium pentoxide
through inhibition of various enzymes involved in DNA synthesis and DNA repair also
contribute to its genotoxicity.

Induction of dominant lethal mutations in mice may result from one, or a combination, of the modes of action mentioned above.

5.   Summary of Data Reported and Evaluation

5.1   Exposure data

Vanadium is widely distributed in the earth's crust in a wide range of minerals and in
fossil fuels. Vanadium pentoxide, the major commercial product of vanadium, is mainly
used in the production of alloys with iron and aluminium. It is also used as an oxidation
catalyst in the chemical industry and in a variety of minor applications. Exposure to vana-

dium pentoxide in the workplace occurs during the refining and processing of vanadium-rich mineral ores, during the burning of fossil fuels, especially petroleum, during the handling of vanadium catalysts in the chemical manufacturing industry and during the cleaning of oil-fired boilers and furnaces. Exposure to vanadium can also occur from ambient air contaminated by the burning of fossil fuels and, at much lower levels, from contaminated food and drinking-water.

5.2   Human carcinogenicity data

No data were available to the Working Group.

5.3   Animal carcinogenicity data

Vanadium pentoxide was tested for carcinogenicity in a single study in mice and rats by inhalation exposure. In both male and female mice, the incidences of alveolar/bronchiolar neoplasms were significantly increased, and there were also increases in male rats. It was uncertain as to whether a marginal increase in alveolar/bronchiolar neoplasms in female rats was related to exposure to vanadium pentoxide.

5.4   Other relevant data

Vanadium pentoxide is rapidly absorbed following inhalation, but poorly through dermal contact or ingestion. Elimination from the lung is initially fast, but complete only after several days. Lung retention can increase due to impaired health status of the lung. Distribution of vanadium pentoxide is mainly to the bone and kidney.

The major non-cancer health effect associated with inhalation exposure to vanadium pentoxide involves acute respiratory irritation, characterized as 'boilermakers bronchitis'. This clinical effect appears to be reversible. Green coloration of the tongue is another frequently observed clinical manifestation of intoxication with vanadium pentoxide.

Vanadium has been recognized as an essential nutritional requirement in animals of high order, but its function is not clear. Vanadium pentoxide has important effects on a broad variety of cellular processes. It stimulates cell differentiation, it causes cell and DNA injury via generation of reactive oxygen species and it alters gene expression. The many biochemical effects induced by vanadium pentoxide, such as the inhibition of a number of different enzymes, can explain many of the metabolic effects observed in experimental animals treated with this compound.

Vanadium pentoxide can pass the blood–placenta barrier. It has been reported to be teratogenic in rodents and it affects sexual development in pre-pubertal animals, the toxicity in males being greater than that in females. The reduced fertility seen in male mice was confirmed by a reduction in sperm motility *in vitro*.

Vanadium pentoxide is mutagenic *in vitro* and possibly *in vivo* in mice. It shows clastogenic and aneugenic activity in cultured mammalian cells, the latter effect probably being due to disturbance of spindle formation and chromosome segregation. Vanadium pentoxide has been reported to inhibit enzymes involved in DNA synthesis and repair of DNA damage. Data on genetic effects in humans exposed to vanadium pentoxide are scarce.

5.5   Evaluation

There is *inadequate evidence* in humans for the carcinogenicity of vanadium pentoxide. There is *sufficient evidence* in experimental animals for the carcinogenicity of vanadium pentoxide.

Overall evaluation

Vanadium pentoxide is *possibly carcinogenic to humans (Group 2B)*.

6.   References

Acevedo-Nava, S., López, I., Bizarro, P., Sánchez, I., Pasos, F., Delgado, V., Vega, M.I., Calderón, N. & Fortoul, T.I. (2001) Alteraciones morfológicas en el hígado de ratón por inhalacion de vanadio. In: *Proceedings of the IV Congreso Mexicano de Toxicología, Mérida, Yucatán, México*

ACGIH Worldwide® (2003) *Documentation of the TLVs® and BEIs® with other Worldwide Occupational Exposure Values — CD-ROM — 2003*, Cincinnati, OH

Al-Bayati, M.A., Giri, S.N., Raabe, O.G., Rosenblatt, L.S. & Shifrine, M. (1989) Time and dose-response study of the effects of vanadate on rats: Morphological and biochemical changes in organs. *J. environ. Pathol. Toxicol. Oncol.*, **9**, 435–455

Altamirano, M., Ayala, M.E., Flores, A., Morales, L. & Dominguez, R. (1991) Sex differences in the effects of vanadium pentoxide administration to prepubertal rats. *Med. Sci. Res.*, **19**, 825–826

Altamirano-Lozano, M.A. & Alvarez-Barrera, L. (1996) Genotoxicity and reprotoxic effects of vanadium and lithium. In: Collery, P., Corbella, J., Domingo, J.L., Etienne, J.C. & Llobert, J.M., eds, *Metal Ions in Biology and Medicine*, Vol. 4, Paris, John Libbey Eurotext, pp. 423–425

Altamirano-Lozano, M., Alvarez-Barrera, L. & Roldán-Reyes, E. (1993) Cytogenetic and teratogenic effects of vanadium pentoxide on mice. *Med. Sci. Res.*, **21**, 711–713

Altamirano-Lozano, M., Alvarez-Barrera, L., Basurto-Alcántara, F., Valverde, M. & Rojas, E. (1996) Reprotoxic and genotoxic studies of vanadium pentoxide in male mice. *Teratog. Carcinog. Mutag.*, **16**, 7–17

Altamirano-Lozano, M., Roldán-Reyes, E., Bonilla, E. & Betancourt, M. (1997) Effect of some metal compounds on sperm motility *in vitro*. *Med. Sci. Res.*, **25**, 147–150

Altamirano-Lozano, M., Roldán, E., Bonilla, E. & Betancourt, M. (1998a) Effect of metal compounds on boar sperm motility in vitro. *Adv. exp. Med. Biol.*, **444**, 105–111

Altamirano-Lozano, M.A., Roldán-Reyes, M.E. & Rojas, E. (1998b) Genetic toxicology of vanadium compounds. In: Nriagu, J.O., ed., *Vanadium in the Environment. Part 2: Health Effects*, Vol. 31, New York, John Wiley & Sons, pp. 159–179

Altamirano-Lozano, M., Valverde, M., Alvarez-Barrera, L., Molina, B. & Rojas, E. (1999) Genotoxic studies of vanadium pentoxide (V_2O_5) in male mice. II. Effects in several mouse tissues. *Teratog. Carcinog. Mutag.*, **19**, 243–255

American Elements (2003) *Vanadium*, Los Angeles, CA (http://www.americanelements.com.vv. html, accessed 19.09.2003)

Arbouine, M. & Smith, N.J. (1991) The determination of vanadium in urine and its application to the biological monitoring of occupationally exposed workers. *Atomic Spectr.*, **12**, 54–58

Atomix (2003) *Product Data Sheet: Vanadium Pentoxide (V_2O_5)*, Hohokus, NJ (http://www. atomixinc.com/vanadiumpentoxide.htm, accessed 19.09.2003)

Auricchio, F., Di Domenico, M., Migliaccio, A., Castoria, G. & Bilancio, A. (1995) The role of estradiol receptor in the proliferative activity of vanadate on MCF-7 cells. *Cell Growth Differ.*, **6**, 105–113

Auricchio, F., Migliaccio, A., Castoria, G., Di Domenico, M., Bilancio, A. & Rotondi, A. (1996) Protein tyrosine phosphorylation and estradiol action. *Ann. N.Y. Acad. Sci.*, **784**, 149–172

AVISMA titanium-magnesium Works (2001) *Vanadium Pentoxide*, Berezniki (http://www. avisma.ru/eng/produkts, accessed 19.09.2003)

Barisione, G., Fontana, L. & Pasquarelli, P. (1993) [Acute exposure to vanadium pentoxide: Critical review and clinical investigation.] *Archiv. Sci. Lav.*, **8**(2), 111–115 (in Italian)

Bauer, G., Güther, V., Hess, H., Otto, A., Roidl, O., Roller, H. & Sattelberger, S. (2003) Vanadium and vanadium compounds. In: Bohnet *et al.*, eds, *Ullmann's Encyclopedia of Industrial Chemistry*, 6th rev. Ed., Vol. 38, Weinheim, Wiley-VCH Verlag GmbH & Co., pp. 1–21

Blotcky, A.J., Duckworth, W.C., Ebrahim, A., Hamel, F.K., Rack, E.P. & Sharma, R.B. (1989) Determination of vanadium in serum by pre-irradiation and post-irradiation chemistry and neutron activation analysis. *J. radioanal. nucl. Chem.*, **134**, 151–160

Bonner, J.C., Rice, A.B., Moomaw, C.R. & Morgan, D.L. (2000) Airway fibrosis in rats induced by vanadium pentoxide. *Am. J. Physiol. Lung Cell mol. Physiol.*, **278**, L209–L216

Bracken, W.N., Sharma, R.P. & Elsner, Y.Y. (1985) Vanadium accumulation and subcellular distribution in relation to vanadate induced cytotoxicity in vitro. *Cell Biol. Toxicol.*, **1**, 259–268

Bräunlich, H., Pfeifer, R., Grau, P. & Reznik, L. (1989) Renal effects of vanadate in rats. *Biomed. biochim. Acta*, **48**, 457–464

Buchet, J.P., Knepper, E. & Lauwerys R. (1982) Determination of vanadium in urine by electrothermal atomic absorption spectrometry. *Anal. chim. Acta*, **136**, 243–248

Buratti, M., Pellegrino, O., Caravelli, G., Calzaferri, G., Bettinelli, M., Colombi, A. & Maroni, M. (1985) Sensitive determination of urinary vanadium by solvent extraction and atomic absorption spectroscopy. *Clin. Chem. Acta*, **150**, 53–58

Byczkowski, J.Z. & Kulkarni, A.P. (1998) Oxidative stress and pro-oxidant biological effects of vanadium. In: Nriagu, J.O., ed., *Vanadium in the Environment. Part 2: Health Effects*, Vol. 31, New York, John Wiley & Sons, pp. 235–264

Byerrum, R.U., Eckardt, R.E., Hopkins, L.L., Libsch, J.F., Rostoker, W., Zenz, C., Gordon, W.A., Mountain, J.T., Hicks, S.P. & Boaz, T.D. (1974) *Vanadium*, Washington DC, National Academy of Sciences, pp. 1–117

Byrne, A.R. (1993) Review of neutron activation analysis in the standardization and study of reference materials, including its application to radionuclide reference materials. *Fresenius J. anal. Chem.*, **345**, 144–151

Byrne, A.R. & Kosta, L. (1978a) Determination of vanadium in biological materials at nanogram level by neutron activation analysis. *J. Radioanal. Chem.*, **44**, 247–264

Byrne, A.R. & Kosta, L. (1978b) Vanadium in foods and in human body fluids and tissues. *Sci. total Environ.*, **10**, 17–30

Byrne, A.R. & Kucera, J. (1991a) New data on levels of vanadium in man and his diet. In: Momcilovic, B., ed., *Proceedings of the 7th International Symposium on Trace Elements in Man and Animals*, pp. 25-18–25-20

Byrne, A.R. & Kucera, J. (1991b) Radiochemical neutron activation analysis of traces of vanadium in biological samples: A comparison of prior dry ashing with post-irradiation wet ashing. *Fresenius J. anal. Chem.*, **340**, 48–52

Byrne, A.R. & Versieck, J. (1990) Vanadium determination in the ultratrace level in biological reference materials and serum by radiochemical neutron activation analysis. *Biol. trace Elem. Res.*, **27**, 529–540

Cantley, L.C., Jr, Josephson, L., Warner, R., Yanagisawa, M., Lechene, C. & Guidotti, G. (1977) Vanadate is a potent (Na,K)-ATPase inhibitor found in ATP derived from muscle. *J. biol. Chem.*, **252**, 7421–7423

Carsey, T.P. (1985) Quantitation of vanadium oxides in airborne dusts by X-ray diffraction. *Anal. Chem.*, **57**, 2125–2130

Chakraborty, D., Bhattacharyya, A., Majumdar, K. & Chatterjee, G.C. (1977) Effects of chronic vanadium pentoxide administration on L-ascorbic acid metabolism in rats: Influence of L-ascorbic acid supplementation. *Int. J. Vitam. Nutr. Res*, **47**, 81–87

Chemical Information Services (2003) *Directory of World Chemical Producers (Version 2003)*, Dallas, TX (http://www.chemical.info.com, accessed 18.09.2003)

Chen, F., Vallyathan, V., Castranova, V. & Shi, X. (2001) Cell apoptosis induced by carcinogenic metals. *Mol. cell. Biochem.*, **222**, 183–188

Chin, L.S., Murray, S.F., Harter, D.H., Doherty, P.F. & Singh, S.K. (1999) Sodium vanadate inhibits apoptosis in malignant glioma cells: A role for Akt/PKB. *J. Biomed. Sci.*, **6**, 213–218

Cohen, M.D., Parsons, E., Schlesinger, R.B. & Zelikoff, J.T. (1993) Immunotoxicity of in vitro vanadium exposures: Effects on interleukin-1, tumor necrosis factor-α, and prostaglandin E_2 production by WEHI-3 macrophages. *Int. J. Immunopharmacol.*, **15**, 437–446

Cohen, M.D., McManus, T.P., Yang, Z., Qu, Q., Schlesinger, R.B. & Zelikoff, J.T. (1996) Vanadium affects macrophage interferon-γ-binding and -inducible responses. *Toxicol. appl. Pharmacol.*, **138**, 110–120

Colin-Barenque, L., Avila-Costa, M.R., Sánchez, I., López, I., Niño-Cabrera, G., Pasos, F., Delgado, V. & Fortul, T.I. (2003) Muerte neuronal en el bulbo olfatorio de ratón inducida por exposición subaguda y crónica de vanadio. In: *Proceedings of the XLVI Congreso Nacional de la Sociedad de Ciencias Fisiológicas, A.C., Aguascalientes, Ags. México*

Conklin, A.W., Skinner, C.S., Felten, T.L. & Sanders, C.L. (1982) Clearance and distribution of intratracheally instilled [48]vanadium compounds in the rat. *Toxicol. Lett.*, **11**, 199–203

Cornelis, R., Versieck, J., Mees, L., Hoste, J. & Barbier, F. (1980) Determination of vanadium in human serum by neutron activation analysis. *J. radioanal. Chem.*, **55**, 35–43

Cornelis, R., Versieck, J., Mees, L., Hoste, J. & Barbier, F. (1981) The ultratrace element vanadium in human serum. *Biol. trace Elem. Res.*, **3**, 257–263

Crans, D.C., Amin, S.S. & Keramidas, A.D.(1998) Chemistry of relevance to vanadium in the environment. In: Nriagu, J.O. ed., *Vanadium in the Environment. Part One. Chemistry and Biochemistry*, New York, John Wiley & Sons, 73–96

Dafnis, E., Spohn, M., Lonis, B., Kurtzman, N.A. & Sabatini, S. (1992) Vanadate causes hypokalemic distal renal tubular acidosis. *Am. J. Physiol.*, **262**, F449–F453

Dai, S., Thompson, K.H., Vera, E. & McNeill, J.H. (1994) Toxicity studies on one-year treatment of non-diabetic and streptozotocin-diabetic rats with vanadyl sulphate. *Pharmacol. Toxicol.*, **75**, 265–273

Deutsche Forschungsgemeinschaft (2002) *List of MAK and BAT Values 2002 — Commission for the Investigation of Health Hazards of Chemical Compounds in the Work Area* (Report No. 38), Weinheim, Wiley-VCH Verlag GmbH, pp. 109, 189

Devereux, T.R., Holliday, W., Anna, C., Ress, N., Roycroft, J. & Sills, R.C. (2002) Map kinase activation correlates with k-*ras* mutation and loss of heterozygosity in chromosome 6 in alveolar bronchiolar carcinomas from B6C3F1 mice exposed to vanadium pentoxide for 2 years. *Carcinogenesis*, **23**, 1737–1743

Dimond, E.G., Caravaca, J. & Benchimol, A. (1963) Vanadium. Excretion, toxicity, lipid effect in man. *Am. J. clin. Nutr.*, **12**, 49–53

Ding, M., Li, J.-J., Leonard, S.S., Ye, J.-P., Shi, X., Colburn, N.H., Castranova, V. & Vallyathan, V. (1999) Vanadate-induce activation of activator protein-1: Role of reactive oxygen species. *Carcinogenesis*, **20**, 663–668

Domingo, J.L. (1994) Metal-induced developmental toxicity in mammals: A review. *J. Toxicol. environ. Health*, **42**, 123–141

Domingo, J.L. (1996) Vanadium: A review of the reproductive and developmental toxicity. *Reprod. Toxicol.*, **10**, 175–182

Dong, W., Simeonova, P.P., Gallucci, R., Matheson, J., Flood, L., Wang, S., Hubbs, A. & Luster, M.I. (1998) Toxic metals stimulate inflammatory cytokines in hepatocytes through oxidative stress mechanisms. *Toxicol. appl. Pharmacol.*, **151**, 359–366

Dreher, K.L., Jaskot, R.H., Lehmann, J.R., Richards, J.H., McGee, J.K., Ghio, A.J. & Costa, D.L. (1997) Soluble transition metals mediate residual oil fly ash induced acute lung injury. *J. Toxicol. environ. Health*, **50**, 285–305

Edel, J. & Sabbioni, E. (1988) Retention of intratracheally instilled and ingested tetravalent and pentavalent vanadium in the rat. *J. trace Elem. Electrolytes Health Dis.*, **2**, 23–30

Edel, J. & Sabbioni, E. (1989) Vanadium transport across placenta and milk of rats to the fetus and newborn. *Biol. trace. Elem. Res.*, **22**, 265–275

Environmental Protection Agency (1977) *Scientific and Technical Assessment Report on Vanadium*, Washington DC, US Environmental Protection Agency (STAR Series, EPA-600/6-77-002)

Environmental Protection Agency (1985) *Health and Environmental Effects Profile for Vanadium Pentoxide* (EPA/600/X-85/114, 53), Cincinnati, OH, US Environmental Protection Agency, Environmental Criteria and Assessment Office

ESPI (1994) *Material Safety Data Sheet: Vanadium Pentoxide*, Ashland, OR (http://www.espimetals.com/msds's/vanadiumoxidev2o5.pdf, accessed 18.09.2003)

Fassett, J.D. & Kingston, H.M. (1985) Determination of nanogram quantities of vanadium in biological material by isotope dilution thermal ionization mass spectrometry with ion counting detection. *Anal. Chem.*, **57**, 2474–2478

Fisher, G.L., McNeill, K.L. & Democko, C.J. (1986) Trace element interactions affecting pulmonary macrophage cytotoxicity. *Environ. Res.*, **39**, 164–171

Fortoul, T.I., Quan-Torres, A., Sánchez, I., Lopez, I.E., Bizarro, P., Mendoza, M.L., Osorio, L.S., Espejel-Maya, G., Avila-Casado M.D.C., Avila-Costa, M.R., Colin-Barenque, L., Villanueva, D.N. & Olaiz-Fernandez, G. (2002) Vanadium in ambient air: Concentrations in lung tissue from autopsies of Mexico City residents in the 1960s and 1990s. *Arch. environ. Health*, **57**, 446–449

French, R.J. & Jones, P.J.H. (1993) Minireview. Role of vanadium in nutrition: Metabolism, essentiality and dietary considerations. *Life Sci.*, **52**, 339–346

Garcia, A.G., Jurkiewicz, A. & Jurkiewicz, N.H. (1981) Contractile effect of vanadate and other vanadium compounds on the rat vas deferens. *Eur. J. Pharmacol.*, **70**, 17–23

GfE mbH (2003) *Products by Product Groups: Vanadium Chemicals*, Nuremburg (http://www.gfe-online.de, accessed 19.09.2003)

Giri, A.K., Sanyal, R., Sharma, A., & Talukder, G. (1979) Cytological and cytochemical changes induced through certain heavy metals in mammalian systems. *Natl. Acad Sci. Lett.*, **2**, 391–394

Graedel, T.E., Hawkins, D.T., & Claxton, L.D. (1986) *Atmospheric Chemical Compounds, Sources, Occurrence and Bioassay*, New York, Academic Press

Greenberg, R.R., Kingston, H.M., Zeisler, R. & Woittiez, J. (1990) Neutron activation analysis of biological samples with a preirradiation separation. *Biol. trace Elem. Res.*, **26/27**, 17–25

Gylseth, B., Leira, H.L., Steinnes, E. & Thomassen, Y. (1979) Vanadium in the blood and urine of workers in a ferroalloy plant. *Scand. J. Work Environ. Health*, **5**, 188–194

Hahn, F.F. (1993) Chronic inhalation bioassays for respiratory tract carcinogenesis. In: Gardner, D.E., Crapo, J.D. & McClellan, R.O., eds, *Target Organ Toxicology Series: Toxicology of the Lung*, 2nd Ed., New York, Raven Press, p. 435

Hamel, F.G. (1998) Endocrine control of vanadium accumulation. In: Nriagu, J.O. ed., *Vanadium in the Environment. Part 2: Health Effects*, New York, John Wiley & Son, pp. 265–276

Hansard, S.L., Ammerman, C.B., Fick, K.R. & Miller, S.M. (1978) Performance and vanadium content of tissues in sheep as influenced by dietary vanadium. *J. Anim. Sci.* **46**, 1091–1095

Hansen, K. & Stern, R.M. (1984) A survey of metal-induced mutagenicity *in vitro* and *in vivo*. *Toxicol. environ. Chem.*, **9**, 87–91

Hauser, R., Elreedy, S., Hoppin, J.A. & Christiani, D.C. (1995a) Airway obstruction in boilermakers exposed to fuel oil ash. A prospective investigation. *Am. J. respir. crit. Care Med.*, **152**, 1478–1484

Hauser, R., Elreedy, S., Hoppin, J.A. & Christiani, D.C. (1995b) Upper airway response in workers exposed to fuel oil ash: Nasal lavage analysis. *Occup. environ. Med.*, **52**, 353–358

Hauser, R., Elreedy, S., Ryan, P.B. & Christiani, D.C. (1998) Urine vanadium concentrations in workers overhauling an oil-fired boiler. *Am. J. ind. Med.*, **33**, 55–60

Hehner, S.P., Hofmann, T.G., Dröge, W. & Schmitz, M.L. (1999) Inhibition of tyrosine phosphatases induces apoptosis independent from the CD95 system. *Cell Death Differ.*, **6**, 833–841

Hery, M., Gerber, J.M., Hecht, G., Hubert, G., Elcabache, J.M. & Honnert, B. (1992) [Handling of catalysts in the chemical industry: Exposure assessment during operations carried out by outside contractors.] *Cahiers de Notes Documentaires*, **149**, 479–486 (in French)

Hery, M., Gerber, J.M., Diebold, F., Honnert, B., Hecht, G. & Hubert, G. (1994) [Exposure to chemical pollutants during the manufacture and reprocessing of inorganic catalysts] *Cahiers de Notes Documentaires*, **155**, 151–156 (in French)

Heydorn, K. (1990) Factors affecting the levels reported for vanadium in human serum. *Biol. trace Elem. Res.*, **27**, 541–551

Highveld Steel & Vanadium Corporation Ltd (2003) *Vanadium Pentoxide Powder*, Witbank, (http://www.highveldsteel.co.za/Marketing/Files/vanadium_pentoxide_powder_R_Grade. htm; http://www.highveldsteel.co.za/Marketing/Files/vanadium_pentoxide_powder_R_ Granular.htm; accessed 19.09.2003)

Hilliard, H.E. (1994) *The Materials Flow of Vanadium in the United States* (Information Circular 9409), Washington DC, US Bureau of Mines

Hoffman, G.L., Duce, R.A. & Zoller, W.H. (1969) Vanadium copper, and aluminum in the lower atmosphere between California and Hawaii. *Environ. Sci. Technol.*, **3**, 1207–1210

Hope, B.K. (1994) A global biogeochemical budget for vanadium. *Sci. tot. Environ.*, **141**, 1–10

Huang, R.-T., Kang, J.F., Jin, H.E., Xue, H. & Wu, B.H. (1989) [Radiological observation on the workers exposed to vanadium]. *Zhonghua Yu Fang Yi Xue Za Zhi* [*Chin. J. Prev. Med.*, **23**, 283–285 (In Chinese)

Huang, C., Zhang, Z., Ding, M., Li, J., Ye, J., Leonard, S.S., Shen, H.-M., Butterworth, L., Lu, Y., Costa, M., Rojanasakul, Y., Castranova, V., Vallyathan, V. & Shi, X. (2000) Vanadate induces p53 transactivation through hydrogen peroxide and causes apoptosis. *J. biol. Chem.*, **275**, 32516–32522

Hudson, T.G.F. (1964) Vanadium: Toxicology and biological significance. In: Browning, E., ed., *Elsevier Monographs on Toxic Agents*, Amsterdam, Elsevier Publishing Company, pp. 67–78

IARC (1989) *IARC Monographs on the Evaluation of Carcinogenic Risks to Humans*, Vol. 45, *Occupational Exposures in Petroleum Refining; Crude Oil and Major Petroleum Fuels*, Lyon, IARC*Press*

INRS (1999) [Threshold limit values for occupational exposure to chemicals in France] In: *Cahiers de Notes Documentaires ND 2098-174-99*, Paris, Hygiène et Sécurité du Travail (in French)

Ingram, J.L., Rice, A.B., Santos, J., Van Houten, B. & Bonner, J.C. (2003) Vanadium-induced HB-EGF expression in human lung fibroblast is oxidant dependent and requires MAP kinases. *Am. J. Physiol. Lung Cell mol. Physiol.*, **284**, L774–L782

Ishida, O., Kihira, K., Tsukamoto, Y. & Marumo, F. (1989) Improved determination of vanadium in biological fluids by electrothermal atomic absorption spectrometry. *Clin. Chem.*, **35**, 127–130

Ivancsists, S., Pilger, A., Diem, E., Schaffer, A. & Rüdiger, H.W. (2002) Vanadate induces DNA strand breaks in cultured human fibroblasts at doses relevant to occupational exposure. *Mutat. Res.*, **519**, 25–35

Johnson, J.L., Cohen, H.J. & Rajagopalan, K.V. (1974) Studies of vanadium toxicity in the rat: Lack of correlation with molybdenum utilization. *Biochem. biophys. Res. Commun.*, **56**, 904–946

Kada, T., Hirano, K. & Shirasu, Y. (1980) Screening of environmental chemical mutagens by the rec-assay system with *Bacillus subtilis*. In: de Serres, F.J. & Hollaender, A., eds, *Chemical Mutagens: Principles and Methods for their Detection*, Vol. 6, New York, Plenum Press, pp. 149–173

Kanematsu, K. & Kada, T. (1978) Mutagenicity of metal compounds. *Mutation. Res.*, **53**, 207–208

Kanematsu, N., Hare, M., & Kada, T. (1980) Rec assay and mutagenicity studies on metal compounds. *Mutat. Res.*, **77**, 109–116

Kawai, T., Seiji, K., Watanabe, T., Nakatsuka, H. & Ikeda, M. (1989) Urinary vanadium as a biological indicator of exposure to vanadium. *Int. Arch. occup. environ. Health*, **61**, 283–287

Kerckaert, G.A., LeBoeuf, R.A. & Isfort, R.J. (1996a) Use of the Syrian hamster embryo cell transformation assay for determining the carcinogenic potential of heavy metal compounds. *Fundam. appl. Toxicol.*, **34**, 67–72

Kerckaert, G.A., Brauninger, R., LeBoeuf, R.A. & Isfort, R.J. (1996b) Use of the Syrian hamster embryo cell transformation assay for carcinogenicity prediction of chemicals currently being tested by the National Toxicology Program in rodent bioassays. *Environ. Health Perspect.*, **104** (Suppl. 5), 1075–1084

Kiviluoto, M., Pyy, L. & Pakarinen, A. (1979) Serum and urinary vanadium of vanadium-exposed workers. *Scand. J. Work Environ. Health*, **5**, 362–367

Kiviluoto, M., Pyy, L. & Pakarinen, A. (1981) Serum and urinary vanadium of workers processing vanadium pentoxide. *Int. Arch. occup. environ. Health*, **48**, 251–256

Knecht, E.A., Moorman, W.J., Clark, J.C., Lynch, D.W. & Lewis, T.R. (1985) Pulmonary effects of acute vanadium pentoxide inhalation in monkeys. *Am. Rev. respir. Dis.*, **132**, 1181–1185

Knecht, E.A., Moorman, W.J., Clark, J.C., Hull, R.D., Biagini, R.E., Lynch, D.W., Boyle, T.J. & Simon, S.D. (1992) Pulmonary reactivity to vanadium pentoxide following subchronic inhalation exposure in a non-human primate animal model. *J. appl. Toxicol.*, **12**, 427–434

Kodavanti, U.P., Hauser, R., Christiani, D.C., Meng, Z.H., McGee, J., Ledbetter, A., Richards, J. & Costa, D.L. (1998) Pulmonary responses to oil fly ash particles in the rat differ by virtue of their specific soluble metals. *Toxicol. Sci.*, **43**, 204–212

Kucera, J. & Sabbioni, E. (1998) Baseline vanadium levels in human blood, serum, and urine. In: Nriagu, J.O., ed., *Vanadium in the Environment. Part 2: Health Effects*, New York, John Wiley & Sons, pp. 75–89

Kucera, J., Byrne, A.R., Mravcová, A. & Lener, J. (1992) Vanadium levels in hair and blood of normal and exposed persons. *Sci. total Environ.*, **15**, 191–205

Kucera, J., Lener, J. & Mnuková, J. (1994) Vanadium levels in urine and cystine levels in fingernails and hair of exposed and normal persons. *Biol. Trace Element Res.*, **43–45**, 327–334

Kucera, J., Lener, J., Mnuková, J. & Bayerová, E. (1998) Vanadium exposure tests in humans: Hair, nails, blood, and urine. In: Nriagu, J.O., ed., *Vanadium in the Environment. Part 2: Health Effects*, New York, John Wiley & Sons, pp. 55–73

Kuzelova, M., Havel, V., Popler, A. & Stepanek, O. (1977) [The problems of occupational medicine in the work of chimney sweeps] *Prac. Lek.*, **29**, 225–228 (in Czech)

Kyono, H., Serita, F., Toya, T., Kubota, H., Arito, H., Takahashi, M., Maruyama, R., Homma, K., Ohta, H., Yamauchi, Y., Nakakita, M., Seki, Y., Ishihara, Y. & Kagawa, J. (1999) A new model rat with acute bronchiolitis and its application to research on the toxicology of inhaled particulate matter. *Ind. Health*, **37**, 47–54

Lagerkvist, B., Nordberg, G.F. & Vouk, V. (1986) Vanadium. In: Friberg, L., Nordberg, G.F. & Vouk, V.B., eds, *Handbook on the Toxicology of Metals, Vol. II, Specific Metals*, 2nd Ed., Amsterdam, Elsevier, pp. 638–663

Lees, R.E.M. (1980) Changes in lung function after exposure to vanadium compounds in fuel oil ash. *Br. J. ind. Med.*, **37**, 253–256

Lener, J., Kucera, J., Kodl, M. & Skokanová, V. (1998) Health effects of environmental exposure to vanadium. In: Nriagu J, ed. *Vanadium in the Environment. Part 2: Health Effects*, New York, John Wiley & Sons, Inc., pp. 1–19

Leonard, A. & Gerber, G.B. (1994) Mutagenicity, carcinogenicity and teratogenicity of vanadium compounds. *Mutat. Res.*, **317**, 81–88

Leonard, A. & Gerber, G.B. (1998) Mutagenicity, carcinogenicity, and teratogenicity of vanadium. In: Nriagu, J.O., ed., *Vanadium in the environment. Part Two: Health Effects*, New York, John Wiley & Sons, Inc., pp. 39–51

Levy, B.S., Hoffman, L. & Gottsegen, S. (1984) Boilermakers' bronchitis. Respiratory tract irritation associated with vanadium pentoxide exposure during oil-to-coal conversion of a power plant. *J. occup. Med.*, **26**, 567–570

Lewis, C.E. (1959) The biological effects of vanadium. II. The signs and symptoms of occupational vanadium exposure. *Arch. ind. Health*, **19**, 497–503

Lewis, R.J., Sr (2000) Sax's Dangerous Properties of Industrial Materials, 10th Ed., New York, John Wiley and Sons, Inc., pp. 3657–3660

Li, S., Zhang, T., Yang, Z. & Gou, X. (1991) [Distribution of vanadium in tissues of nonpregnant and pregnant Wistar rats]. *Hua Xi Yi Ke Da Xue Xue Bao*, **22**, 196–200 (in Chinese)

Li, H., Yang, Z. & Zhang, T. (1995) [Toxicity of vanadium to Leydig cells in vitro]. *Hua Xi Yi Ke Da Xue Xue Bao*, **26**, 433–435 (in Chinese)

Lide, D.R., ed. (2003) *CRC Handbook of Chemistry and Physics*, 84th Ed., Boca Raton, FL, CRC Press LLC, p. 4-93

Liochev, S. & Fridovich, I. (1988) Superoxide is responsible for the vanadate stimulation of NAD(P)H oxidation by biological membranes. *Arch. Biochem. Biophys.*, **263**, 299–304

Liu, G. *et al.* (1992) The measure of cytogenic effects of vanadium pentaoxide on pregnant mice and embryo. *J. environ. Health*, **9**, 119 (cited in Sun, 1996)

Magyar, M.J. (2002) *Minerals Yearbook: Vanadium*, Reston, VA, US Geological Survey (http://minerals.usgs.gov/minerals/pubs/commodity/vanadium/index.html, accessed 19.09.2003)

Mamane, Y. & Pirrone, N. (1998) Vanadium in the atmosphere. In: Nriagu, J.O., ed., *Vanadium in the Environment. Part 1: Chemistry and Biochemistry*, New York, John Wiley & Sons, pp. 37–71

Martens, C.S., Wesolowski, J.J., Kaifer, R., John, W. & Harriss, R.C. (1973) Sources of vanadium in Puerto Rican and San Francisco Bay area aerosols. *Environ. Sci. Technol.*, **7**, 817–820

McKee, J.M. (1998) Vanadium. In: Wexler, P., ed., *Encyclopedia of Toxicology*, Vol. 3, San Diego, CA, Academic Press, pp. 386–387

Meza, P., Colin-Barenque, L., Mondragón, A., Avila-Costa, M.R., Sánchez, I., López, I., Niño-Cabrera, G., Pasos, F., Delgado, V. & Fortoul, T. (2003) Efecto de la inhalación subaguda y crónica de vanadio sobre la ultraestructura del cerebelo de ratón. In: *Proceedings of the XLVI Congreso Nacional de la Sociedad de Ciencias Fisiológicas, A.C., Aguascalientes, Ags. México*

Minoia, C., Sabbioni, E., Apostoli, P., Pietra, R., Pozzoli, L., Gallorini, M., Nicolaou, G., Alessio, L. & Capodaglio, E. (1990) Trace element reference values in tissues from inhabitants of the European community. I. A study of 46 elements in urine, blood and serum of Italian subjects. *Sci. total Environ.*, **95**, 89–105

Minoia, C., Pietra, R., Sabbioni, E., Ronchi, A., Gatti, A., Cavalleri, A. & Manzo, L. (1992) Trace element reference values in tissues from inhabitants of the European Community. III. The control of preanalytical factors in the biomonitoring of trace elements in biological fluids. *Sci. total Environ.*, **120**, 63–79

Minoia, C., Sabbioni, E., Ronchi, A., Gatti, A., Pietra, R., Nicolotti, A., Fortaner, S., Balducci, C., Fonte, A. & Roggi, C. (1994) Trace element reference values in tissues from inhabitants of the European Community. IV. Influence of dietary factors. *Sci. total Environ.*, **141**, 181–195

Moens, L. & Dams, R. (1995) NAA and ICP-MS: A comparison between two methods for trace and ultra-trace element analysis. *J. radioanal. nucl. Chem.*, **192**, 29–38

Moens, L., Verrept, P., Dams, R., Greb, U., Jung, G. & Laser, B. (1994) New high-resolution inductively coupled plasma mass spectrometry technology applied for the determination of V, Fe, Cu, Zn and Ag in human serum. *J. anal. Atom. Spectr.*, **9**, 1075–1078

Mondragón, A., Colin-Barenque, L., Meza, P., Avila-Costa, M.R., Sánchez, I., López, I., Niño-Cabrera, G., Pasos, F., Delgado, V., Ordóñez, J., Gutiérrez, A., Aley, P. & Fortoul, T. (2003) Efectos de la inhibición crónica de vanadio sobre la citología del bulbo olfatorio de ratón. In: *Proceedings of the XLVI Congreso Nacional de la Sociedad de Ciencias Fisiológicas, A.C., Aguascalientes, Ags. México*

Mongold, J.J., Cros, G.H., Vian, L., Tep, A., Ramanadham, S., Siou, G., Diaz, J., McNeill, J.H. & Serrano, J.J. (1990) Toxicological aspects of vanadyl sulphate on diabetic rats: Effects on vanadium levels and pancreatic B-cell morphology. *Pharmacol. Toxicol.*, **67**, 192–198

Montiel-Flores, E., Avila-Costa, M., Aley, P., López, I., Acevedo, S., Saldivar, L., Espejel, G., González, A., Avila-Casado, M., Niño, G., Bizarro, P., Mussali, P., García-Morales, D., Colin-Barenque, L., Delgado, V. & Fortoul, T. (2003) Vanadium inhalation induces neuronal alterations in *corpus striatum*. An experimental model in mice. In: *Proceedings of the 42nd Annual Meeting of the Society of Toxicology, Salt Lake City, Utah, USA*

Motolese, A., Truzzi, M., Giannini, A. & Seidenari, S. (1993) Contact dermatitis and contact sensitization among enamellers and decorators in the ceramics industry. *Contact Derm.*, **28**, 59–62

Mravcová, A., Jírová, D., Janci, H. & Lener, J. (1993) Effects of orally administered vanadium on the immune system and bone metabolism in experimental animals. *Sci total Environ.*, **Suppl.**, 663–669

Musk, A.W. & Tees, J.G. (1982) Asthma caused by occupational exposure to vanadium compounds. *Med. J. Aust.*, **1**, 183–184

Myron, D.R., Gvand, S.H. & Nielsen, F.H. (1977) Vanadium content of selected foods as determined by flameless atomic absorption spectroscopy. *J. agric. food Chem.*, **25**, 297–300

Myron, D.R., Zimmerman, T.J., Shuler, T.R., Klevay, L.M., Lee, D.E. & Nielsen, F.H. (1978) Intake of nickel and vanadium by humans. A survey of selected diets. *Am. J. clin. Nutr.*, **31**, 527–531

National Institute for Occupational Safety and Health (1976) *National Occupational Hazard Survey (1970) Database*, Cincinnati, OH

National Institute for Occupational Safety and Health (1984) *National Occupational Exposure Survey (1980–1983) Database*, Cincinnati, OH

National Institute for Occupational Safety and Health (1994) Vanadium oxides — Method 7504, Issue 2. In: *NIOSH Manual of Analytical Methods (NMAM)*, 4th Ed., dated August 15, 1994

National Institute for Occupational Safety and Health (2005) Vanadium dust. In: *NIOSH Pocket Guide to Chemical Hazard* (DHHS (NIOSH) Publication No. 2005-149), Cincinnati, OH, Department of Health and Human Services, p. 328

National Library of Medicine (2003) ChemIDplus (http://chem2.sis.nlm.nih.gov/chemidplus/chemidlite.jsp, accessed 11.10.2003)

National Toxicology Program (2002) *Toxicology and Carcinogenesis Studies of Vanadium Pentoxide (CAS No. 1314-62-1) in F3344/N rats and B6C3F1 mice (Inhalation Studies)* (Technical Report series No. 507; NIH Publication No. 03-4441) Research Triangle Park, NC

Nechay, B.R. (1984) Mechanisms of action of vanadium. *Am. Rev. Pharmacol. Toxicol.*, **24**, 501–524

Nielsen, F.H. (1987) Vanadium. In: Mertz, W., ed., *Trace Elements in Human and Animal Nutrition*, 5th Ed., Vol. 1, San Diego, Academic Press, pp. 275–300

Nielsen, F.H. (1991) Nutritional requirements for boron, silicon, vanadium, nickel, and arsenic: Current knowledge and speculation. *FASEB J.*, **5**, 2661–2667

Nriagu, J.O. (1998) History, occurrence, and uses of vanadium. In: Nriagu, J.O., ed., *Vanadium in the Environment. Part 1: Chemistry and Biochemistry*, John Wiley & Sons, Inc., pp. 1–24

Nriagu, J.O. & Pirrone, N. (1998) Emission of vanadium into the atmosphere. In: Nriagu, J.O., ed., *Vanadium in the Environment. Part 1: Chemistry and Biochemistry*, John Wiley & Sons, Inc., pp. 25–36

Oberg, S.G., Parker, R.D.R. & Sharma, R.P. (1978) Distribution and elimination of an intratracheally administered vanadium compound in the rat. *Toxicology*, **11**, 315–323

O'Neil, M.J., ed. (2001) *The Merck Index*, 13th Ed., Whitehouse Station, NJ, Merck & Co., Inc., p. 1767

Parfett, C.L.J. & Pilon, R. (1995) Oxidative stress-regulated gene expression and promotion of morphological transformation induced in C3H/10T1/2 cells by ammonium metavanadate. *Food chem. Toxicol.*, **33**, 301–308

Pazhynich, V.M. (1966) Maximum permissible concentration of vanadium pentoxide in the atmosphere. *Hyg. Sanit.*, **31**, 6–11

Pazhynich, V.M. (1967) [Experimental basis for the determination of maximum allowable concentrations of vanadium pentoxide in atmospheric air]. In: Rjazanov, V.A., ed. [*The Biological Effect and Hygienic Importance of Atmospheric Pollutants*], Moscow, Medicina Publishing House, pp. 201–217 (in Russian)

Perron, L. (1994) *Canadian Minerals Yearbook: Vanadium*, Natural Resources Canada

Perron, L. (2001) *Canadian Minerals Yearbook: Vanadium*, Natural Resources Canada

Pierce, L.M., Alessandrini, F., Godleski, J.J. & Paulauskis, J.D. (1996) Vanadium-induced chemokine mRNA expression and pulmonary inflammation. *Toxicol. appl. Pharmacol.*, **138**, 1–11

Pistelli, R., Pupp, N., Forastiere, F., Agabiti, N., Corbo, G.M., Tidei, F. & Perucci, C.A. (1991) [Increase of non-specific bronchial responsiveness following occupational exposure to vanadium]. *Med. Lav.*, **82**, 270–275 (in Italian)

Plunkett, E.R. (1987) *Handbook of Industrial Toxicology*, 3rd Ed., New York, Chemical Publishing Co., pp. 563–564

Pyy, L., Hakala, E. & Lajunen, L.J. (1984) Screening for vanadium in urine and blood serum by electrothermal atomic absorption spectrometry and d.c. plasma atomic emission spectrometry. *Anal. chim. Acta*, **158**, 297–303

Ramírez, P., Eastmond, D.A., Laclette, J.P. & Ostrosky-Wegman, P. (1997) Disruption of microtubule assembly and spindle formation as a mechanism for the induction of aneuploid cells by sodium arsenite and vanadium pentoxide. *Mutat. Res.*, **386**, 291–298

Reade Advanced Materials (1997) *Vanadium Oxide Powder*, Providence, RI (http://www.reade.com/Products/Oxides/vanadium_oxide.htm, accessed 19.09.2003)

Rehder, D. (1991) The bioinorganic chemistry of vanadium. *Angew. Chem. Int. Ed. Engl.*, **30**, 148–167

Ress, N.B., Chou, B.J., Renne, R.A., Dill, J.A., Miller, R.A., Roycroft, J.H., Hailey, J.R., Haseman, J.K. & Bucher, J.R. (2003) Carcinogenicity of inhaled vanadium pentoxide in F344/N rats and B6C3F$_1$ mice. *Toxicol. Sci.*, **74**, 287–296

Rhoads, K. & Sanders, C.L. (1985) Lung clearance, translocation, and acute toxicity of arsenic, beryllium, cadmium, cobalt, lead, selenium, vanadium, and ytterbium oxides following deposition in rat lung. *Environ. Res.*, **36**, 359–378

Rice, A.B., Moomaw, C.R., Morgan, D.L. & Bonner, J.C. (1999) Specific inhibitors of platelet-derived growth factor or epidermal growth factor receptor tyrosine kinase reduce pulmonary fibrosis in rats. *Am. J. Phatol.*, **155**, 213–221

Rivedal, E., Roseng, L.E. & Sanner, T. (1990) Vanadium compounds promote the induction of morphological transformation of hamster embryo cells with no effect on gap junctional cell communication. *Cell Biol. Toxicol.*, **6**, 303–314

Rojas, E., Valverde, M., Sordo, M., Altamirano-Lozano, M. & Ostrosky-Wegman, P. (1996a) Single cell gel electrophoresis assay in the evaluation of metal carcinogenicity. In: Collery, P., Corbella, J., Domingo, J.L., Elienne, J.-C. & Llobet, J.M. eds, *Metal Ions in Biology and Medicine*, Vol. 4, Paris, John Libbey Eurotext, pp. 375–377

Rojas, E., Valverde, M., Herrera, L.A., Altamirano-Lozano, M. & Ostrosky-Wegman, P. (1996b) Genotoxicity of vanadium pentoxide evaluated by the single cell gel electrophoresis assay in human lymphocytes. *Mutat. Res.*, **359**, 77–84

Roldán, R.E. & Altamirano, L.M.A. (1990) Chromosomal aberrations, sister-chromatid exchanges, cell-cycle kinetics and satellite associations in human lymphocyte cultures exposed to vanadium pentoxide. *Mutat. Res.*, **245**, 61–65

Roldán-Reyes, E., Aguilar-Morales, C., Frías-Vásquez, S. & Altamirano-Lozano, M. (1997) Induction of sister chromatid exchanges in human lymphocytes by vanadium pentoxide in combination with caffeine. *Med. Sci. Res.*, **25**, 501–504

Roshchin, A.V. (1967a) Toxicology of vanadium compounds used in modern industry. *Hyg. Sanit.*, **32**, 345–352

Roshchin, A.V. (1967b) Vanadium. In: Izraelson, Z.I., ed., *Toxicology of the Rare Metals (AEC-tr-6710)*, Jerusalem, Israel Program for Scientific Translations, Washington, DC, pp. 52–59

Roschin, A.V. (1968) [*Vanadium and its Compounds*], Moscow, Medicina Publishing House (in Russian)

Ryan, R. P., Terry, C. E., & Leffingwell, S. S., eds (1999) *Toxicology Desk Reference: The Toxic Exposure and Medical Monitoring Index*, Taylor and Francis, pp. 1211–1217

Sabbioni, E., Clerici, L. & Brazzelli, A. (1983) Different effects of vanadium ions on some DNA-metabolizing enzymes. *J. Toxicol. environ. Health*, **12**, 737–748

Sabbioni, E., Pozzi, G., Pintar, A., Casella, L. & Garattini, S. (1991) Cellular retention, cytotoxicity and morphological transformation by vanadium(IV) and vanadium(V) in BALB/3T3 cell lines. *Carcinogenesis*, **12**, 47–52

Sabbioni, E., Kucera, J., Pietra, R. & Vesterberg, O. (1996) A critical review on normal concentrations of vanadium in human blood, serum, and urine. *Sci. total Environ.*, **188**, 49–58

Sánchez, G.I., López, I.E., Acevedo, S., Bizarro, P., Pasos, F., Delgado, V., Rondán, A. & Fortoul, T.I. (2001) Cambios ultraestructurales en pulmón causados por la inhalación aguda de V_2O_5: Modelo en ratón. In: *Proceedings of the IV Congreso Mexicano de Toxicología, Mérida, Yucatán, México*

Sánchez, G.I., López, I., Mussali, P., Bizarro, N.P., Niño, G., Saldivar, L., Espejel, G., Avila, M., Morales, D., Colin, L., Delgado, V., Acevedo, S., González, A., Avila-Costa, M.R. & Fortoul, T.I. (2003) Vanadium concentrations in lung, liver, kidney, testes and brain, after the inhalation of 0.02M of V_2O_5. An experimental model in mice. In: *Proceedings of the 42nd Annual Meeting of the Society of Toxicology, Lake City, Utah, USA*

Sato, K., Kusaka, Y., Akino, H., Kanamaru, H. & Okada, K. (2002) Direct effect of vanadium on citrate uptake by rat renal brush border membrane vesicles (BBMV). *Ind. Health*, **40**, 278–281

Schiff, L.J. & Graham, J.A. (1984) Cytotoxic effect of vanadium and oil-fired fly ash on hamster tracheal epithelium. *Environ. Res.*, **34**, 390–402

Schroeder, H.A., Balassa, J.J. & Tipton, I.H. (1963) Abnormal trace metals in man — Vanadium. *J. chron. Dis.*, **16**, 1047–1071

Seiler, H.G. (1995) Analytical procedures for the determination of vanadium in biological materials. In: Sigel, H. & Sigel, A., eds, *Metal Ions in Biological Systems, Vol. 31, Vanadium and its Role in Life*, New York, Marcel Dekker, pp. 671–688

Seljankina, K.P. (1961) [Data for determining the maximum permisible content of vanadium in water basins.] *Gig. Sanit.*, **26**, 6–12 (in Russian)

Sharma, R.P., Flora, S.J.S., Drown, D.B. & Oberg, S.G. (1987) Persistence of vanadium compounds in lungs after intratracheal instillation in rats. *Toxicol. ind. Health*, **3**, 321–329

Shi, X. & Dalal, N.S. (1992) Hydroxyl radical generation in the NADH/microsomal reduction of vanadate. *Free Rad. Res. Commun.*, **17**, 369–376

Shi, X., Sun, X. & Dalal, N.S. (1990) Reaction of vanadium(V) with thiols generates vanadium(IV) and thiyl radicals. *FEBS Lett.*, **271**, 185–188

Shi, X., Flynn, D.C., Liu, K. & Dalal, N. (1997) Vanadium (IV) formation in the reduction of vanadate by glutathione reductase/NADPH and the role of molecular oxygen. *Ann. clin. Lab. Sci.*, **27**, 422–427

Si, R. *et al.* (1982) Studies on the mutagenicity and teratogenicity of vanadium pentaoxide. Weisheng Zhuanye Cankao Ziliao. School of Public Health, Sichuan Medical College, **22**, 36 [cited in Sun, 1996]

Silbajoris, R., Ghio, A.J., Samet, J.M., Jaskot, R., Dreher, K.L. & Brighton, L.E. (2000) *In vivo* and *in vitro* correlation of pulmonary MAP kinase activation following metallic exposure. *Inhalat. Toxicol.*, **12**, 453–468

Sit, K.H., Paramanantham, R., Bay, B.H., Wong, K.P., Thong, P. & Watt, F. (1996) Induction of vanadium accumulation and nuclear sequestration causing cell suicide in human Chang liver cells. *Experientia*, **52**, 778–785

Sjöberg, S.G. (1950) Vanadium pentoxide dust: A clinical and experimental investigation on its effect after inhalation. *Acta med. scand.*, **138** (Suppl. 238), 1–188

Smith, M.M., White, M.A. & Ide, C. (1992) *The Biological Monitoring of Occupational Exposure to Vanadium in Boiler Cleaners*. Health and Safety Executive (Research Report IR/L/TM/92/1), Sudbury, Suffolk

Sokolov, S.M. (1981) Experimental data for establishing time differentiated maximum permissible concentrations of vanadium pentoxide in the atmosphere. *Gig. Sanit.*, 17–19

Stern, A., Yin, X., Tsang, S.-S., Davison, A. & Moon, J. (1993) Vanadium as modulator of cellular regulatory cascades and oncogene expression. *Biochem. Cell Biol.*, **71**, 103–112

STN International (2003) STN International Registry file (http://stnweb.cas.org/html/english/a_index.html, accessed 19.09.2003)

Strategic Minerals Corp. (2003) *Vanadium pentoxide (V_2O_5)*, Danbury, CT (http://www.stratcor.com/products/vanadium_pentoxide.html, accessed 19.09.2003)

Sun, M. *et al.* (1989) *Study on the Maximum Allowable Concentration of Vanadium in Surface Water, Report on the National Technical Committee of Environmental Health Standard* [reference cited in Sun, 1996]

Sun, M.-L. (1996) Toxicity of vanadium and its environmental health standard. *J. Health Toxicol.*, **10**, 25–31

Suva (2003) *Grenzwerte am Arbeitsplatz 2003*, Luzern [Swiss OELs]

Thompson, K.H. & McNeill, J.H. (1993) Effect of vanadyl sulfate feeding on susceptibility to peroxidative change in diabetic rats. *Res. Comm. chem. Pathol. Pharmacol.*, **80**, 187–200

Thompson, K.H., Battell, M. & McNeill, J.H. (1998) Toxicology of vanadium in mammals. In: Nriagu, J.O., ed., *Vanadium in the Environment. Part Two: Health Effects*, New York, John Wiley & Sons, pp. 21–37

Todaro, A., Bronzato, R., Buratti, M. & Colombi, A. (1991) [Acute exposure to vanadium-containing dusts: Health effects and biological monitoring in a group of boiler maintenance workers.] *Med. Lav.*, **82**, 142–147 (in Italian)

Toya, T., Fukuda, K., Takaya, M. & Arito, H. (2001) Lung lesions induced by intratracheal instillation of vanadium pentoxide powder in rats. *Ind. Health*, **39**, 8–15

TRI (1987–2001) *Toxic Release Inventory Database*, Bethesda, MD, National Library of Medicine (http://www.epa.gov/triexplorer/chemical.htm, accessed 19.09.2003)

Tsukamoto, Y., Saka, S., Kumano, K., Iwanami, S., Ishida, O. & Marumo, F. (1990) Abnormal accumulation of vanadium in patients on chronic hemodialysis therapy. *Nephron*, **56**, 368–373

Työsuojelusäädöksiä (2002) *HTP arvot 2002*, Sosiali-ja terveysministeriö [Finnish OELs]

Usutani S., Nishiama, K., Sato, I., Matsura, K., Sawada, Y., Hosokawa, Y. & Izumi, S. (1979) [An investigation of the environment in a certain vanadium refinery]. *Jpn. J. ind. Health*, **21**, 11–20 (in Japanese)

Van Vleet, J.F., Boon, G.D. & Ferrans, V.J. (1981) Induction of lesions of selenium-vitamin E deficiency in weanling swine fed silver, cobalt, tellurium, zinc, cadmium, and vanadium. *Am. J. vet. Res.*, **42**, 789–799

Venugopal, B. & Luckey, T.D. (1978) *Chemical Toxicity of Metals and Metalloids*, Vol. 2, *Metal Toxicity in Mammals*, New York, Plenum Press

Versieck, J. & Cornelis, R. (1980) Normal levels of trace elements in human blood plasma or serum. *Anal. chim. Acta*, **116**, 217–254

Versieck, J., Vanballenberghe, L., de Kesel, A., Hoste, J., Wallaeys, B., Vandenhaute, J., Baeck, N., Steyaert, H., Byrne, A.R. & Sunderman, F.W., Jr (1988) Certification of a second-generation biological reference material (freeze-dried human serum) for trace element determinations. *Anal. Chim. Acta*, **204**, 63–75

Wang, Y.-Z. & Bonner, J.C. (2000) Mechanism of extracellular signal-regulated kinase (ERK)-1 and ERK-2 activation by vanadium pentoxide in rat pulmonary myofibroblasts. *Am. J. resp. Cell mol. Biol.*, **22**, 590–596

Waters, M.D. (1977) Toxicology of vanadium. In: Goyer R.A. & Melhman M.A., eds, *Advances in Modern Toxicology*, Vol. 2, *Toxicology of Trace Elements*, New York, Halsted Press, pp. 147–189

Waters, M.D., Gardner, D.E. & Coffin, D.L. (1974) Cytotoxic effects of vanadium on rabbit alveolar macrophages in vitro. *Toxicol. appl. Pharmacol.*, **28**, 253–263

White, M.A., Reeves, G.D., Moore, S., Chandler, H.A. & Holden, H.J. (1987) Sensitive determination of urinary vanadium as a measure of occupational exposure during cleaning of oil fired boilers. *Ann. occup. Hyg.*, **31**, 339–343

WHO (1988) *Vanadium* (Environmental Health Criteria 81), Geneva, International Programme on Chemical Safety

WHO (2000) *Air Quality Guidelines for Europe* (European Series No. 91), 2nd Ed., Copenhagen, Regional Office for Europe

WHO (2001) *Vanadium Pentoxide and Other Inorganic Vanadium Compounds (Concise International Chemical Assessment Document 29)*, Geneva, International Programme on Chemical Safety

Wide, M. (1984) Effect of short-term exposure to five industrial metals on the embryonic and fetal development of the mouse. *Environ. Res.*, **33**, 47–53

Williams, N. (1952) Vanadium poisoning from cleaning oil-fired boilers. *Br. J. ind. Med.*, **9**, 50–55

Woodin, M.A., Hauser, R., Liu, Y., Smith, T.J., Siegel, P.D., Lewis, D.M., Tollerud, D.J. & Christiani, D.C. (1998) Molecular markers of acute upper airway inflammation in workers exposed to fuel-oil ash. *Am. J. respir. crit. Care Med.*, **158**, 182–187

Woodin, M.A., Liu, Y., Hauser, R., Smith, T.J. & Christiani, D.C. (1999) Pulmonary function in workers exposed to low levels of fuel-oil ash. *J. occup. environ. Med.*, **41**, 973–980

Woodin, M.A., Liu, Y., Neuberg, D., Hauser, R., Smith, T.J. & Christiani, D.C. (2000) Acute respiratory symptoms in workers exposed to vanadium-rich fuel-oil ash. *Am. J. Ind. Med.*, **37**, 353–363

Woolery, M. (1997) Vanadium compounds. In: Kroschwitz, J.I. & Howe-Grant, M., eds, *Kirk-Othmer Encyclopedia of Chemical Technology*, 4th Ed., Vol. 24, New York, John Wiley & Sons, pp. 797–811

Wrbitzky, R., Göen, T., Letzel, S., Frank, F. & Angerer, J. (1995) Internal exposure of waste incineration workers to organic and inorganic substances. *Intern. Arch. occup. environ. Health*, **68**, 13–21

Wyers, H. (1946) Some toxic effects of vanadium pentoxide. *Br. J. ind. Med.*, **3**, 177–182

Yamaguchi, M., Oishi, H. & Suketa, Y. (1989) Effect of vanadium on bone metabolism in weanling rats: Zinc prevents the toxic effect of vanadium. *Res. exp. Med.*, **198**, 47–53

Yuen, V.G., Orvig, C., Thompson, K.H. & McNeill, J.H. (1993) Improvement in cardiac dysfunction in streptozotocin-induced diabetic rats following chronic oral administration of bis(maltolato)oxovanadium(IV). *Can. J. Physiol. Pharmacol.*, **71**, 270–276

Zeiger, E., Anderson, B., Haworth, S., Lawlor, T. & Mortelmans, K. (1992) Salmonella mutagenicity tests: V. Results from the testing of 311 chemicals. *Environ. mol. Mutagen.*, **19** (Suppl. 21), 2–141

Zen, X. *et al.* (1988) Study on the mutagenicity of vanadium Pentaoxide, *Report on the National Symposium of Trace Element and Health* [cited in Sun, 1996]

Zenz, C., ed. (1994) *Occupational Medicine*, 3rd Ed., St Louis, MO, Mosby, pp. 584–594

Zenz, C. & Berg, B.A. (1967) Human responses to controlled vanadium pentoxide exposure. *Arch. Environ. Health*, **14**, 709–712

Zenz, C., Bartlett, J.P. & Thiede, W. (1962) Acute vanadium pentoxide intoxication. *Arch. environ. Health*, **5**, 542–546

Zhang, T., Yang, Z., Li, S. & Go, X. (1991a) [Transplacental passage of vanadium after treatment with vanadium pentoxide in Wistar rat]. *Hua Xi Yi Ke Da Xue Xue Bao*, **22**, 296–299 (in Chinese)

Zhang, T., Gou, X. & Yang, Z. (1991b) [A study on developmental toxicity of vanadium pentoxide in NIH mice]. *Hua Xi Yi Ke Da Xue Xue Bao*, **22**, 192–195 (in Chinese)

Zhang, T., Yang, Z., Zeng, C. & Gou, X. (1993a) [A study on developmental toxicity of vanadium pentoxide in Wistar rats]. *Hua Xi Yi Ke Da Xue Xue Bao*, **24**, 92–96 (in Chinese)

Zhang, T., Gou, X. & Yang, Z. (1993b) [Study of teratogenicity and sensitive period of vanadium pentoxide in Wistar rats]. *Hua Xi Yi Ke Da Xue Xue Bao*, **24**, 202–205 (in Chinese)

Zhang, L., Rice, A.B., Adler, K., Sannes, P., Martin, L., Gladwell, L., Koo, J.-S., Gray, T.E. & Bonner, J.C. (2001a) Vanadium stimulates human bronchial epithelial cells to produce heparin-binding epidermal growth factor-like growth factor: A mitogen for lung fibroblasts. *Am. J. respir. Cell mol. Biol.*, **24**, 123–131

Zhang, Z., Wang, Y., Vikis, H.G., Johnson, L., Liu, G., Li, J., Anderson, M.W., Sills, R.C., Hong, H.L., Devereux, T.R., Jacks, T., Guan, K.-L. & You, M. (2001b) Wildtype *Kras2* can inhibit lung carcinogenesis in mice. *Nature Genetics*, **29**, 25–33

Zhong, B.-Z., Gu, Z.-W., Wallace, W.E., Whong, W.-Z. & Ong, T. (1994) Genotoxicity of vanadium pentoxide in Chinese hamster V79 cells. *Mutat. Res.*, **321**, 35–42

van Zinderen Bakker, X. & Jaworski, J.F. (1980) *Effects of Vanadium in the Canadian Environment*, Ottawa, National Research Council Canada, Associate Committee Scientific Criteria for Environmental Quality, pp. 1–94

Zoller, W.H., Gordon, G.E., Gladney, E.S. & Jones, A.G. (1973) The sources and distribution of vanadium in the atmosphere. In: Kothy, E.L., ed., *Trace Elements in the Environment: Symposium* (Advances in Chemistry Series No. 123), Washington DC, American Chemical Society, pp. 31–47

Zoller, W.H., Gladney, E.S. & Duce, R.A. (1974) Atmospheric concentrations and sources of trace metals at the South Pole. *Science*, **183**, 198–200

Zychlinski, L. & Byczkowski, J.Z. (1990) Inhibitory effects of vanadium pentoxide on respiration of rat liver mitochondria. *Arch. environ. Contam. Toxicol.*, **19**, 138–142

Zychlinski, L., Byczkowski, J.Z. & Kulkarni, A.P. (1991) Toxic effects of long-term intratracheal administration of vanadium pentoxide in rats. *Arch. environ. Contam. Toxicol.*, **20**, 295–298

LIST OF ABBREVIATIONS USED IN THIS VOLUME

8-OHdG: 8-hydroxydeoxyguanosine
AAS: atomic absorption spectrometry
AFS: atomic fluorescence spectrometry
ALA: δ-aminolevulinic acid
ALAD: δ-aminolevulinic acid dehydratase
ATPases: adenosinetriphosphatases
AUC: area under the curve
BALF: bronchoalveolar lavage fluid
BEI: biological exposure index
CI: confidence interval
Co-FAP: cobalt-labelled fused aluminosilicate
D$_{50}$: median diameter
DMA: dimethylarsinic acid
DMPO: 5,5-dimethyl-1-pyrroline *N*-oxide
ECP: eosinophilic cationic protein
EDTA: ethylenediaminetetraacetic acid
ERK-1/2: extracellular signal-regulated kinases 1 and 2
ESR: electron spin resonance
F-AAS: flame atomic absorption spectrometry
FPG: formamido-pyrimidine DNA glycosylase
G6PD: glucose-6-phosphate dehydrogenase
GF-AAS: graphite furnace atomic absorption spectrometry
GIP: giant-cell interstitial pneumonia
GSD: geometric standard deviation
GSH: glutathione
HB: horizontal Bridgeman
HB-EGF: heparin-binding epidermal growth factor
HFC: high-frequency cell
HIP: hot isostatic pressing
HPLC: high-performance liquid chromatography
HSE: heat shock element
IC: integrated circuit
ICP-AAS: inductively coupled atomic absorption spectrometry
ICP-MS: inductively coupled plasma mass spectrometry
IDMS: isotope dilution mass spectrometry

IEC: ion-exchange chromatography
IL-1: interleukin-1
IL-6: interleukin-6
IL-8: interleukin-8
INAA: instrumental neutron activation analysis
LDH: lactate dehydrogenase
LED: light-emitting diode
MIP-2: mitogen-activated protein
MMAD: mass median aerodynamic diameter
MMA^{III}: monomethylarsonous acid
MMA^V: monomethylarsonic acid
MMS: methyl methanesulfonate
MOCVD: metal-organic chemical vapour deposition
MPO: myeloperoxidase
MRE: metal response element
MTT: dimethylthiazol diphenyl tetrazolium
NAA: neutron activation analysis
PBMC: peripheral blood mononucleated cells
PGE$_2$: prostaglandin E$_2$
PM$_{10}$: particulate matter with aerodynamic diameter $\leq$ 10 μm
RNAA: radiochemical neutron activation analysis
ROS: reactive oxygen species
SHE: Syrian hamster embryo
SI: semi-insulating
SIR: standardized incidence ratio
SMR: standardized mortality ratio
SOD: superoxide dismutase
SRR: standardized registration ratio
T$_{1/2}$: half-life
TGF-β1: transforming growth factor β1
TLV: threshold limit value
TNFα: tumour necrosis factor α
UV: ultraviolet

SUPPLEMENTARY CORRIGENDA TO VOLUMES 1–85

Volume 69

p. 5, *After* S. Reynaud, *add*:
Post-meeting scientific assistance:
Yann Grosse.

Volume 84

p. 3, *Members*
After:
Anthony B. DeAngelo, Environmental Carcinogenesis Division, US Environmental Protection Agency, National Health and Environmental Effects Research Laboratory, MD-68 ERC, 86 TW Alexander Drive, Research Triangle Park, NC 27711, USA
delete: (Subgroup Chair: Cancer in Experimental Animals)

After:
David M. DeMarini, Environmental Carcinogenesis Division (MD-68), US Environmental Protection Agency, 86 Alexander Drive, Research Triangle Park, NC 27711, USA
add: (Subgroup Chair: Cancer in Experimental Animals)

p. 6, *Post-meeting scientific assistance*:
After Catherine Cohet, *add* Fatiha El Ghissassi.

CUMULATIVE CROSS INDEX TO *IARC MONOGRAPHS ON THE EVALUATION OF CARCINOGENIC RISKS TO HUMANS*

The volume, page and year of publication are given. References to corrigenda are given in parentheses.

-

A

A-α-C	*40*, 245 (1986); *Suppl. 7*, 56 (1987)
Acetaldehyde	*36*, 101 (1985) (*corr. 42*, 263); *Suppl. 7*, 77 (1987); *71*, 319 (1999)
Acetaldehyde formylmethylhydrazone (*see* Gyromitrin)	
Acetamide	*7*, 197 (1974); *Suppl. 7*, 56, 389 (1987); *71*, 1211 (1999)
Acetaminophen (*see* Paracetamol)	
Aciclovir	*76*, 47 (2000)
Acid mists (*see* Sulfuric acid and other strong inorganic acids, occupational exposures to mists and vapours from)	
Acridine orange	*16*, 145 (1978); *Suppl. 7*, 56 (1987)
Acriflavinium chloride	*13*, 31 (1977); *Suppl. 7*, 56 (1987)
Acrolein	*19*, 479 (1979); *36*, 133 (1985); *Suppl. 7*, 78 (1987); *63*, 337 (1995) (*corr. 65*, 549)
Acrylamide	*39*, 41 (1986); *Suppl. 7*, 56 (1987); *60*, 389 (1994)
Acrylic acid	*19*, 47 (1979); *Suppl. 7*, 56 (1987); *71*, 1223 (1999)
Acrylic fibres	*19*, 86 (1979); *Suppl. 7*, 56 (1987)
Acrylonitrile	*19*, 73 (1979); *Suppl. 7*, 79 (1987); *71*, 43 (1999)
Acrylonitrile-butadiene-styrene copolymers	*19*, 91 (1979); *Suppl. 7*, 56 (1987)
Actinolite (*see* Asbestos)	
Actinomycin D (*see also* Actinomycins)	*Suppl. 7*, 80 (1987)
Actinomycins	*10*, 29 (1976) (*corr. 42*, 255)
Adriamycin	*10*, 43 (1976); *Suppl. 7*, 82 (1987)
AF-2	*31*, 47 (1983); *Suppl. 7*, 56 (1987)
Aflatoxins	*1*, 145 (1972) (*corr. 42*, 251); *10*, 51 (1976); *Suppl. 7*, 83 (1987); *56*, 245 (1993); *82*, 171 (2002)
Aflatoxin B₁ (*see* Aflatoxins)	
Aflatoxin B₂ (*see* Aflatoxins)	
Aflatoxin G₁ (*see* Aflatoxins)	
Aflatoxin G₂ (*see* Aflatoxins)	
Aflatoxin M₁ (*see* Aflatoxins)	
Agaritine	*31*, 63 (1983); *Suppl. 7*, 56 (1987)
Alcohol drinking	*44* (1988)
Aldicarb	*53*, 93 (1991)

Aldrin	*5*, 25 (1974); *Suppl. 7*, 88 (1987)
Allyl chloride	*36*, 39 (1985); *Suppl. 7*, 56 (1987); *71*, 1231 (1999)
Allyl isothiocyanate	*36*, 55 (1985); *Suppl. 7*, 56 (1987); *73*, 37 (1999)
Allyl isovalerate	*36*, 69 (1985); *Suppl. 7*, 56 (1987); *71*, 1241 (1999)
Aluminium production	*34*, 37 (1984); *Suppl. 7*, 89 (1987)
Amaranth	*8*, 41 (1975); *Suppl. 7*, 56 (1987)
5-Aminoacenaphthene	*16*, 243 (1978); *Suppl. 7*, 56 (1987)
2-Aminoanthraquinone	*27*, 191 (1982); *Suppl. 7*, 56 (1987)
para-Aminoazobenzene	*8*, 53 (1975); *Suppl. 7*, 56, 390 (1987)
ortho-Aminoazotoluene	*8*, 61 (1975) (*corr. 42*, 254); *Suppl. 7*, 56 (1987)
para-Aminobenzoic acid	*16*, 249 (1978); *Suppl. 7*, 56 (1987)
4-Aminobiphenyl	*1*, 74 (1972) (*corr. 42*, 251); *Suppl. 7*, 91 (1987)
2-Amino-3,4-dimethylimidazo[4,5-*f*]quinoline (*see* MeIQ)	
2-Amino-3,8-dimethylimidazo[4,5-*f*]quinoxaline (*see* MeIQx)	
3-Amino-1,4-dimethyl-5*H*-pyrido[4,3-*b*]indole (*see* Trp-P-1)	
2-Aminodipyrido[1,2-*a*:3′,2′-*d*]imidazole (*see* Glu-P-2)	
1-Amino-2-methylanthraquinone	*27*, 199 (1982); *Suppl. 7*, 57 (1987)
2-Amino-3-methylimidazo[4,5-*f*]quinoline (*see* IQ)	
2-Amino-6-methyldipyrido[1,2-*a*:3′,2′-*d*]imidazole (*see* Glu-P-1)	
2-Amino-1-methyl-6-phenylimidazo[4,5-*b*]pyridine (*see* PhIP)	
2-Amino-2-methyl-9*H*-pyrido[2,3-*b*]indole (*see* MeA-α-C)	
3-Amino-1-methyl-5*H*-pyrido[4,3-*b*]indole (*see* Trp-P-2)	
2-Amino-5-(5-nitro-2-furyl)-1,3,4-thiadiazole	*7*, 143 (1974); *Suppl. 7*, 57 (1987)
2-Amino-4-nitrophenol	*57*, 167 (1993)
2-Amino-5-nitrophenol	*57*, 177 (1993)
4-Amino-2-nitrophenol	*16*, 43 (1978); *Suppl. 7*, 57 (1987)
2-Amino-5-nitrothiazole	*31*, 71 (1983); *Suppl. 7*, 57 (1987)
2-Amino-9*H*-pyrido[2,3-*b*]indole (*see* A-α-C)	
11-Aminoundecanoic acid	*39*, 239 (1986); *Suppl. 7*, 57 (1987)
Amitrole	*7*, 31 (1974); *41*, 293 (1986) (*corr. 52*, 513; *Suppl. 7*, 92 (1987); *79*, 381 (2001)
Ammonium potassium selenide (*see* Selenium and selenium compounds)	
Amorphous silica (*see also* Silica)	*42*, 39 (1987); *Suppl. 7*, 341 (1987); *68*, 41 (1997) (*corr. 81*, 383)
Amosite (*see* Asbestos)	
Ampicillin	*50*, 153 (1990)
Amsacrine	*76*, 317 (2000)
Anabolic steroids (*see* Androgenic (anabolic) steroids)	
Anaesthetics, volatile	*11*, 285 (1976); *Suppl. 7*, 93 (1987)
Analgesic mixtures containing phenacetin (*see also* Phenacetin)	*Suppl. 7*, 310 (1987)
Androgenic (anabolic) steroids	*Suppl. 7*, 96 (1987)
Angelicin and some synthetic derivatives (*see also* Angelicins)	*40*, 291 (1986)
Angelicin plus ultraviolet radiation (*see also* Angelicin and some synthetic derivatives)	*Suppl. 7*, 57 (1987)
Angelicins	*Suppl. 7*, 57 (1987)
Aniline	*4*, 27 (1974) (*corr. 42*, 252); *27*, 39 (1982); *Suppl. 7*, 99 (1987)

B

Benz[c]acridine	3, 241 (1973); 32, 129 (1983); Suppl. 7, 58 (1987)
Benzal chloride (see also α-Chlorinated toluenes and benzoyl chloride)	29, 65 (1982); Suppl. 7, 148 (1987); 71, 453 (1999)
Benz[a]anthracene	3, 45 (1973); 32, 135 (1983); Suppl. 7, 58 (1987)
Benzene	7, 203 (1974) (corr. 42, 254); 29, 93, 391 (1982); Suppl. 7, 120 (1987)
Benzidine	1, 80 (1972); 29, 149, 391 (1982); Suppl. 7, 123 (1987)
Benzidine-based dyes	Suppl. 7, 125 (1987)
Benzo[b]fluoranthene	3, 69 (1973); 32, 147 (1983); Suppl. 7, 58 (1987)
Benzo[j]fluoranthene	3, 82 (1973); 32, 155 (1983); Suppl. 7, 58 (1987)
Benzo[k]fluoranthene	32, 163 (1983); Suppl. 7, 58 (1987)
Benzo[ghi]fluoranthene	32, 171 (1983); Suppl. 7, 58 (1987)
Benzo[a]fluorene	32, 177 (1983); Suppl. 7, 58 (1987)
Benzo[b]fluorene	32, 183 (1983); Suppl. 7, 58 (1987)
Benzo[c]fluorene	32, 189 (1983); Suppl. 7, 58 (1987)
Benzofuran	63, 431 (1995)
Benzo[ghi]perylene	32, 195 (1983); Suppl. 7, 58 (1987)
Benzo[c]phenanthrene	32, 205 (1983); Suppl. 7, 58 (1987)
Benzo[a]pyrene	3, 91 (1973); 32, 211 (1983) (corr. 68, 477); Suppl. 7, 58 (1987)
Benzo[e]pyrene	3, 137 (1973); 32, 225 (1983); Suppl. 7, 58 (1987)
1,4-Benzoquinone (see para-Quinone)	
1,4-Benzoquinone dioxime	29, 185 (1982); Suppl. 7, 58 (1987); 71, 1251 (1999)
Benzotrichloride (see also α-Chlorinated toluenes and benzoyl chloride)	29, 73 (1982); Suppl. 7, 148 (1987); 71, 453 (1999)
Benzoyl chloride (see also α-Chlorinated toluenes and benzoyl chloride)	29, 83 (1982) (corr. 42, 261); Suppl. 7, 126 (1987); 71, 453 (1999)
Benzoyl peroxide	36, 267 (1985); Suppl. 7, 58 (1987); 71, 345 (1999)
Benzyl acetate	40, 109 (1986); Suppl. 7, 58 (1987); 71, 1255 (1999)
Benzyl chloride (see also α-Chlorinated toluenes and benzoyl chloride)	11, 217 (1976) (corr. 42, 256); 29, 49 (1982); Suppl. 7, 148 (1987); 71, 453 (1999)
Benzyl violet 4B	16, 153 (1978); Suppl. 7, 58 (1987)
Bertrandite (see Beryllium and beryllium compounds)	
Beryllium and beryllium compounds	1, 17 (1972); 23, 143 (1980) (corr. 42, 260); Suppl. 7, 127 (1987); 58, 41 (1993)
Beryllium acetate (see Beryllium and beryllium compounds)	
Beryllium acetate, basic (see Beryllium and beryllium compounds)	
Beryllium-aluminium alloy (see Beryllium and beryllium compounds)	
Beryllium carbonate (see Beryllium and beryllium compounds)	
Beryllium chloride (see Beryllium and beryllium compounds)	
Beryllium-copper alloy (see Beryllium and beryllium compounds)	
Beryllium-copper-cobalt alloy (see Beryllium and beryllium compounds)	

Butyl benzyl phthalate	*29*, 193 (1982) (*corr. 42*, 261);
	Suppl. 7, 59 (1987); *73*, 115 (1999)
β-Butyrolactone	*11*, 225 (1976); *Suppl. 7*, 59
	(1987); *71*, 1317 (1999)
γ-Butyrolactone	*11*, 231 (1976); *Suppl. 7*, 59
	(1987); *71*, 367 (1999)

C

Cabinet-making (*see* Furniture and cabinet-making)	
Cadmium acetate (*see* Cadmium and cadmium compounds)	
Cadmium and cadmium compounds	*2*, 74 (1973); *11*, 39 (1976)
	(*corr. 42*, 255); *Suppl. 7*, 139
	(1987); *58*, 119 (1993)
Cadmium chloride (*see* Cadmium and cadmium compounds)	
Cadmium oxide (*see* Cadmium and cadmium compounds)	
Cadmium sulfate (*see* Cadmium and cadmium compounds)	
Cadmium sulfide (*see* Cadmium and cadmium compounds)	
Caffeic acid	*56*, 115 (1993)
Caffeine	*51*, 291 (1991)
Calcium arsenate (*see* Arsenic in drinking-water)	
Calcium chromate (*see* Chromium and chromium compounds)	
Calcium cyclamate (*see* Cyclamates)	
Calcium saccharin (*see* Saccharin)	
Cantharidin	*10*, 79 (1976); *Suppl. 7*, 59 (1987)
Caprolactam	*19*, 115 (1979) (*corr. 42*, 258);
	39, 247 (1986) (*corr. 42*, 264);
	Suppl. 7, 59, 390 (1987); *71*, 383
	(1999)
Captafol	*53*, 353 (1991)
Captan	*30*, 295 (1983); *Suppl. 7*, 59 (1987)
Carbaryl	*12*, 37 (1976); *Suppl. 7*, 59 (1987)
Carbazole	*32*, 239 (1983); *Suppl. 7*, 59
	(1987); *71*, 1319 (1999)
3-Carbethoxypsoralen	*40*, 317 (1986); *Suppl. 7*, 59 (1987)
Carbon black	*3*, 22 (1973); *33*, 35 (1984);
	Suppl. 7, 142 (1987); *65*, 149
	(1996)
Carbon tetrachloride	*1*, 53 (1972); *20*, 371 (1979);
	Suppl. 7, 143 (1987); *71*, 401
	(1999)
Carmoisine	*8*, 83 (1975); *Suppl. 7*, 59 (1987)
Carpentry and joinery	*25*, 139 (1981); *Suppl. 7*, 378
	(1987)
Carrageenan	*10*, 181 (1976) (*corr. 42*, 255); *31*,
	79 (1983); *Suppl. 7*, 59 (1987)
Cassia occidentalis (*see* Traditional herbal medicines)	
Catechol	*15*, 155 (1977); *Suppl. 7*, 59
	(1987); *71*, 433 (1999)
CCNU (*see* 1-(2-Chloroethyl)-3-cyclohexyl-1-nitrosourea)	
Ceramic fibres (*see* Man-made vitreous fibres)	
Chemotherapy, combined, including alkylating agents (*see* MOPP and	
other combined chemotherapy including alkylating agents)	

Chlorophenoxy herbicides (occupational exposures to)	*41*, 357 (1986)
4-Chloro-*ortho*-phenylenediamine	*27*, 81 (1982); *Suppl. 7*, 60 (1987)
4-Chloro-*meta*-phenylenediamine	*27*, 82 (1982); *Suppl. 7*, 60 (1987)
Chloroprene	*19*, 131 (1979); *Suppl. 7*, 160 (1987); *71*, 227 (1999)
Chloropropham	*12*, 55 (1976); *Suppl. 7*, 60 (1987)
Chloroquine	*13*, 47 (1977); *Suppl. 7*, 60 (1987)
Chlorothalonil	*30*, 319 (1983); *Suppl. 7*, 60 (1987); *73*, 183 (1999)
para-Chloro-*ortho*-toluidine and its strong acid salts (*see also* Chlordimeform)	*16*, 277 (1978); *30*, 65 (1983); *Suppl. 7*, 60 (1987); *48*, 123 (1990); *77*, 323 (2000)
4-Chloro-*ortho*-toluidine (see *para*-chloro-*ortho*-toluidine)	
5-Chloro-*ortho*-toluidine	*77*, 341 (2000)
Chlorotrianisene (*see also* Nonsteroidal oestrogens)	*21*, 139 (1979); *Suppl. 7*, 280 (1987)
2-Chloro-1,1,1-trifluoroethane	*41*, 253 (1986); *Suppl. 7*, 60 (1987); *71*, 1355 (1999)
Chlorozotocin	*50*, 65 (1990)
Cholesterol	*10*, 99 (1976); *31*, 95 (1983); *Suppl. 7*, 161 (1987)
Chromic acetate (*see* Chromium and chromium compounds)	
Chromic chloride (*see* Chromium and chromium compounds)	
Chromic oxide (*see* Chromium and chromium compounds)	
Chromic phosphate (*see* Chromium and chromium compounds)	
Chromite ore (*see* Chromium and chromium compounds)	
Chromium and chromium compounds (*see also* Implants, surgical)	*2*, 100 (1973); *23*, 205 (1980); *Suppl. 7*, 165 (1987); *49*, 49 (1990) (*corr. 51*, 483)
Chromium carbonyl (*see* Chromium and chromium compounds)	
Chromium potassium sulfate (*see* Chromium and chromium compounds)	
Chromium sulfate (*see* Chromium and chromium compounds)	
Chromium trioxide (*see* Chromium and chromium compounds)	
Chrysazin (*see* Dantron)	
Chrysene	*3*, 159 (1973); *32*, 247 (1983); *Suppl. 7*, 60 (1987)
Chrysoidine	*8*, 91 (1975); *Suppl. 7*, 169 (1987)
Chrysotile (*see* Asbestos)	
CI Acid Orange 3	*57*, 121 (1993)
CI Acid Red 114	*57*, 247 (1993)
CI Basic Red 9 (*see also* Magenta)	*57*, 215 (1993)
Ciclosporin	*50*, 77 (1990)
CI Direct Blue 15	*57*, 235 (1993)
CI Disperse Yellow 3 (see Disperse Yellow 3)	
Cimetidine	*50*, 235 (1990)
Cinnamyl anthranilate	*16*, 287 (1978); *31*, 133 (1983); *Suppl. 7*, 60 (1987); *77*, 177 (2000)
CI Pigment Red 3	*57*, 259 (1993)
CI Pigment Red 53:1 (*see* D&C Red No. 9)	
Cisplatin (*see also* Etoposide)	*26*, 151 (1981); *Suppl. 7*, 170 (1987)
Citrinin	*40*, 67 (1986); *Suppl. 7*, 60 (1987)
Citrus Red No. 2	*8*, 101 (1975) (*corr. 42*, 254); *Suppl. 7*, 60 (1987)

Cyclohexanone	*47*, 157 (1989); *71*, 1359 (1999)
Cyclohexylamine (*see* Cyclamates)	
Cyclopenta[*cd*]pyrene	*32*, 269 (1983); *Suppl. 7*, 61 (1987)
Cyclopropane (*see* Anaesthetics, volatile)	
Cyclophosphamide	*9*, 135 (1975); *26*, 165 (1981); *Suppl. 7*, 182 (1987)
Cyproterone acetate	*72*, 49 (1999)

D

2,4-D (*see also* Chlorophenoxy herbicides; Chlorophenoxy herbicides, occupational exposures to)	*15*, 111 (1977)
Dacarbazine	*26*, 203 (1981); *Suppl. 7*, 184 (1987)
Dantron	*50*, 265 (1990) (*corr. 59*, 257)
D&C Red No. 9	*8*, 107 (1975); *Suppl. 7*, 61 (1987); *57*, 203 (1993)
Dapsone	*24*, 59 (1980); *Suppl. 7*, 185 (1987)
Daunomycin	*10*, 145 (1976); *Suppl. 7*, 61 (1987)
DDD (*see* DDT)	
DDE (*see* DDT)	
DDT	*5*, 83 (1974) (*corr. 42*, 253); *Suppl. 7*, 186 (1987); *53*, 179 (1991)
Decabromodiphenyl oxide	*48*, 73 (1990); *71*, 1365 (1999)
Deltamethrin	*53*, 251 (1991)
Deoxynivalenol (*see* Toxins derived from *Fusarium graminearum*, *F. culmorum* and *F. crookwellense*)	
Diacetylaminoazotoluene	*8*, 113 (1975); *Suppl. 7*, 61 (1987)
N,N-Diacetylbenzidine	*16*, 293 (1978); *Suppl. 7*, 61 (1987)
Diallate	*12*, 69 (1976); *30*, 235 (1983); *Suppl. 7*, 61 (1987)
2,4-Diaminoanisole and its salts	*16*, 51 (1978); *27*, 103 (1982); *Suppl. 7*, 61 (1987); *79*, 619 (2001)
4,4′-Diaminodiphenyl ether	*16*, 301 (1978); *29*, 203 (1982); *Suppl. 7*, 61 (1987)
1,2-Diamino-4-nitrobenzene	*16*, 63 (1978); *Suppl. 7*, 61 (1987)
1,4-Diamino-2-nitrobenzene	*16*, 73 (1978); *Suppl. 7*, 61 (1987); *57*, 185 (1993)
2,6-Diamino-3-(phenylazo)pyridine (*see* Phenazopyridine hydrochloride)	
2,4-Diaminotoluene (*see also* Toluene diisocyanates)	*16*, 83 (1978); *Suppl. 7*, 61 (1987)
2,5-Diaminotoluene (*see also* Toluene diisocyanates)	*16*, 97 (1978); *Suppl. 7*, 61 (1987)
ortho-Dianisidine (*see* 3,3′-Dimethoxybenzidine)	
Diatomaceous earth, uncalcined (*see* Amorphous silica)	
Diazepam	*13*, 57 (1977); *Suppl. 7*, 189 (1987); *66*, 37 (1996)
Diazomethane	*7*, 223 (1974); *Suppl. 7*, 61 (1987)
Dibenz[*a,h*]acridine	*3*, 247 (1973); *32*, 277 (1983); *Suppl. 7*, 61 (1987)
Dibenz[*a,j*]acridine	*3*, 254 (1973); *32*, 283 (1983); *Suppl. 7*, 61 (1987)
Dibenz[*a,c*]anthracene	*32*, 289 (1983) (*corr. 42*, 262); *Suppl. 7*, 61 (1987)

H

Haematite	*1*, 29 (1972); *Suppl. 7*, 216 (1987)
Haematite and ferric oxide	*Suppl. 7*, 216 (1987)
Haematite mining, underground, with exposure to radon	*1*, 29 (1972); *Suppl. 7*, 216 (1987)
Hairdressers and barbers (occupational exposure as)	*57*, 43 (1993)
Hair dyes, epidemiology of	*16*, 29 (1978); *27*, 307 (1982);
Halogenated acetonitriles	*52*, 269 (1991); *71*, 1325, 1369, 1375, 1533 (1999)
Halothane (*see* Anaesthetics, volatile)	
HC Blue No. 1	*57*, 129 (1993)
HC Blue No. 2	*57*, 143 (1993)
α-HCH (*see* Hexachlorocyclohexanes)	
β-HCH (*see* Hexachlorocyclohexanes)	
γ-HCH (*see* Hexachlorocyclohexanes)	
HC Red No. 3	*57*, 153 (1993)
HC Yellow No. 4	*57*, 159 (1993)
Heating oils (*see* Fuel oils)	
Helicobacter pylori (infection with)	*61*, 177 (1994)
Hepatitis B virus	*59*, 45 (1994)
Hepatitis C virus	*59*, 165 (1994)
Hepatitis D virus	*59*, 223 (1994)
Heptachlor (*see also* Chlordane/Heptachlor)	*5*, 173 (1974); *20*, 129 (1979)
Hexachlorobenzene	*20*, 155 (1979); *Suppl. 7*, 219 (1987); *79*, 493 (2001)
Hexachlorobutadiene	*20*, 179 (1979); *Suppl. 7*, 64 (1987); *73*, 277 (1999)
Hexachlorocyclohexanes	*5*, 47 (1974); *20*, 195 (1979) (*corr. 42*, 258); *Suppl. 7*, 220 (1987)
Hexachlorocyclohexane, technical-grade (*see* Hexachlorocyclohexanes)	
Hexachloroethane	*20*, 467 (1979); *Suppl. 7*, 64 (1987); *73*, 295 (1999)
Hexachlorophene	*20*, 241 (1979); *Suppl. 7*, 64 (1987)
Hexamethylphosphoramide	*15*, 211 (1977); *Suppl. 7*, 64 (1987); *71*, 1465 (1999)
Hexoestrol (*see also* Nonsteroidal oestrogens)	*Suppl. 7*, 279 (1987)
Hormonal contraceptives, progestogens only	*72*, 339 (1999)
Human herpesvirus 8	*70*, 375 (1997)
Human immunodeficiency viruses	*67*, 31 (1996)
Human papillomaviruses	*64* (1995) (*corr. 66*, 485)
Human T-cell lymphotropic viruses	*67*, 261 (1996)
Hycanthone mesylate	*13*, 91 (1977); *Suppl. 7*, 64 (1987)
Hydralazine	*24*, 85 (1980); *Suppl. 7*, 222 (1987)
Hydrazine	*4*, 127 (1974); *Suppl. 7*, 223 (1987); *71*, 991 (1999)
Hydrochloric acid	*54*, 189 (1992)
Hydrochlorothiazide	*50*, 293 (1990)
Hydrogen peroxide	*36*, 285 (1985); *Suppl. 7*, 64 (1987); *71*, 671 (1999)
Hydroquinone	*15*, 155 (1977); *Suppl. 7*, 64 (1987); *71*, 691 (1999)
1-Hydroxyanthraquinone	*82*, 129 (2002)
4-Hydroxyazobenzene	*8*, 157 (1975); *Suppl. 7*, 64 (1987)

K

Kaempferol	*31*, 171 (1983); *Suppl. 7*, 65 (1987)
Kaposi's sarcoma herpesvirus	*70*, 375 (1997)
Kepone (*see* Chlordecone)	
Kojic acid	*79*, 605 (2001)

L

Lasiocarpine	*10*, 281 (1976); *Suppl. 7*, 65 (1987)
Lauroyl peroxide	*36*, 315 (1985); *Suppl. 7*, 65 (1987); *71*, 1485 (1999)
Lead acetate (*see* Lead and lead compounds)	
Lead and lead compounds (*see also* Foreign bodies)	*1*, 40 (1972) (*corr. 42*, 251); *2*, 52, 150 (1973); *12*, 131 (1976); *23*, 40, 208, 209, 325 (1980); *Suppl. 7*, 230 (1987)
Lead arsenate (*see* Arsenic and arsenic compounds)	
Lead carbonate (*see* Lead and lead compounds)	
Lead chloride (*see* Lead and lead compounds)	
Lead chromate (*see* Chromium and chromium compounds)	
Lead chromate oxide (*see* Chromium and chromium compounds)	
Lead naphthenate (*see* Lead and lead compounds)	
Lead nitrate (*see* Lead and lead compounds)	
Lead oxide (*see* Lead and lead compounds)	
Lead phosphate (*see* Lead and lead compounds)	
Lead subacetate (*see* Lead and lead compounds)	
Lead tetroxide (*see* Lead and lead compounds)	
Leather goods manufacture	*25*, 279 (1981); *Suppl. 7*, 235 (1987)
Leather industries	*25*, 199 (1981); *Suppl. 7*, 232 (1987)
Leather tanning and processing	*25*, 201 (1981); *Suppl. 7*, 236 (1987)
Ledate (*see also* Lead and lead compounds)	*12*, 131 (1976)
Levonorgestrel	*72*, 49 (1999)
Light Green SF	*16*, 209 (1978); *Suppl. 7*, 65 (1987)
d-Limonene	*56*, 135 (1993); *73*, 307 (1999)
Lindane (*see* Hexachlorocyclohexanes)	
Liver flukes (*see Clonorchis sinensis*, *Opisthorchis felineus* and *Opisthorchis viverrini*)	
Lucidin (*see* 1,3-Dihydro-2-hydroxymethylanthraquinone)	
Lumber and sawmill industries (including logging)	*25*, 49 (1981); *Suppl. 7*, 383 (1987)
Luteoskyrin	*10*, 163 (1976); *Suppl. 7*, 65 (1987)
Lynoestrenol	*21*, 407 (1979); *Suppl. 7*, 293 (1987); *72*, 49 (1999)

M

Madder root (*see also Rubia tinctorum*)	*82*, 129 (2002)

N

1-Naphthylthiourea	*30*, 347 (1983); *Suppl. 7*, 263 (1987)
Neutrons	*75*, 361 (2000)
Nickel acetate (*see* Nickel and nickel compounds)	
Nickel ammonium sulfate (*see* Nickel and nickel compounds)	
Nickel and nickel compounds (*see also* Implants, surgical)	*2*, 126 (1973) (*corr. 42*, 252); *11*, 75 (1976); *Suppl. 7*, 264 (1987) (*corr. 45*, 283); *49*, 257 (1990) (*corr. 67*, 395)
Nickel carbonate (*see* Nickel and nickel compounds)	
Nickel carbonyl (*see* Nickel and nickel compounds)	
Nickel chloride (*see* Nickel and nickel compounds)	
Nickel-gallium alloy (*see* Nickel and nickel compounds)	
Nickel hydroxide (*see* Nickel and nickel compounds)	
Nickelocene (*see* Nickel and nickel compounds)	
Nickel oxide (*see* Nickel and nickel compounds)	
Nickel subsulfide (*see* Nickel and nickel compounds)	
Nickel sulfate (*see* Nickel and nickel compounds)	
Niridazole	*13*, 123 (1977); *Suppl. 7*, 67 (1987)
Nithiazide	*31*, 179 (1983); *Suppl. 7*, 67 (1987)
Nitrilotriacetic acid and its salts	*48*, 181 (1990); *73*, 385 (1999)
5-Nitroacenaphthene	*16*, 319 (1978); *Suppl. 7*, 67 (1987)
5-Nitro-*ortho*-anisidine	*27*, 133 (1982); *Suppl. 7*, 67 (1987)
2-Nitroanisole	*65*, 369 (1996)
9-Nitroanthracene	*33*, 179 (1984); *Suppl. 7*, 67 (1987)
7-Nitrobenz[*a*]anthracene	*46*, 247 (1989)
Nitrobenzene	*65*, 381 (1996)
6-Nitrobenzo[*a*]pyrene	*33*, 187 (1984); *Suppl. 7*, 67 (1987); *46*, 255 (1989)
4-Nitrobiphenyl	*4*, 113 (1974); *Suppl. 7*, 67 (1987)
6-Nitrochrysene	*33*, 195 (1984); *Suppl. 7*, 67 (1987); *46*, 267 (1989)
Nitrofen (technical-grade)	*30*, 271 (1983); *Suppl. 7*, 67 (1987)
3-Nitrofluoranthene	*33*, 201 (1984); *Suppl. 7*, 67 (1987)
2-Nitrofluorene	*46*, 277 (1989)
Nitrofural	*7*, 171 (1974); *Suppl. 7*, 67 (1987); *50*, 195 (1990)
5-Nitro-2-furaldehyde semicarbazone (*see* Nitrofural)	
Nitrofurantoin	*50*, 211 (1990)
Nitrofurazone (*see* Nitrofural)	
1-[(5-Nitrofurfurylidene)amino]-2-imidazolidinone	*7*, 181 (1974); *Suppl. 7*, 67 (1987)
N-[4-(5-Nitro-2-furyl)-2-thiazolyl]acetamide	*1*, 181 (1972); *7*, 185 (1974); *Suppl. 7*, 67 (1987)
Nitrogen mustard	*9*, 193 (1975); *Suppl. 7*, 269 (1987)
Nitrogen mustard *N*-oxide	*9*, 209 (1975); *Suppl. 7*, 67 (1987)
Nitromethane	*77*, 487 (2000)
1-Nitronaphthalene	*46*, 291 (1989)
2-Nitronaphthalene	*46*, 303 (1989)
3-Nitroperylene	*46*, 313 (1989)
2-Nitro-*para*-phenylenediamine (*see* 1,4-Diamino-2-nitrobenzene)	
2-Nitropropane	*29*, 331 (1982); *Suppl. 7*, 67 (1987); *71*, 1079 (1999)
1-Nitropyrene	*33*, 209 (1984); *Suppl. 7*, 67 (1987); *46*, 321 (1989)

Nonsteroidal oestrogens   *Suppl. 7*, 273 (1987)
Norethisterone   *6*, 179 (1974); *21*, 461 (1979);
                                            Suppl. 7, 294 (1987); *72*, 49
                                            (1999)
Norethisterone acetate   *72*, 49 (1999)
Norethynodrel   *6*, 191 (1974); *21*, 461 (1979)
                                            (*corr. 42*, 259); *Suppl. 7*, 295
                                            (1987); *72*, 49 (1999)
Norgestrel   *6*, 201 (1974); *21*, 479 (1979);
                                            Suppl. 7, 295 (1987); *72*, 49 (1999)
Nylon 6   *19*, 120 (1979); *Suppl. 7*, 68 (1987)

O

Ochratoxin A   *10*, 191 (1976); *31*, 191 (1983)
                                            (*corr. 42*, 262); *Suppl. 7*, 271
                                            (1987); *56*, 489 (1993)
Oestradiol   *6*, 99 (1974); *21*, 279 (1979);
                                            Suppl. 7, 284 (1987); *72*, 399
                                            (1999)

Oestradiol-17β (*see* Oestradiol)
Oestradiol 3-benzoate (*see* Oestradiol)
Oestradiol dipropionate (*see* Oestradiol)
Oestradiol mustard   *9*, 217 (1975); *Suppl. 7*, 68 (1987)
Oestradiol valerate (*see* Oestradiol)
Oestriol   *6*, 117 (1974); *21*, 327 (1979);
                                            Suppl. 7, 285 (1987); *72*, 399
                                            (1999)

Oestrogen-progestin combinations (*see* Oestrogens,
    progestins (progestogens) and combinations)
Oestrogen-progestin replacement therapy (*see* Post-menopausal
    oestrogen-progestogen therapy)
Oestrogen replacement therapy (*see* Post-menopausal oestrogen
    therapy)
Oestrogens (*see* Oestrogens, progestins and combinations)
Oestrogens, conjugated (*see* Conjugated oestrogens)
Oestrogens, nonsteroidal (*see* Nonsteroidal oestrogens)
Oestrogens, progestins (progestogens) and combinations   *6* (1974); *21* (1979); *Suppl. 7*, 272
                                            (1987); *72*, 49, 339, 399, 531
                                            (1999)
Oestrogens, steroidal (*see* Steroidal oestrogens)
Oestrone   *6*, 123 (1974); *21*, 343 (1979)
                                            (*corr. 42*, 259); *Suppl. 7*, 286
                                            (1987); *72*, 399 (1999)

Oestrone benzoate (*see* Oestrone)
Oil Orange SS   *8*, 165 (1975); *Suppl. 7*, 69 (1987)
Opisthorchis felineus (infection with)   *61*, 121 (1994)
Opisthorchis viverrini (infection with)   *61*, 121 (1994)
Oral contraceptives, combined   *Suppl. 7*, 297 (1987); *72*, 49 (1999)
Oral contraceptives, sequential (*see* Sequential oral contraceptives)
Orange I   *8*, 173 (1975); *Suppl. 7*, 69 (1987)
Orange G   *8*, 181 (1975); *Suppl. 7*, 69 (1987)
Organolead compounds (*see also* Lead and lead compounds)   *Suppl. 7*, 230 (1987)

N-Phenyl-2-naphthylamine	*16*, 325 (1978) (*corr. 42*, 257); *Suppl. 7*, 318 (1987)
ortho-Phenylphenol	*30*, 329 (1983); *Suppl. 7*, 70 (1987); *73*, 451 (1999)
Phenytoin	*13*, 201 (1977); *Suppl. 7*, 319 (1987); *66*, 175 (1996)
Phillipsite (*see* Zeolites)	
PhIP	*56*, 229 (1993)
Pickled vegetables	*56*, 83 (1993)
Picloram	*53*, 481 (1991)
Piperazine oestrone sulfate (*see* Conjugated oestrogens)	
Piperonyl butoxide	*30*, 183 (1983); *Suppl. 7*, 70 (1987)
Pitches, coal-tar (*see* Coal-tar pitches)	
Polyacrylic acid	*19*, 62 (1979); *Suppl. 7*, 70 (1987)
Polybrominated biphenyls	*18*, 107 (1978); *41*, 261 (1986); *Suppl. 7*, 321 (1987)
Polychlorinated biphenyls	*7*, 261 (1974); *18*, 43 (1978) (*corr. 42*, 258); *Suppl. 7*, 322 (1987)
Polychlorinated camphenes (*see* Toxaphene)	
Polychlorinated dibenzo-*para*-dioxins (other than 2,3,7,8-tetrachlorodibenzodioxin)	*69*, 33 (1997)
Polychlorinated dibenzofurans	*69*, 345 (1997)
Polychlorophenols and their sodium salts	*71*, 769 (1999)
Polychloroprene	*19*, 141 (1979); *Suppl. 7*, 70 (1987)
Polyethylene (*see also* Implants, surgical)	*19*, 164 (1979); *Suppl. 7*, 70 (1987)
Poly(glycolic acid) (*see* Implants, surgical)	
Polymethylene polyphenyl isocyanate (*see also* 4,4′-Methylenediphenyl diisocyanate)	*19*, 314 (1979); *Suppl. 7*, 70 (1987)
Polymethyl methacrylate (*see also* Implants, surgical)	*19*, 195 (1979); *Suppl. 7*, 70 (1987)
Polyoestradiol phosphate (*see* Oestradiol-17β)	
Polypropylene (*see also* Implants, surgical)	*19*, 218 (1979); *Suppl. 7*, 70 (1987)
Polystyrene (*see also* Implants, surgical)	*19*, 245 (1979); *Suppl. 7*, 70 (1987)
Polytetrafluoroethylene (*see also* Implants, surgical)	*19*, 288 (1979); *Suppl. 7*, 70 (1987)
Polyurethane foams (*see also* Implants, surgical)	*19*, 320 (1979); *Suppl. 7*, 70 (1987)
Polyvinyl acetate (*see also* Implants, surgical)	*19*, 346 (1979); *Suppl. 7*, 70 (1987)
Polyvinyl alcohol (*see also* Implants, surgical)	*19*, 351 (1979); *Suppl. 7*, 70 (1987)
Polyvinyl chloride (*see also* Implants, surgical)	*7*, 306 (1974); *19*, 402 (1979); *Suppl. 7*, 70 (1987)
Polyvinyl pyrrolidone	*19*, 463 (1979); *Suppl. 7*, 70 (1987); *71*, 1181 (1999)
Ponceau MX	*8*, 189 (1975); *Suppl. 7*, 70 (1987)
Ponceau 3R	*8*, 199 (1975); *Suppl. 7*, 70 (1987)
Ponceau SX	*8*, 207 (1975); *Suppl. 7*, 70 (1987)
Post-menopausal oestrogen therapy	*Suppl. 7*, 280 (1987); *72*, 399 (1999)
Post-menopausal oestrogen-progestogen therapy	*Suppl. 7*, 308 (1987); *72*, 531 (1999)
Potassium arsenate (*see* Arsenic and arsenic compounds)	
Potassium arsenite (*see* Arsenic and arsenic compounds)	
Potassium bis(2-hydroxyethyl)dithiocarbamate	*12*, 183 (1976); *Suppl. 7*, 70 (1987)
Potassium bromate	*40*, 207 (1986); *Suppl. 7*, 70 (1987); *73*, 481 (1999)
Potassium chromate (*see* Chromium and chromium compounds)	

Q

R

Radiation (*see* gamma-radiation, neutrons, ultraviolet radiation, X-radiation)	
Radionuclides, internally deposited	*78* (2001)
Radon	*43*, 173 (1988) (*corr. 45*, 283)
Refractory ceramic fibres (*see* Man-made vitreous fibres)	
Reserpine	*10*, 217 (1976); *24*, 211 (1980) (*corr. 42*, 260); *Suppl. 7*, 330 (1987)
Resorcinol	*15*, 155 (1977); *Suppl. 7*, 71 (1987); *71*, 1119 (1990)
Retrorsine	*10*, 303 (1976); *Suppl. 7*, 71 (1987)
Rhodamine B	*16*, 221 (1978); *Suppl. 7*, 71 (1987)
Rhodamine 6G	*16*, 233 (1978); *Suppl. 7*, 71 (1987)
Riddelliine	*10*, 313 (1976); *Suppl. 7*, 71 (1987); *82*, 153 (2002)
Rifampicin	*24*, 243 (1980); *Suppl. 7*, 71 (1987)
Ripazepam	*66*, 157 (1996)
Rock (stone) wool (*see* Man-made vitreous fibres)	
Rubber industry	*28* (1982) (*corr. 42*, 261); *Suppl. 7*, 332 (1987)
Rubia tinctorum (*see also* Madder root, Traditional herbal medicines)	*82*, 129 (2002)
Rugulosin	*40*, 99 (1986); *Suppl. 7*, 71 (1987)

S

Saccharated iron oxide	*2*, 161 (1973); *Suppl. 7*, 71 (1987)
Saccharin and its salts	*22*, 111 (1980) (*corr. 42*, 259); *Suppl. 7*, 334 (1987); *73*, 517 (1999)
Safrole	*1*, 169 (1972); *10*, 231 (1976); *Suppl. 7*, 71 (1987)
Salted fish	*56*, 41 (1993)
Sawmill industry (including logging) (*see* Lumber and sawmill industry (including logging))	
Scarlet Red	*8*, 217 (1975); *Suppl. 7*, 71 (1987)
Schistosoma haematobium (infection with)	*61*, 45 (1994)
Schistosoma japonicum (infection with)	*61*, 45 (1994)
Schistosoma mansoni (infection with)	*61*, 45 (1994)
Selenium and selenium compounds	*9*, 245 (1975) (*corr. 42*, 255); *Suppl. 7*, 71 (1987)
Selenium dioxide (*see* Selenium and selenium compounds)	
Selenium oxide (*see* Selenium and selenium compounds)	
Semicarbazide hydrochloride	*12*, 209 (1976) (*corr. 42*, 256); *Suppl. 7*, 71 (1987)
Senecio jacobaea L. (*see also* Pyrrolizidine alkaloids)	*10*, 333 (1976)
Senecio longilobus (*see also* Pyrrolizidine alkaloids, Traditional herbal medicines)	*10*, 334 (1976); *82*, 153 (2002)
Senecio riddellii (*see also* Traditional herbal medicines)	*82*, 153 (1982)
Seneciphylline	*10*, 319, 335 (1976); *Suppl. 7*, 71 (1987)
Senkirkine	*10*, 327 (1976); *31*, 231 (1983); *Suppl. 7*, 71 (1987)

T

T-2 Toxin (*see* Toxins derived from *Fusarium sporotrichioides*)

Toxins derived from *Fusarium graminearum, F. culmorum* and F. crookwellense	*11*, 169 (1976); *31*, 153, 279 (1983); *Suppl. 7*, 64, 74 (1987); *56*, 397 (1993)
Toxins derived from *Fusarium moniliforme*	*56*, 445 (1993)
Toxins derived from *Fusarium sporotrichioides*	*31*, 265 (1983); *Suppl. 7*, 73 (1987); *56*, 467 (1993)
Traditional herbal medicines	*82*, 41 (2002)
Tremolite (*see* Asbestos)	
Treosulfan	*26*, 341 (1981); *Suppl. 7*, 363 (1987)
Triaziquone (*see* Tris(aziridinyl)-*para*-benzoquinone)	
Trichlorfon	*30*, 207 (1983); *Suppl. 7*, 73 (1987)
Trichlormethine	*9*, 229 (1975); *Suppl. 7*, 73 (1987); *50*, 143 (1990)
Trichloroacetic acid	*63*, 291 (1995) (*corr. 65*, 549); *84* (2004)
Trichloroacetonitrile (*see also* Halogenated acetonitriles)	*71*, 1533 (1999)
1,1,1-Trichloroethane	*20*, 515 (1979); *Suppl. 7*, 73 (1987); *71*, 881 (1999)
1,1,2-Trichloroethane	*20*, 533 (1979); *Suppl. 7*, 73 (1987); *52*, 337 (1991); *71*, 1153 (1999)
Trichloroethylene	*11*, 263 (1976); *20*, 545 (1979); *Suppl. 7*, 364 (1987); *63*, 75 (1995) (*corr. 65*, 549)
2,4,5-Trichlorophenol (*see also* Chlorophenols; Chlorophenols, occupational exposures to; Polychlorophenols and their sodium salts)	*20*, 349 (1979)
2,4,6-Trichlorophenol (*see also* Chlorophenols; Chlorophenols, occupational exposures to; Polychlorophenols and their sodium salts)	*20*, 349 (1979)
(2,4,5-Trichlorophenoxy)acetic acid (*see* 2,4,5-T)	
1,2,3-Trichloropropane	*63*, 223 (1995)
Trichlorotriethylamine-hydrochloride (*see* Trichlormethine)	
T$_2$-Trichothecene (*see* Toxins derived from *Fusarium sporotrichioides*)	
Tridymite (*see* Crystalline silica)	
Triethanolamine	*77*, 381 (2000)
Triethylene glycol diglycidyl ether	*11*, 209 (1976); *Suppl. 7*, 73 (1987); *71*, 1539 (1999)
Trifluralin	*53*, 515 (1991)
4,4′,6-Trimethylangelicin plus ultraviolet radiation (*see also* Angelicin and some synthetic derivatives)	*Suppl. 7*, 57 (1987)
2,4,5-Trimethylaniline	*27*, 177 (1982); *Suppl. 7*, 73 (1987)
2,4,6-Trimethylaniline	*27*, 178 (1982); *Suppl. 7*, 73 (1987)
4,5′,8-Trimethylpsoralen	*40*, 357 (1986); *Suppl. 7*, 366 (1987)
Trimustine hydrochloride (*see* Trichlormethine)	
2,4,6-Trinitrotoluene	*65*, 449 (1996)
Triphenylene	*32*, 447 (1983); *Suppl. 7*, 73 (1987)
Tris(aziridinyl)-*para*-benzoquinone	*9*, 67 (1975); *Suppl. 7*, 367 (1987)
Tris(1-aziridinyl)phosphine-oxide	*9*, 75 (1975); *Suppl. 7*, 73 (1987)
Tris(1-aziridinyl)phosphine-sulphide (*see* Thiotepa)	
2,4,6-Tris(1-aziridinyl)-*s*-triazine	*9*, 95 (1975); *Suppl. 7*, 73 (1987)
Tris(2-chloroethyl) phosphate	*48*, 109 (1990); *71*, 1543 (1999)

W

Welding	*49*, 447 (1990) (*corr. 52*, 513)
Wollastonite	*42*, 145 (1987); *Suppl. 7*, 377 (1987); *68*, 283 (1997)
Wood dust	*62*, 35 (1995)
Wood industries	*25* (1981); *Suppl. 7*, 378 (1987)

X

X-radiation	*75*, 121 (2000)
Xylenes	*47*, 125 (1989); *71*, 1189 (1999)
2,4-Xylidine	*16*, 367 (1978); *Suppl. 7*, 74 (1987)
2,5-Xylidine	*16*, 377 (1978); *Suppl. 7*, 74 (1987)
2,6-Xylidine (*see* 2,6-Dimethylaniline)	

Y

Yellow AB	*8*, 279 (1975); *Suppl. 7*, 74 (1987)
Yellow OB	*8*, 287 (1975); *Suppl. 7*, 74 (1987)

Z

Zalcitabine	*76*, 129 (2000)
Zearalenone (*see* Toxins derived from *Fusarium graminearum*, *F. culmorum* and *F. crookwellense*)	
Zectran	*12*, 237 (1976); *Suppl. 7*, 74 (1987)
Zeolites other than erionite	*68*, 307 (1997)
Zidovudine	*76*, 73 (2000)
Zinc beryllium silicate (*see* Beryllium and beryllium compounds)	
Zinc chromate (*see* Chromium and chromium compounds)	
Zinc chromate hydroxide (*see* Chromium and chromium compounds)	
Zinc potassium chromate (*see* Chromium and chromium compounds)	
Zinc yellow (*see* Chromium and chromium compounds)	
Zineb	*12*, 245 (1976); *Suppl. 7*, 74 (1987)
Ziram	*12*, 259 (1976); *Suppl. 7*, 74 (1987); *53, 423* (1991)

List of IARC Monographs on the Evaluation of Carcinogenic Risks to Humans*

Volume 1
Some Inorganic Substances, Chlorinated Hydrocarbons, Aromatic Amines, *N*-Nitroso Compounds, and Natural Products
1972; 184 pages (out-of-print)

Volume 2
Some Inorganic and Organo-metallic Compounds
1973; 181 pages (out-of-print)

Volume 3
Certain Polycyclic Aromatic Hydrocarbons and Heterocyclic Compounds
1973; 271 pages (out-of-print)

Volume 4
Some Aromatic Amines, Hydra-zine and Related Substances, *N*-Nitroso Compounds and Miscellaneous Alkylating Agents
1974; 286 pages (out-of-print)

Volume 5
Some Organochlorine Pesticides
1974; 241 pages (out-of-print)

Volume 6
Sex Hormones
1974; 243 pages (out-of-print)

Volume 7
Some Anti-Thyroid and Related Substances, Nitrofurans and Industrial Chemicals
1974; 326 pages (out-of-print)

Volume 8
Some Aromatic Azo Compounds
1975; 357 pages (out-of-print)

Volume 9
Some Aziridines, *N*-, *S*- and *O*-Mustards and Selenium
1975; 268 pages (out-of-print)

Volume 10
Some Naturally Occurring Substances
1976; 353 pages (out-of-print)

Volume 11
Cadmium, Nickel, Some Epoxides, Miscellaneous Industrial Chemicals and General Considerations on Volatile Anaesthetics
1976; 306 pages (out-of-print)

Volume 12
Some Carbamates, Thio-carbamates and Carbazides
1976; 282 pages (out-of-print)

Volume 13
Some Miscellaneous Pharmaceutical Substances
1977; 255 pages

Volume 14
Asbestos
1977; 106 pages (out-of-print)

Volume 15
Some Fumigants, the Herbicides 2,4-D and 2,4,5-T, Chlorinated Dibenzodioxins and Miscella-neous Industrial Chemicals
1977; 354 pages (out-of-print)

Volume 16
Some Aromatic Amines and Related Nitro Compounds—Hair Dyes, Colouring Agents and Miscellaneous Industrial Chemicals
1978; 400 pages

Volume 17
Some *N*-Nitroso Compounds
1978; 365 pages

Volume 18
Polychlorinated Biphenyls and Polybrominated Biphenyls
1978; 140 pages (out-of-print)

Volume 19
Some Monomers, Plastics and Synthetic Elastomers, and Acrolein
1979; 513 pages (out-of-print)

Volume 20
Some Halogenated Hydrocarbons
1979; 609 pages (out-of-print)

Volume 21
Sex Hormones (II)
1979; 583 pages

Volume 22
Some Non-Nutritive Sweetening Agents
1980; 208 pages

Volume 23
Some Metals and Metallic Compounds
1980; 438 pages (out-of-print)

Volume 24
Some Pharmaceutical Drugs
1980; 337 pages

Volume 25
Wood, Leather and Some Associated Industries
1981; 412 pages

Volume 26
Some Antineoplastic and Immunosuppressive Agents
1981; 411 pages (out-of-print)

Volume 27
Some Aromatic Amines, Anthraquinones and Nitroso Compounds, and Inorganic Fluorides Used in Drinking-water and Dental Preparations
1982; 341 pages (out-of-print)

Volume 28
The Rubber Industry
1982; 486 pages (out-of-print)

Volume 29
Some Industrial Chemicals and Dyestuffs
1982; 416 pages (out-of-print)

Volume 30
Miscellaneous Pesticides
1983; 424 pages (out-of-print)

*High-quality photocopies of all out-of-print volumes may be purchased from University Microfilms International, 300 North Zeeb Road, Ann Arbor, MI 48106-1346, USA (Tel.: +1 313-761-4700, +1 800-521-0600).

Supplement No. 6
**Genetic and Related Effects:
An Updating of Selected *IARC
Monographs* from Volumes 1
to 42**
1987; 729 pages (out-of-print)

Supplement No. 7
**Overall Evaluations of Carcino-
genicity: An Updating of
IARC Monographs Volumes 1–42**
1987; 440 pages (out-of-print)

Supplement No. 8
**Cross Index of Synonyms and
Trade Names in Volumes 1 to 46
of the *IARC Monographs***
1990; 346 pages (out-of-print)